Biomass Processing Technologies for Sustainable Energy and Material Production

This book summarizes the state-of-the-art approaches and novel technologies developed for the sustainable production of energy and structural and functional materials. Each chapter includes a detailed review of current aspects pertaining to the energy and material system supported by sustainability assessment tools. It covers sustainable utilization of biomass for energy in the form of green hydrogen, bio-oil, biodiesel, algal fuel, biogas, bioethanol, and electrochemical conversion using microbial fuel cells and materials in the form of bioadsorbents and so forth.

- Presents an up-to-date review of biomass processing technologies
- Offers insights into the use of sustainability assessment tools for the design of material and energy systems
- Highlights the sustainable production of energy from biomass, including algae and fungal processing
- Reviews pertinent sustainability assessment tools
- Covers the production of sustainable structural and functional material from biomass

This book is aimed at graduate students and researchers in chemical engineering, bioenergy, bioprocessing, and green technology.

Biomass Processing Technologies for Sustainable Energy and Material Production

Edited by
V. Sivasubramanian
M. Yogesh Kumar

CRC Press
Taylor & Francis Group
Boca Raton New York London

CRC Press is an imprint of the
Taylor & Francis Group, an **informa** business

Designed cover image: Shutterstock

First edition published 2026
by CRC Press
2385 NW Executive Center Drive, Suite 320, Boca Raton FL 33431

and by CRC Press
4 Park Square, Milton Park, Abingdon, Oxon, OX14 4RN

CRC Press is an imprint of Taylor & Francis Group, LLC

ISBN: 978-1-032-68936-4 (hbk)
ISBN: 978-1-041-22559-1 (pbk)
ISBN: 978-1-003-73290-7 (ebk)

DOI: 10.1201/9781003732907

Typeset in Times
by SPi Technologies India Pvt Ltd (Straive)

Contents

Preface

Biomass refers to renewable organic material derived from plants, animals, or microorganisms intentionally grown for human use or waste and by-products of industrial processes that use bioresources as raw materials. Human use of biomass as a source of energy dates back to antiquity with the burning of wood to cook and to keep warm. The weaving and dyeing of plant-derived fibers for clothing and reinforcement of clay using straw in construction materials represent early use of biomass-derived functional and structural materials, respectively. The progress in science and technology has led to innovations that have diversified our choice of energy and material resources and technologies that process them far more efficiently. However, climate change driven by human activity and loss of biodiversity caused by environmental pollution and overexploitation of natural resources have obliged us to rethink the way we process and utilize resources. Owing to their renewable and biodegradable nature and carbon neutrality potential, structural and functional materials derived from biomass and innovative processing technologies that can extract energy from biomass are crucial for advancing sustainability and achieving a circular bioeconomy.

This book is a collection of chapters authored by eminent researchers from leading technical institutes and research organizations that summarizes the state-of-the-art approaches and novel technologies developed for the sustainable production of biofuels, structural and functional materials using biomass as raw material. We aim to provide up-to-date knowledge on sustainable production of bio-hydrogen, bio-oil, and biodiesel from lignocellulosic biomass, algal fuel, and materials in the form of bioadsorbents and biosurfactants for bioremediation, bioconcrete for construction, biodegradable bioplastics, and antimicrobial smart packaging materials and catalysts based on plant, animal, and microbial biomass. Each chapter includes a detailed review of current knowledge pertaining to the energy and material system.

Biomass, being renewable and biodegradable, are excellent source of structural and functional materials. Lignocellulosic biomass abundantly produced as a by-product of agricultural processing and forestry can be effectively valorized through lignin-first approach to obtain high-value bioaromatics in addition to better-quality biofuels. Chemical methods for valorization of lignocellulosic biomass are reviewed in Chapter 1. Over 40% of the global CO_2 emissions are attributed to the burning of fossil fuels for electricity generation. India is the second largest producer of coal and ranks third in global emissions from coal mining. Carbon-neutral biomass briquettes and pellets as a substitute for coal can mitigate emissions associated with coal mining and promote effective, sustainable disposal of agricultural residues. A case study on policies and technologies for partial replacement of coal with biomass through cofiring in the Indian context is presented in Chapter 2. Hydrothermal processing techniques of biomass for bioenergy can employ wet biomass, eliminating the energy-intensive pretreatment step, thus reducing the overall carbon footprint of the biomass-derived fuel. Hydrothermal processing technologies for conversion of lignocellulosic biomass to biofuel are outlined in Chapter 3. The other strategies for biomass conversion, such as combustion and thermal and biochemical methods, are outlined in Chapter 4. Sustainability assessment of pyrolysis and integrated energy systems for energy production from wood waste through life cycle sustainability analysis is presented in Chapter 5. A review on cultivation and processing for microalgae for energy and food applications is presented in Chapter 6. Hydrogen is of paramount importance in the transition to a sustainable energy system, offering a clean, versatile, and scalable solution to meet growing energy demand while addressing climate change and environmental concerns. Photobiochemical processes for green hydrogen production are discussed in Chapter 7.

Biomass, being renewable and biodegradable, are excellent source of sustainable structural and functional materials. Chapters 8–15 deal with processing biomass for material applications. Biomass-derived adsorbents are efficient, economical, and eco-friendly materials with enormous potential for environmental and industrial applications. The application of biomass-derived absorbents for the removal of dyes from industrial wastewater is discussed in Chapter 8. The conventional cement manufacturing is energy intensive and contributes to about 8% of anthropogenic CO_2 emissions. The ability of certain species of microorganisms to precipitate calcite has been engineered to produce biocement and bioconcrete through processes that are energy efficient and clean. A review of microbial biocement and incorporation of biomaterials such as plant fibers and biosilica to improve the mechanical properties is presented in Chapter 9. Production of bioplastic from agricultural waste is discussed in Chapter 10. Chapter 11 reviews biomass-derived catalysts for biodiesel production. Mycelium technology that uses fungal materials to produce sustainable packaging and construction materials is discussed in Chapter 12. Synthesis of chitosan nanomaterials and their application in the biomedical field for drug delivery applications is explained in Chapter 13. Design of antimicrobial food packaging materials that incorporates cellulose fibers from plant sources as reinforcement and essential oils to impart antimicrobial activity is discussed in Chapter 14. Biosurfactants are organic molecules produced by microorganisms that can modify surface wettability and interfacial tension. The applications of these molecules for applications in the food industry and in healthcare are reviewed in Chapter 15. Microalgae serve as a source of lipids for both functional materials and biodiesel, and Chapter 16. Summarises the key technologies for lipid extraction and biodiesel conversion.

The book presents a multidisciplinary approach to reduce greenhouse gas emissions, mitigate environmental and ecological damage, and achieve carbon neutrality goals through energy-efficient process for energy and material applications.

Dr. **V. Sivasubramanian, M.Tech., MBA, Ph.D.**
Professor
Department of Chemical Engineering
National Institute of Technology Calicut
Kozhikode
Kerala-673601
India

Dr. M Yogesh Kumar
Assistant Professor
Department of Chemical Engineering
National Institute of Technology Calicut
Kozhikode
Kerala-673601
India

Editor's Biography

Dr. V. Sivasubramanian is a Professor of Chemical Engineering at the National Institute of Technology Calicut, Kozhikode, Kerala, India. With more than 25 years of teaching and research experience and 5 years of industrial experience, he has made significant contributions in the areas of CO_2 sequestration, biofuels, fluidization engineering—hydrodynamics and mass transfer, biochemical engineering, environmental engineering, and energy engineering. He has published over 200 peer-reviewed international journal and conference papers, and has authored/edited seven books. Under his guidance, 20 Ph.D. scholars have successfully completed their research, alongside 55 undergraduate and 35 postgraduate projects.

Dr. Yogesh Kumar Murugesan is an Assistant Professor in the Department of Chemical Engineering at the National Institute of Technology Calicut, Kozhikode, India. He received his Ph.D. from McGill University, Canada, where his research in the Materials Modelling Research Group focused on the investigation of the structure of biological fibrous composites and their potential for biomimetic materials engineering. His current research interests include sustainable materials for construction applications, energy-efficient processing, and biomass valorization. He also serves as the Vice Chairperson of the Centre for Sustainable Technologies at NIT Calicut.

Contributors

Umaiyambika Neduvel Annal
Department of Chemical Engineering
Saveetha Engineering College
Chennai, India

Seung-Yeop Baek
School of Applied Chemical Engineering
Kyungpook National University
Daegu, South Korea

Kasturi Bidkar
Department of Biotechnology
National Institute of Technology Warangal
Warangal, India

V. Dhineshkumar
Department of Food Technology
Mahendra Engineering College
Namakkal, India

R. Divya
Department of Biotechnology
National Institute of Technology Warangal
Warangal, India

Praveen Kumar Ghodke
Department of Chemical Engineering
National Institute of Technology Calicut
Kozhikode, India

S. Anuradha Jabasingh
School of Chemical and Bio Engineering
Addis Ababa Institute of Technology
Addis Ababa University
Addis Ababa, Ethiopia

Sowmiya Kadalmani
Biofuels and Bionanomaterial Laboratory
Centre for Biotechnology
A.C.Tech Campus
Anna University
Chennai, India

Habtamu Shebabaw Kassa
School of Chemical and Bio-Engineering
Addis Ababa Institute of Technology
Addis Ababa University
Addis Ababa, Ethiopia

Uma Krishnakumar
Department of Biotechnology
Sahrdaya College of Engineering and Technology
Kodakara, India

J. Aravind Kumar
Institute of Energy and Environmental Engineering
Saveetha School of Engineering
SIMATS
Chennai, India

Poonam Kumari
Bioengineering and Environmental Science Lab
Department of Energy and Environmental Engineering
CSIR-Indian Institute of Chemical Technology (CSIR-IICT)
Hyderabad, India

S. Lalithakumari
School of Electrical and Electronics
Sathyabama Institute of Science and Technology
Chennai, India

Jerold Manuel
Department of Biotechnology
National Institute of Technology Warangal
Warangal, India

Lalehvash Moghaddam
School of Chemistry and Physics
Centre for Agriculture and the Bioeconomy
Queensland University of Technology
Brisbane, Queensland, Australia

Shegaw Ahmed Mohammed
School of Chemical and Bio-Engineering
Addis Ababa Institute of Technology

Addis Ababa University
Addis Ababa, Ethiopia

S. Venkata Mohan
CSIR-National Environmental Engineering Research Institute (CSIR-NEERI)
Nagpur, India

I. Ganesh Moorthy
School of Chemical Engineering
Vellore Institute of Technology
Vellore, India

L. Muruganandam
School of Chemical Engineering
Vellore Institute of Technology
Vellore, India

Yogesh Kumar Murugesan
National Institute of Technology Calicut
Calicut, India

Arunodhaya Natarajan
Department of Chemical Engineering
Agni College of Technology
Chennai, India

Gazliya Nazimudheen
FibscapeAgro Products
Thrissur, India

G. S. Nirmala
School of Chemical Engineering
Vellore Institute of Technology
Vellore, India

Sreeda Pallipad
CSIR-National Metallurgical Laboratory
Jamshedpur, India

Soo-Young Park
Polymeric Nano Materials Laboratory (PNML)
Department of Polymer S&E
Kyungpook National University
Buk-gu, Daegu, South Korea

Nithya V. Pillai
National Institute of Technology Calicut
Kozhikode, India

Akshay Prakash
Department of Biotechnology
National Institute of Technology Warangal
Warangal, India

Aneesh Raj
Department of Chemical Engineering
National Institute of Technology Calicut
Kozhikode, India

Geethalakshmi Ramakrishnan
Research scholar, Biofuels and Bionanomaterial laboratory
Centre for Biotechnology
A.C.Tech Campus
Anna University
Chennai, India

A. Annam Renita
Department of Chemical Engineering
Sathyabama Institute of Science and Technology
Chennai, India

Renganathan Sahadevan
Biofuels and Bionanomaterial laboratory
Centre for Biotechnology
A.C.Tech Campus
Anna University
Chennai, India

S. Sathish
Department of Chemical Engineering
Sathyabama Institute of Science and Technology
Chennai, India

Nimmi C. Sekhar
Department of Chemical Engineering
Government Engineering College Thrissur
Kerala, India

P. Sinu
Department of Electrical Engineering
Indian Institute of Technology
Hyderabad, India

V. Sivasubramanian
National Institute of Technology Calicut
Kozhikode, India

Michael Rahul Soosai
Department of Chemical Engineering
Sathyabama Institute of Science and Technology
Chennai, India

Vishal Bellie Subramani
Centre for Agriculture and the Bioeconomy
School of Chemistry & Physics
Queensland University of Technology
Brisbane, Queensland, Australia

C. Suganya
Department of Pharmaceutical Technology
Mahendra Engineering College
Namakkal, India

Feba C. Varghese
Department of Biotechnology
Sahrdaya College of Engineering and Technology
Kodakara, India

C. Vigneshwaran
Department of Pharmaceutical Technology
Mahendra Engineering College
Namakkal, India

1 Lignin-First Approach to Lignocellulosic Biomass Valorisation

Vishal Bellie Subramani and Lalehvash Moghaddam

1.1 INTRODUCTION

Lignocellulosic biomass (LCB) encompasses the organic matter on Earth that has captured and stored sunlight as chemical energy. It is broadly characterised as organic material of plant origin, whether naturally occurring or cultivated by human intervention. It comprises a range of sources, including forest residues, agricultural wastes, energy crops, algal biomass, sewage sludge and various other organic materials. These are complex materials and are an abundant source of organic carbon containing three highly functional biopolymers, namely cellulose (35–50%), hemicellulose (25–30%) and lignin (15–30%) [1–3]. Moreover, biomass also contains low amounts of proteins, ash, lipids and other lipophilic compounds [4, 5]. The extent of each component varies with the type of biomass, i.e., plant species, tissue type and cell wall development [6]. An estimated 181.5 billion tons of LCB are produced globally every year, and its disposal causes significant concern. It is a vital renewable feedstock for modern biorefineries owing to its cost-effectiveness, widespread availability and carbon neutrality, which can facilitate the progression towards a bio-based economy and is compatible with the existing energy infrastructure [7]. However, converting LCB into biofuels or value-added products poses significant challenges, primarily stemming from variations in biomass quality, inherent variability and compositional disparities of the resulting bio-oil compared to crude oil.

1.1.1 COMPONENTS OF LIGNOCELLULOSIC BIOMASS

Cellulose is one of the abundant biopolymers comprising linear β-1, 4-glucose chain links. It is commonly regarded as a polysaccharide with the chemical formula $(C_6H_{10}O_5)_X$ [8]. Cellulose can easily be disentangled into various valuable products such as biofuels (bioethanol, biobutanol, biohydrogen and biodiesel) and platform chemicals (i.e. levulinic, formic acids and other derived products). Hemicellulose is the second major component of plant biomass. It connects the cellulose fibres and the lignin and gives the cellulose-hemicellulose-lignin network rigidity [9]. Hemicellulose is a heterogeneous polymer composed of C5 and C6 polysaccharides. It is built up of short chains with important applications for producing biofuels and alternatively for generating valuable chemical intermediates like furfural. Lignin is a heteropolymer of phenolic alcohols that shields the polysaccharide microstructure from external physical, chemical and biological attacks. These biopolymers are organised into a complex and highly recalcitrant lignocellulosic structure.

Lignin accounts for almost 30% of organic carbon content occurring in the biosphere. It is an alkyl aromatic polymer and is biosynthesised through enzymatic dehydrogenation of three phenylpropanoid monolignols: p-coumaryl, coniferyl and sinapyl alcohols. They differ based on the number of methoxy groups attached to an aromatic ring [10, 11]. Lignin represents a highly branched structure that contains various functional groups like hydroxyl (OH), methoxyl, carbonyl (C=O) and

DOI: 10.1201/9781003732907-1

carboxyl (COOH) moieties [12]. It is presently the second most rich renewable natural resource next to cellulose, and thus, extracting value-added products from lignin plays an essential role in the economics of biorefineries. Therefore, before converting lignin into valuable chemicals, the detailed structure of lignin and its properties should be understood. The chemical structure of lignin is strongly related to the plant species and the extraction process. Recent advancements in analytical techniques assist in understanding and analysing the chemical bonds. Recent studies [13] have explained that lignin structure can also be altered due to the biosynthesis pathway's presence of various aromatic rings and side chains.

1.1.2 Structure of Lignin

Lignin contains a complex mixture of high molecular weight aromatic polymers that are highly insoluble and the most underutilised fraction of the lignocellulosic biomass [14, 15]. Lignin in the plant structure helps in water transport due to its hydrophobic nature, provides water impermeability functions to the cell wall, and imparts stiffness for external resistance against physical forces like wind and biological attacks. The lignin structures of lignocellulosic biomass can be broadly categorised based on their botanical origin, viz., hardwood (angiosperm/dicot) lignin, softwood (gymnosperm) lignin and herbaceous (monocot) lignin. It is understood that the molecular weight, composition and amount of lignin vary among plant species. However, the lignin abundance substantially decreases in the order of softwoods (25–31%) > hardwoods (16–24%) > herbaceous crops (16–21%). Although advanced Nuclear Magnetic Resonance (NMR) techniques are available, previous investigations [5] have proven that it is challenging to identify and quantify various structures and linkages in lignin due to the complex nature of the lignin structure. Hardwood lignin is an ideal feedstock for producing high bio-aromatic monomers owing to its high proportion of syringyl (S) units and β-O-4 linkages [10, 16, 17]. Softwood lignin usually constitutes approximately 90% of coniferyl alcohols, whereas hardwood lignin usually contains 25–50% coniferyl alcohols and around 50–75% sinapyl alcohol. The lignin from grass typically includes all three monomer alcohols. Lignin structure is formed through several different linkages between the hydroxy (-OH) and methoxy (-OCH_3) substituted phenylpropane monolignol units [18–20]. However, the lignin matrix can be created by these moieties through different linkages (aryl- or alkyl-ether, ester and C-C) with various functional groups (-OH, -OCH_3, C=O, -C -OH, propyl moiety, etc.), resulting in highly tangled lignin structures [21]. These monomers are interconnected by several inter-unit bonds, including many types of ethers (e.g. β-O-4, α-O-4 and 4-O-5) and C-C linkages [22] as presented in Figure 1.1. These monolignols are integrated into the lignin structure in the form of phenyl propanoids, namely p-hydroxyphenyl (H), guaiacyl (G) and syringyl (S) units. Good knowledge of the structure, chemical composition and characteristics of cellulose, hemicellulose and lignin is essential (Table 1.1) to understand their interrelationships, which will help to depict the chemical reaction synthesis pathway derived materials.

1.1.3 Native and Technical Lignin

Lignin can be classified as native lignin and technical lignin. Native lignin as a natural polymer refers to the original lignin structure in the lignocellulose without any modifications. On the contrary, technical lignin can be further classified into two main types based on the fractionation methods. Lignin extracted from biomass such as kraft lignin (KL), organosolv lignin (OL), pyrolytic lignin (PL), sulphite lignin/lignosulphate (SL), etc., is referred to as extracted lignin, while the lignin obtained from residues after the removal of carbohydrates via hydrolysis is hydrolytic lignin (HL). The composition and molecular weight of these technical lignins vary depending on their sources and extraction methods [26]. Lignin depolymerisation is often targeted towards the cleavage of β-O-4 (also called β-ether bonds) inter-unit linkages in native lignin, and the extent of monomer yields strongly depends on the number of β-O-4 linkages in the lignin used for depolymerisation [27]. Lignin model

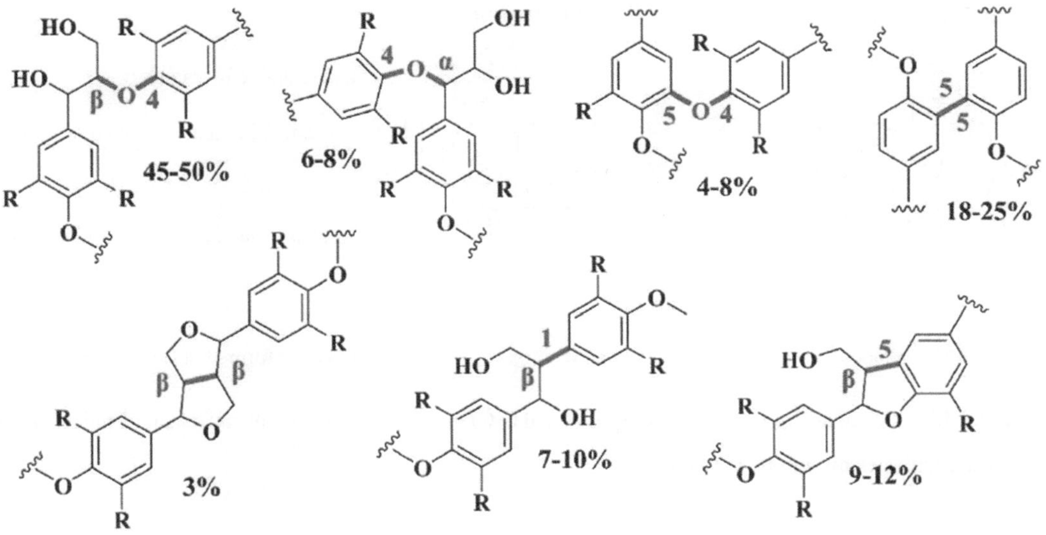

FIGURE 1.1 The structural representation of the common lignin linkages and their abundance in lignocellulosic biomass.

(Adapted with permission from [23]).

TABLE 1.1
Structure and Chemical Composition of Cellulose, Hemicellulose and Lignin [24, 25]

Component	Lignin	Hemicellulose	Cellulose
Monomers	**1,2,3**[a]	D-xylose, mannose, L-arabinose, galactose, glucuronic acid	D-pyran glucose monomers
Bonds between the monomers	various C-O and C-C bonds, various ether bonds, mainly β-O-4 ether bond	β-1,4-glycosidic bonds in main chains; β-1,2-, β -1,3-, β-1,6-glycosidic bonds in side chains	β-1,4-glycosidic bonds
Degree of polymerisation	4,000	<200	500–10,000
Types of polymers	**G, GS, GSH**[b]	polyxylose, Gal-Glu-Man[c], Glu-Man[d]	β-Glucan
Composition	amorphous, inhomogeneous, nonlinear three-dimensional polymer	three-dimensional inhomogeneous molecular with a small crystalline region	three-dimensional linear molecular composed of the crystalline region and the amorphous region
Dominant bonding type between main components	covalent with hemicellulose	covalent with lignin	hydrogen bonds with other components

[a] **1**: p-coumaryl alcohol, **2**: coniferyl alcohol, **3**: sinapyl alcohol.

[b] **H**: p-hydroxyl phenyl, **G**: guaiacyl, **S**: syringyl.

[c] Gal-glu-Man: galactoglucomannan.

[d] Glu-Man: glucomannan.

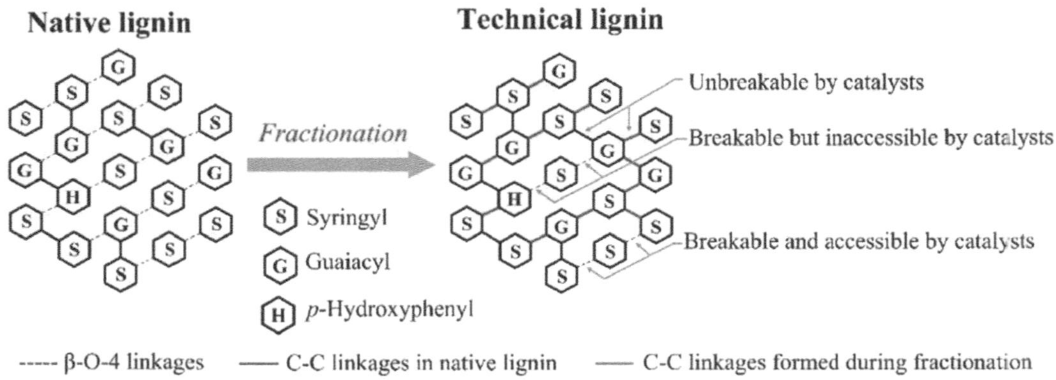

FIGURE 1.2 Schematic representation of (β-O-4 and C-C) linkage distributions in native and technical lignin. **(Adapted with permission from [27]).**

compounds featuring β-O-4 linkages can be effectively depolymerised with noble or non-noble catalysts, e.g., RANEY® Ni, Ni/C, Ru/C, Pd/C, etc., at mild temperatures as low as 120°C to produce monomers in high yields. However, poor yields (<10%) were obtained when technical lignins were depolymerised with similar conditions and even at elevated temperatures [28]. Henceforth, it infers that low monomer yields are due to the low quality of technical lignin rather than the catalyst activities and reaction temperatures.

During biosynthesis and chemical fractionation, lignin undergoes a sequence of chemical reactions, namely polymerisation (coupling) [29], depolymerisation (cleavage) and repolymerisation (condensation) [30]. These reactions significantly affect monomer yields as they cause varied β-O-4 and C-C contents in native and technical lignins [31]. The extent of monomer yield is attributed to four reasons (Figure 1.2). Firstly, lignin is formed by the oxidative coupling of the three major monolignols: p-coumaryl alcohol, coniferyl alcohol and sinapyl alcohol. The coupling reaction forms both β-O-4 and C-C inter-unit linkages. While the β-O-4 linkages can be cleaved by hydrolysis (catalysed by acid or alkali), oxidation methods, hydrogenolysis employing metal catalysts, cleaving of the robust C-C linkages are cumbersome and rarely reported. Secondly, using an acid or alkali catalyst increases the formation of additional robust C-C inter-unit linkages, further reducing the monomers from the technical lignin. Thirdly, a condensed lignin structure can trap the β-O-4 linkages so that they cannot be accessed by heterogenous catalysts. Lastly, the mass transfer between the lignin and the catalyst can be influenced by the poor solubility of the solvent and its processing efficiency.

1.2 TECHNICAL LIGNIN DEPOLYMERISATION TO AROMATIC MONOMERS

Depolymerisation of lignin into aromatic monomers is challenging owing to its intricate structure. Nevertheless, extensive research has been conducted on the depolymerisation of technical lignin by many researchers primarily under four specific depolymerisation reaction mechanisms: (i) acid and base-catalysed, (ii) reductive, (iii) oxidative and (iv) thermal depolymerisation [23, 32–34]. Often, depolymerisation occurs through cleavage of inter-unit linkages, modifying the lignin structure. However, these investigations offer valuable insights into the reaction mechanisms, thereby aiding in developing appropriate methodologies for tailored products.

A standard lignin depolymerisation strategy includes a base-catalysed mechanism using soluble NaOH, KOH, K_2CO_3, or solid MgO, CaO, etc., to cleave lignin monomer linkages and other functional groups attached to the aromatic rings. This fragmentation generates phenolic monomers, while repolymerisation between the reactive species occurs to form the oligomer fraction [35].

The most commonly used homogeneous base catalyst is aqueous NaOH at high temperatures between 240 and 340°C. Moreover, the isolation of lignin monomers is greatly facilitated in basic solution and proportional to the base concentration [36]. As an example, aromatic products like syringol, syringyl aldehyde, guaiacol, vanillin, o-methoxy catechol, etc., are the main compounds obtained at 300°C and 250 bar pressure at 4 min residence time when 2% (w/w) NaOH was reacted with 5% (w/w) starting lignin material [37]. On the other hand, the acid-catalysed mechanism using acids like HCl, H_2SO_4, etc., has been extensively studied with lignin model compounds, particularly abundant with β-O-4 linkages. Initially, dehydration of the hydroxyl groups at the C_α-position occurs, forming quinone methide intermediates on the aliphatic chain side. Subsequently, the β-aryl ether bond undergoes reductive cleavage through an allylic rearrangement, forming intermediate compounds that are then hydrogenated to yield monomeric ketone and phenol structures [38]. The oligomers are generally formed through cleavage of preferred sites (typically ether bonds influenced by acid or base hydrolysis), and then the oligomers are further converted into respective monomers. Due to high operating temperature and pressure, the formation of by-products is inevitable, making the final mixture challenging to purify. To overcome this issue, lignin depolymerisation may be carried out under milder conditions using acid catalysts [13]. However, both base- and acid-catalysed lignin depolymerisation encounter challenges with repolymerisation of product intermediates, resulting in low monomer yields.

In the process of reductive depolymerisation, lignin is hydrocracked in the presence of redox catalysts and H_2, which can be supplied from an external H_2 source or generated by hydrogen-donating solvents (*in-situ* hydrogen source) such as tetralin, isopropanol and formic acid. Native lignin (softwood lignin) depolymerisation under mild conditions of 34 bar and 195°C using different catalysts gave a high conversion yield (78 wt%) of monomeric products such as 4-propylguaiacol and 3-(4-guaiacol) propanol [39, 40]. Contrary to the above-mentioned mild conditions, most of the lignin disassembles at harsher conditions (temperatures of 320–450°C and pressure of 35 bar) without the need for a solvent, which indicates that the reaction occurs between the lignin substrate and the solid catalyst.

The most common catalysts used for hydrodeoxygenation are sulphided CoMo and NiMo; however, high selectivity can be achieved under low severity conditions (20 bar and 150°C) using a noble metal in conjunction with a base metal catalyst (e.g., PtRh, NiRe, PtSn, ZnP) [41–44]. By employing bi-functional catalysts with acid and metal sites, lignin can be converted into cycloalkanes. Heterogeneous catalysts, particularly transition metals supported on various substrates, highly favoured lignin conversion to monomers, proving their ability to break C-O and/or C-C linkages. Previous study [45] showed that technical lignins were selectively degraded to alkylated phenols over Pt/C, Pd/C, Ru/C and Ni/C in supercritical *t*-BuOH increased lignin-oil and less char formation. Zakzeski et al. employed Pt/Al_2O_3, Pd/C and Ru/C catalysts in ethanol/water mixtures for the Kraft lignin depolymerisation process and achieved a maximum yield of 17.6 wt% of monomeric phenols [1]. In another study, Song et al. [17] reported that Ni/C in alcohol solvent exhibited greater efficiency in the formation of monomeric phenols with high selectivity >90% and lignin conversion of 50% compared to other transition metals such as Pd, Ru, Pt, etc. Lignin hydrogenolysis over bimetallic catalysts (NiRu, NiRh and NiPd) at 100°C, at 1 bar H_2 pressure, led to the formation of phenol derivatives, however, with yields lower than 10% [46]. Furthermore, processes utilising homogeneous catalysts necessitate intricate separation procedures.

The oxidative lignin cracking process interests researchers due to the many hydroxyl groups in lignin. During the oxidative degradation of lignin, aromatic aldehydes and carboxylic acids are formed as the main products [47]. The main oxidising agents used are metal oxides such as chromium (VI) oxides, Dess–Martin periodinane and hydrogen peroxide. During the oxidative deconstruction of lignin, vanillin is obtained as a major product with yields of about 5–15 wt% depending on the lignin source. Most of the procedures and studies involve alkali or mineral acids combined with hydrogen peroxide; however, such processes may not be ideal as they produce radicals that cause partial repolymerisation and formation of complex lignin structures [47].

Thermal depolymerisation through fast pyrolysis of lignin can produce various volatile compounds like benzene, alkyl-substituted derivatives, alcohols, ketones, aldehydes, monomeric phenols like guaiacol, syringol, catechol and other polysubstituted phenols [48]. However, product distribution from pyrolysis of lignin is highly complex and is influenced by several factors like feedstock type, reaction temperature, heating rate, etc. [49] and increases the energy costs of the process. However, the formation of a complex reaction mixture hinders all the above processes, either demanding harsh reaction conditions, yielding a wide range of products, or resulting in very low aromatic yields. Collectively, these factors pose challenges to economic viability and the successful commercialisation of the process. To further enhance lignin conversion and product yield, researchers in this field are actively exploring alternative lignin depolymerisation approaches and strategies.

1.3 PARADIGM SHIFT TOWARDS EFFECTIVE UTILISATION OF NATIVE LIGNIN

In the conventional pulping process, lignin is removed prior to the manufacture of pulp and paper. Residual lignin will impact the pulp quality and/or the cost of bleaching. Alternatively, to produce 2G ethanol, delignification is a common pretreatment approach that improves saccharification and fermentation efficiencies by removing the lignin barrier that blocks the cellulase enzyme access to cellulose and reducing the non-productive binding of the enzymes to lignin [50]. During the lignocellulose biomass processing, lignin is modified, and the extent of modification depends on the extent of the severity of the fractionation/pretreatment conditions [51, 52]. As these approaches are "cellulose-first", considerable cleavage is done to the lignin ether bonds (especially the β-O-4 bonds), and recalcitrant C-C bonds are formed [52, 53]. Such lignin modifications lead to ineffective depolymerisation with low aromatic monomer yields and the formation of many phenolic derivatives, even when working under relatively harsh conditions [13, 46, 54]. This necessitates alternative approaches to mitigate the impact of lignin modification on depolymerised products.

1.3.1 Conversion of Lignin to Aromatic Chemicals: Impact of Lignin Structure

The type of lignin feedstocks, whether derived from lignocellulosic materials or obtained as lignin extracts, significantly impacts the lignin valorisation process [55, 56]. The variation in the lignin structure can hinder the yield of aromatic monomers and the selectivity of target products. Hence, structural analysis of lignin is important for advancement in lignin valorisation technology [57].

Generally, pretreatments involving harsh conditions to extract high-purity lignin fractions lead to altered lignin structures. This affects factors such as solubility, average molecular weight and the nature of intra-unit and inter-unit lignin linkages [58, 59]. As a result, the conversion from lignin extracts has resulted in a reduced yield of monomers compared to those achieved through the direct conversion of native lignin [60]. Gaining insights into the formation of unreactive condensed lignin structures during the extraction process and understanding how different lignin structures influence the yield of aromatic monomers when utilising various lignin feedstocks holds significant importance in achieving efficient lignin conversion [61, 62].

Hydrothermal conversion of carbohydrates is extensively studied and has been incorporated into industrial processes to generate organic acids and furan-based chemicals [63]. However, the limited dissolution of lignin in water hinders its viable utilisation in industry. To overcome this limitation, the addition of acids/bases (e.g., H_2SO_4, H_3PO_4, NaOH and KOH) has been used to enhance lignin-water solubility [64, 65]. In addition, organic solvents with low boiling points, such as ethanol, methanol, tetrahydrofuran, etc., which are easy to recover, have been employed as co-solvents for effective delignification in HTL processes [66, 67]. Furthermore, pure organic solvents have been utilised in various studies [68–70] for lignin depolymerisation to produce cycloalkanes and phenols in the presence of H_2 or N_2 atmosphere. In addition, catalysts are employed to obtain chemicals and fuels from lignin and have been well studied [23]. Despite significant efforts to enhance the effective utilisation of lignin through different reaction systems using different acids or bases, solvents, gases

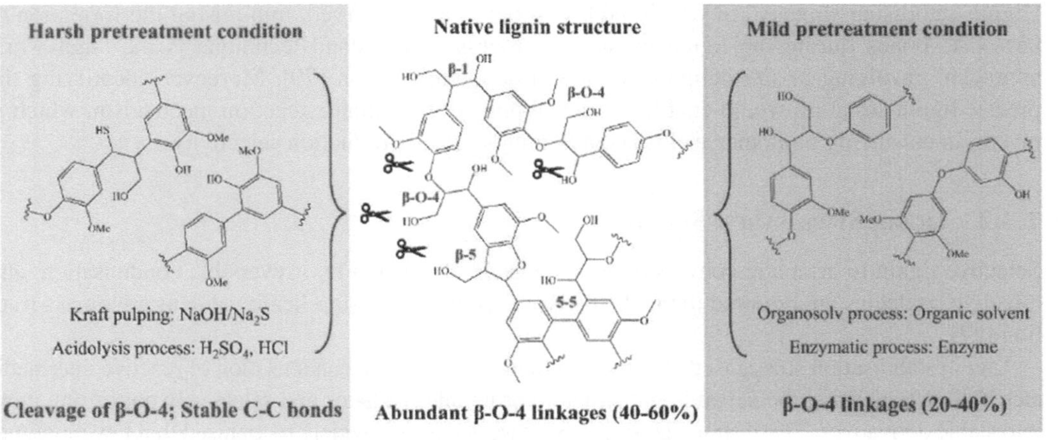

FIGURE 1.3 Representative structure of lignin extracts.

(Reprinted with permission from [57]).

and catalysts, it is imperative to have a thorough understanding of the structure–activity relationship, given the wide range of lignin feedstocks utilised. The structure of the extracted lignin primarily relies on the specific extraction method used in processing the LCB. The representative lignin structure based on specific extraction methods is presented in Figure 1.3.

The kraft pulping technique utilising $NaOH/Na_2S$ or the soda pulping process employing NaOH brings about significant alterations in the structure of lignin (referred to as kraft lignin and soda lignin, respectively). Notably, this involves the cleavage of C–O bonds followed by subsequent repolymerisation, resulting in the formation of highly stable C–C bonds [71]. Other drawbacks include a high concentration of bases and sulphides in the kraft lignin and partial solubilisation of (hemi) cellulose [72] that affects the lignin purity. Employing the acidolysis process can enhance lignin removal and the hemicellulose fraction through hydrolysis of the glycosidic linkages. This process yields a high degree of purified cellulose in the solid residues [65]. In general, under harsh pretreatment conditions, lignin extracts tend to expose an increased number of hydroxyl (OH) groups, promoting water solubility arising from the breakdown of the β–O–4 linkages. The solubilised lignin in the liquid phase becomes more amenable to conversion, but the aromatic monomer yields are restricted due to the irreversible lignin repolymerisation during this process [73, 74].

The use of acids and bases can be minimised by using water-organic solvents at mild conditions to dissolve lignin into the liquid phase [75]. Alcohol as an organic solvent is also preferred due to its low cost and eco-friendliness. The organosolv lignin with mildly altered structure containing 20–40% of β–O–4 linkages can be obtained utilising a slightly acidic water-organic solvent mixture. Subsequently, a remarkable pure lignin fraction, closely resembling its native form, can be extracted while the carbohydrates are hydrolysed in the water phase. Furthermore, the hydrolysis of the cellulose-lignin mixtures can be promoted using cellulolytic enzymes, improving the lignin purity and retaining considerable amounts of β–O–4 linkages [66, 76].

To maintain the native lignin structure, researchers suggest that future biorefineries employ a one-pot process where lignin extraction and conversion happens simultaneously [55]. This process retains 40–60% of β–O–4 linkages, significantly improving the potential for lignin valorisation and inhibiting the repolymerisation of the lignin products. Furthermore, the impact of varying biomass sources with distinct S/G/H ratios on the production of aromatic chemicals should be considered. Hence, it is crucial to comprehend how the lignin structure influences the product yield and conversion efficiency.

Understanding these concepts has led to the development of two relatively new approaches to minimise the effect of lignin modification on depolymerisation products. The first involves the "lignin stabilising strategy" in which the β-O-4 bonds are preserved, minimising the formation of new C-C bonds during the lignin extraction process. The second technique is the "lignin-first approach" involving *in-situ* delignification and depolymerisation [70]. Moreover, identifying the precise lignin structure would enable the development of a suitable reaction mechanism, which is pivotal in enhancing monomer yields and attaining targeted production selectivity.

1.3.2 Lignin Stabilisation Strategies

Selective lignin to aromatic conversion is immensely obstructed by irreversible condensation, and effective strategies are required to tackle the issue, as lignin has significant value as a biomass fraction [27].

Lignin stabilisation strategies distinctively aim to overcome the condensation of reactive intermediates by functionalising the native lignin structure or its intermediates and selectively converting them into stable derivatives. These strategies have been targeted as it is widely recognised that low monomer yields result from lignin condensation occurring during biomass fractionation. Further, the theoretical monomer yield can be related to the degree of condensation, and it is roughly proportional to the square of the fraction of cleavable inter-unit ether (β-O-4) bonds [70]. During biomass fractionation, the cleavage of lignin side chains catalysed by acid or base generates formaldehyde, which induces lignin's repolymerisation. Subsequently, formaldehyde reacts with aromatic rings to form hydroxymethyls [77].

Similarly, α-hydroxyls in lignin also initiate substantial condensation. Therefore, the addition of α-hydroxyls and formaldehyde scavenging reagents such as phenol, guaiacol, anisole and diols can minimise lignin condensation [27]. There are a few proposed reactive reagents that preserve the lignin from condensation by blocking the electron-rich sites on the aromatic ring and the benzylic cation on the side chain, as shown in Figure 1.4. Nevertheless, it is acknowledged that alternative reactive reagents also exist and hold promising potential for lignin depolymerisation. Moreover, lignin isolation techniques that preserve native lignin are recommended to promote depolymerisation with better selectivity and yield. This paves the way towards innovative fractionation techniques that include ammonia-based fractionation [78–81], ionic liquid-assisted fractionation [82–85], γ-valerolactone-assisted hydrolysis [86] and mild organosolv methods [87–89]. The addition of an alternative media enables biomass fractionation to occur under mild processing conditions, reducing the extent of ether bond cleavage and improving selective downstream depolymerisation to low molecular weight chemicals. The different schemes and overviews are represented in Figure 1.5.

Shuai et al. [52] used formaldehyde and HCl to stabilise the lignin by forming acetals with the lignin's α- and γ-OH side chain groups. The pathway for lignin condensation, along with the stabilisation strategy, is outlined in [52]. The recovered lignin was then liquefied in the presence of

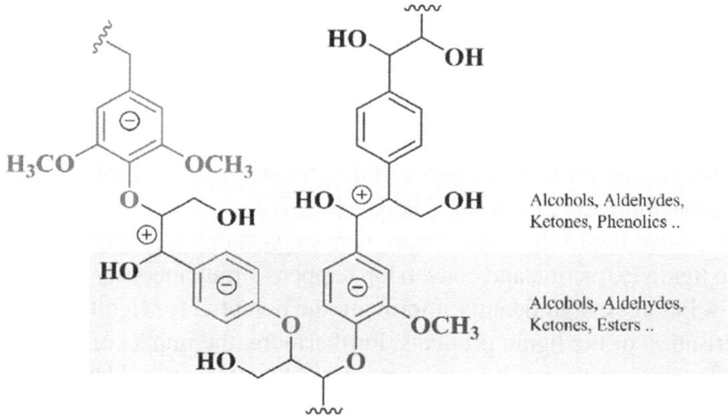

FIGURE 1.4 Schematic of proposed reactive reagents for reducing lignin condensation.

(Adapted with permission from [27]).

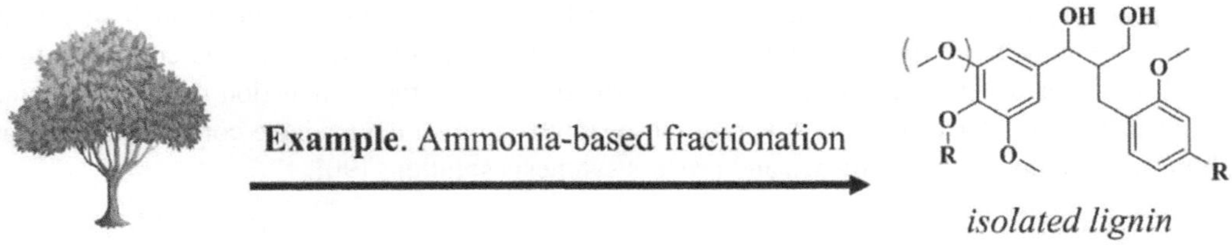

Scheme 1: Passive Preservation – Retention of β-O-4 through mild fractionation

Advantages
+ Minor condensation
+ Decoupling from depolymerisation
+ Practical methodologies

Limitations
- Condensation is not fully excluded
- Limited to mild process conditions
- Trade-off : degradation-lignin yield

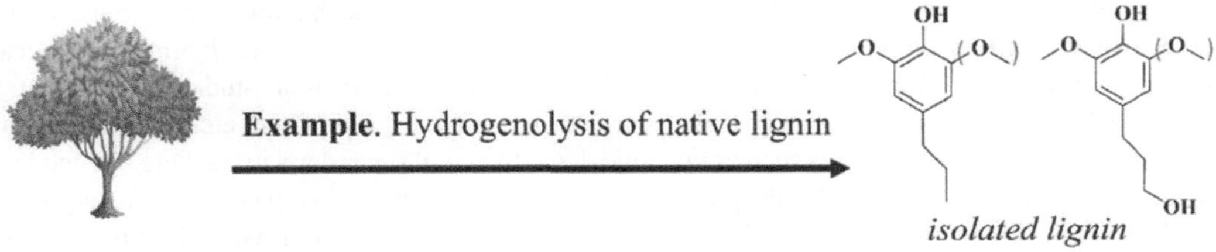

Scheme 2: Active stabilisation, approach 1 – Tandem depolymerisation stabilisation

Advantages
+ Insignificant condensation
+ High monomer yields
+ High selectivity

Limitations
- Compatibility
- Fine-tuning of process conditions
- Catalyst recuperation from pulp

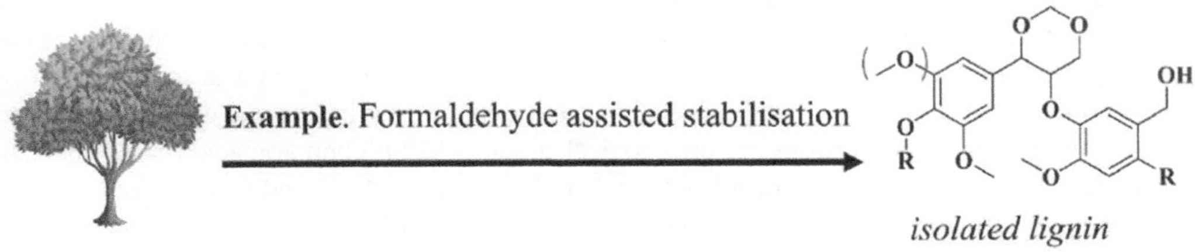

Scheme 3: Active stabilisation, approach 2 – Forced preservation of β-O-4 linkages

Advantages
+ Insignificant condensation
+ Decoupling from depolymerisation
+ High monomer yields

Limitations
- Additional (sacrificial) reagent
- Possible side reactions (e.g. pulp)
- Lower lignin product selectivity

FIGURE 1.5 Overview of (β-O-4) passive preservation and active stabilisation strategies.

(Adapted with permission from [70]).

Ru/C and an external H_2 source at 40 bar to obtain a close-to-theoretical yield of bio-aromatic monomers. The disadvantage of this strategy is that it is a two-step approach and uses a toxic chemical, formaldehyde and an external H_2 source.

Current advances [81] in ammonia-based fractionation of lignocellulosic biomass use liquid ammonia to cleave lignin-carbohydrate complexes and selectively extract lignin (92% lignin) as a major product stream. The (Extractive-Ammonia) EA-derived lignins retained much of lignin's functionality and the *β*-aryl ether linkages. However, the lignin product streams contained up to 4 wt% of nitrogen, which could lead to NO_X formation during the combustion of the compounds produced from the derived lignin. Overall, these stabilisation strategies have contributed to establishing suitable upgrading pathways and process synthesis solutions [90].

1.3.3 Lignin-First Approach

The lignin-first approach has garnered significant attention, mainly focusing on using highly selective catalysts to effectively prevent lignin repolymerisation, optimise the yields of aromatic monomers and achieve complete biomass valorisation [70].

The lignin-first approach, also known as Early-Stage Catalytic Conversion of Lignin (ECCL), is a one-pot process in which lignin is simultaneously extracted from lignocellulosic biomass and catalytically converted into lignin-derived aromatics and carbohydrate pulp.

The ECCL approach is undoubtedly the backbone of emerging technologies that influence the choice of the current biorefinery scheme (Figure 1.6). Valorisation of isolated lignin to chemicals becomes difficult due to the formation of stable C-C bonds during extraction. Studies on liquefaction of whole biomass such as sugarcane bagasse, corn stover, softwood, hardwood, etc., have shown that the complex mixtures of products such as phenolic derivatives and sugar derivatives were obtained at a temperature range of 280–330°C in the presence of acids, bases, water and with or without catalysts [2].

The lignin-first approach (Figure 1.7) involves *in-situ* delignification of native lignin in concert with complete fractionation of cellulose and hemicellulose components to achieve high product yield and selectivity. However, this approach requires an external H_2 source with high initial pressure to promote catalytic hydrogenolysis of lignin and inhibit condensation reactions via the hydrogenation of the reactive intermediates. Subsequently, this reduces the char formation and elevates selective product yields. Furthermore, to produce high bio-aromatic monomer yields using this approach, predominantly, hardwood is chosen as the ideal feedstock rather than softwood or grass lignocellulose due to the higher proportion of syringol (S) unit and β-O-4 linkages [10, 16, 17]. As an alternative to hardwood, grasses such as bagasse, switchgrass and cotton gin trash can be utilised, offering several advantages. It is abundant, conveniently located in centralised areas, and can be easily transported in large quantities. This minimises both harvesting and transportation costs. Furthermore, grasses typically require minimal or no milling or grinding, unlike hardwood. Grasses also contain reasonable proportions of noncanonical subunits like *p*-coumaric acid (*p*CA) and ferulic acid (FA), along with a mixture of *p*-hydroxyphenyl (H), guaiacol (G) and S units; however, they have a complex structure compared to hardwood [23]. Consequently, sustainable reaction pathways need to be targeted to simultaneously capture these advantages and overcome the challenges to enhance the monomer yield from grasses. The bio-aromatic esters and monomers produced from the

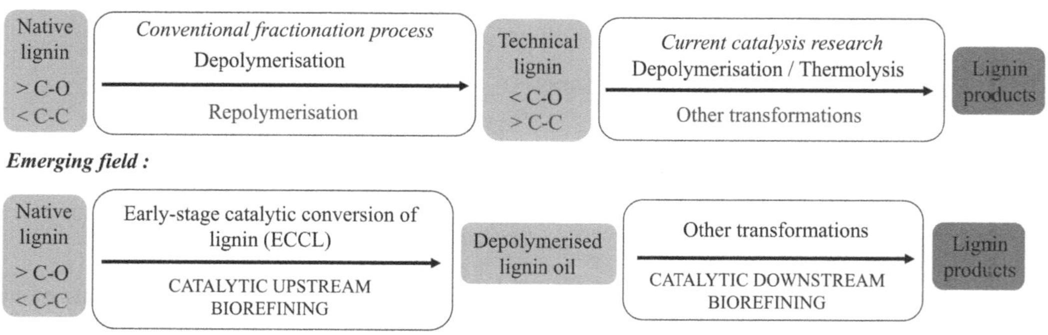

FIGURE 1.6 A comparison of lignin valorisation between conventional fractionation processes and emerging catalytic upstream biorefining processes.

(Adapted with permission from [2]).

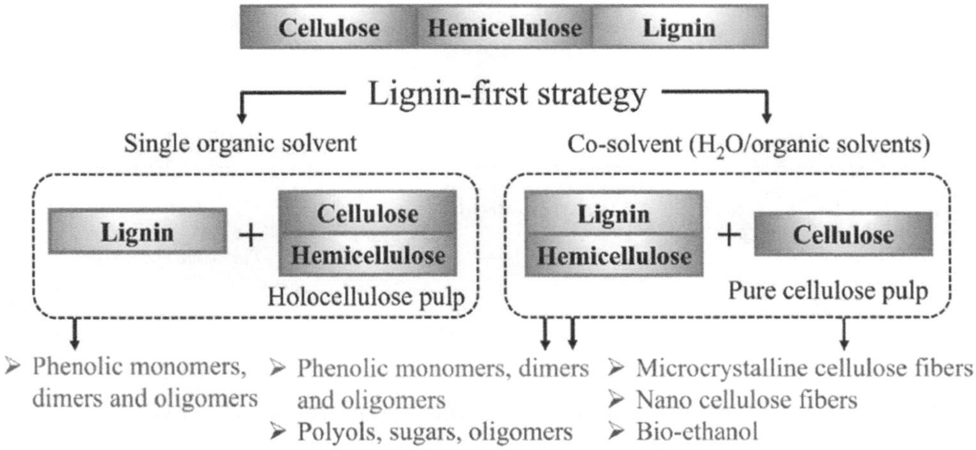

FIGURE 1.7 Schematic representation of "lignin-first" biorefinery approaches.

(Adapted with permission from [80]).

lignin noncanonical units and monolignols can be used as macromolecular building blocks in manufacturing biopolymers [91, 92]. This could open up new avenues for commercialisation in future biorefineries centred around grass lignocellulose.

Studies [93–97] indicate that the extensive decomposition of the propyl side chain can be prevented by varying the conditions of Catalytic Upstream Biorefining (CUB) to give ortho-(di) methoxy-4-propylphenol derivatives. Therefore, CUB provides a highly aromatic lignin-derived stream by minimising condensed lignin structures when they are extracted from the lignocellulosic matrix. In this process, the most reactive functional groups are rendered less reactive through catalytic reduction. For instance, aldehydic intermediates are converted into alcohols, and ketones are subjected to hydrodeoxygenation to form methylene groups. Li et al. [16] reported one-pot catalytic hydrocracking of raw woody biomass (birch) into monophenols (from lignin) and diols (from carbohydrate) at 235°C using Ni-W_2C/AC catalyst. The yield of total monophenols from the birch reached 46.5 wt% (based on the biomass lignin content). Although this study used a low-cost catalyst Ni-W_2C/AC, a high initial H_2 pressure (6 MPa) was required to achieve a high product yield. In the depolymerisation of bagasse and pinewood using identical reaction conditions, the total phenolic yields decreased to 23.4% and 10.1%, respectively. This highlights that achieving a high phenolic yield is contingent on both the hardwood species employed and the pressure of H_2 utilised. In another study [17], the depolymerisation of birch native lignin with Ni loaded on carbon and Al_2O_3 in pure alcohol as a solvent successfully cleaved birch native lignin propyl-guaiacol and propyl-syringol with more than 90% of total selectivity and 50 wt% lignin conversion.

Parsell et al. [95] reported that a bimetallic Zn/Pd/C catalyst converted native lignin in intact wild-type poplar (hardwood) directly to two methoxyphenols, propyl-guaiacol and propyl-syringol, in the solvent methanol, leaving behind the carbohydrates as a solid residue (Figure 1.8). The carbohydrate residue was subjected to enzymatic hydrolysis using cellulases, giving 95 wt% of the theoretical glucose yield. The results validated that H_2 (with approximately 3.4 MPa of initial pressure) was required to achieve up to 54% of phenolic yield. In other studies [10, 98], the same reductive products (4-n-propyl-guaiacol and 4-n-propyl syringol) were formed with high yields using Ru/C and Ni/Al_2O_3 catalysts and high-pressure H_2 (3 MPa) for delignification of raw birch sawdust, achieving over 90% of delignification with ~50% of phenolic monomers and 20% of phenolic dimers.

In a previous study [99], a novel one-pot catalytic oxidation-hydrogenation process for the valorisation of sugarcane bagasse was investigated, as shown in Figure 1.9. Herein, sugarcane bagasse was subjected to complete depolymerisation using Pd/C in atmospheric pressure O_2 and methanol as solvent. At optimised conditions of 250°C and 8 h, 31.7 wt% of ferulic acid (FA) and p-coumaric

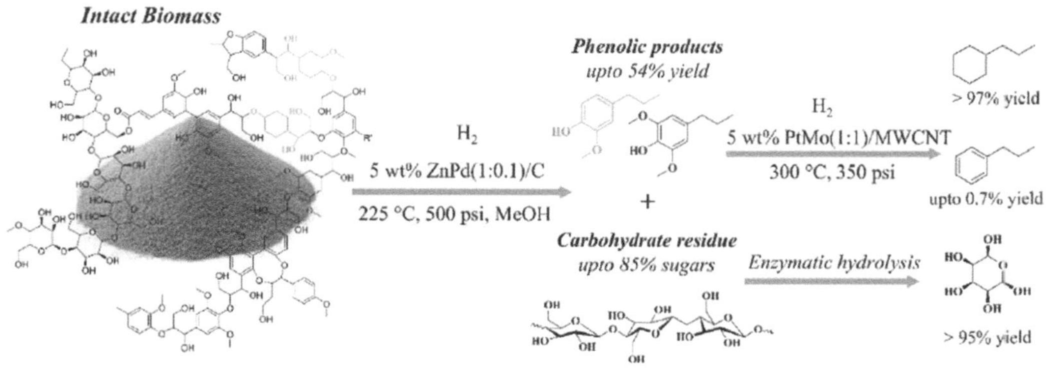

FIGURE 1.8 Scheme of lignin-first approach for hydrodeoxygenation of wood biomass.

(Adapted with permission from [95]).

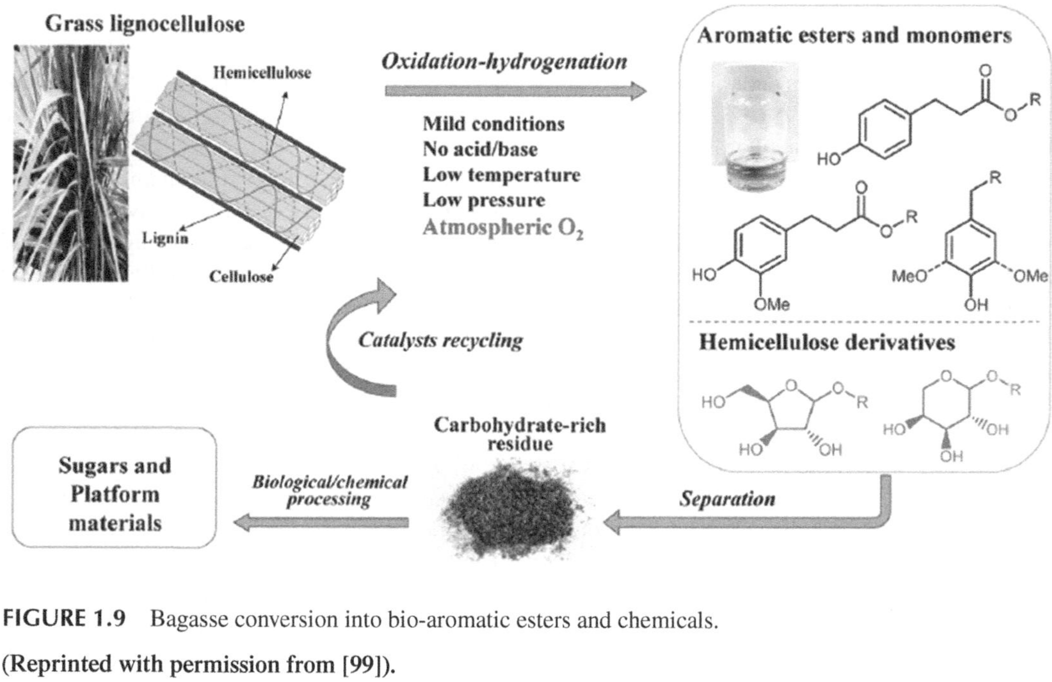

FIGURE 1.9 Bagasse conversion into bio-aromatic esters and chemicals.

(Reprinted with permission from [99]).

acid (pCA) derived aromatic esters and 25.8 wt% aromatic monomers were obtained. Conversely, this study revealed that the extent of bio-aromatics yields, and selectivity depends on the concentration of lignin noncanonical subunits, the proportion of β-O-4 linkages, and G, H and S units. Additionally, the carbohydrate-rich pulp obtained after the catalytic depolymerisation was converted to glucose (65% yield) through enzymatic hydrolysis. However, this required extended processing times of close to 72 h. This study showed that high bio-aromatics yield could be achieved without employing either an acid or base treatment or elevated H_2 gas pressures, potentially reducing operational costs. Despite these advantages, significant safety concern is associated with this process on a pilot scale. The use of methanol solvent along with Pd/C catalyst can lead to the unintended formation of H_2, even at room temperature. This hydrogen generation triggers a highly exothermic reaction when combined with air (or O_2). This reaction can potentially result in the ignition of the solvent, leading to a catastrophic scenario. Therefore, modifications to the process are necessary to address this safety implication.

Recently [100], lignin in cotton gin trash (CGT), a waste product of cotton processing, was depolymerised using argon gas and ethanol solvent over Pd/C. The results show that 50% of the bio-aromatics is *trans*-isoeugenol, a precursor to produce vanillin. The fractionated cellulose-rich portion of CGT was converted to 39 mol% (80% purity) of 5-chloromethylfurfural (CMF), and a solid fuel which has a higher heating value of 20 MJ kg^{-1}. The liquid extract, removed prior to CGT processing, was hydrogenated to produce C_{13}–C_{17} liquid hydrocarbons, and the solid extract rich in nutrients is suitable for crop and plant growth. The study reported 23% aromatic monomer yield when 300 mg of extractive-free CGT was subjected to hydrogenation with 30 mg of 10 wt% Pd/C catalyst and 10 mL of ethanol, bubbled with argon at atmospheric pressure at a reaction temperature of 250°C for 7 hours of reaction time. The study showed that successive processing of cotton gin trash with water/ethanol extraction, followed by "lignin-first" biomass depolymerisation, and then hydrolysis of the carbohydrate-rich solid residue to produce CMF and hydrochar could be a smart choice since these resources could be sufficiently utilised. A CGT-waste valorisation approach was proposed (Figure 1.10) and the main products are CMF, bio-aromatics, liquid hydrocarbons, and solid fuels. However, CGT impurities were shown to affect the reusability of the Pd/C catalyst and hinder its performance.

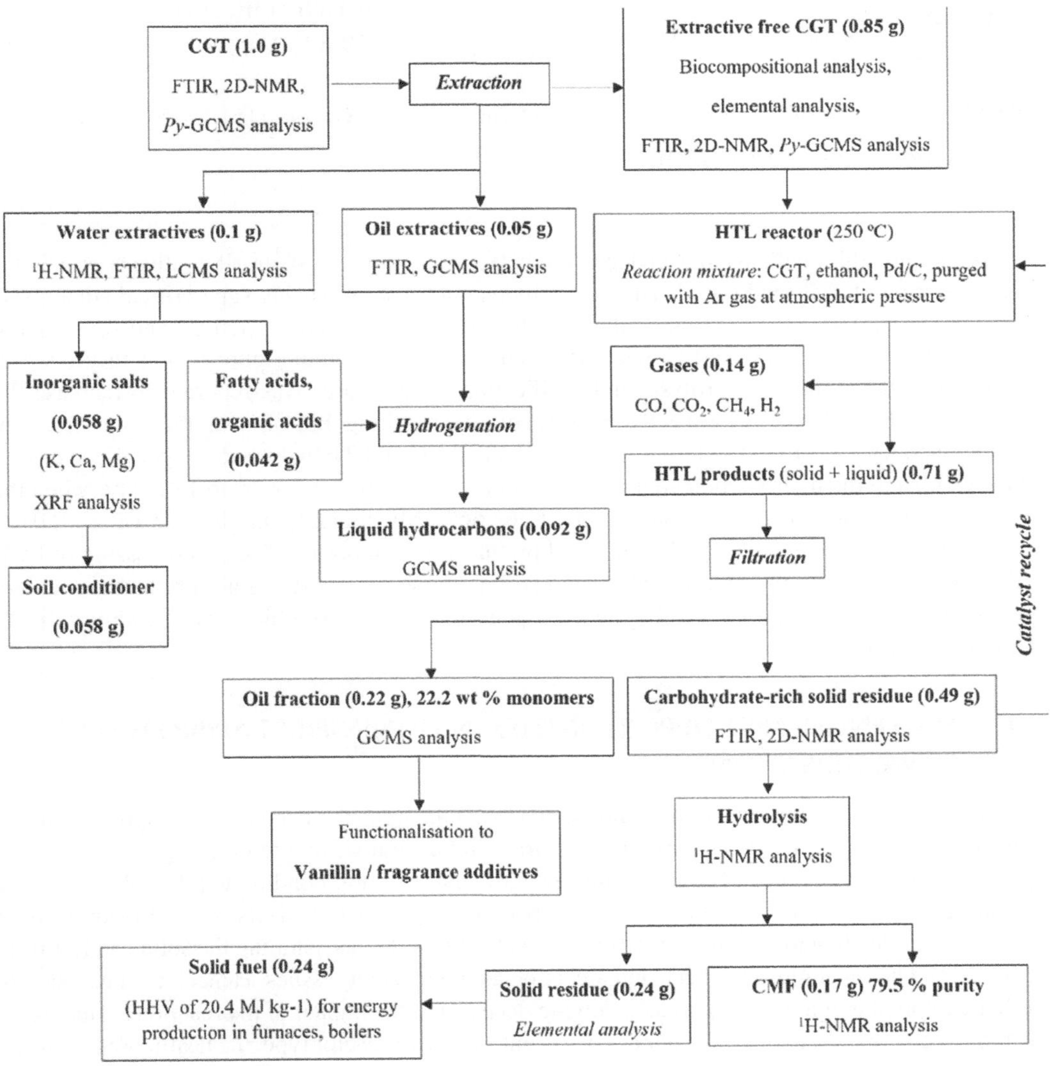

FIGURE 1.10 Flowsheet of the proposed CGT-waste valorisation approach.

(Adapted with permission from [100]).

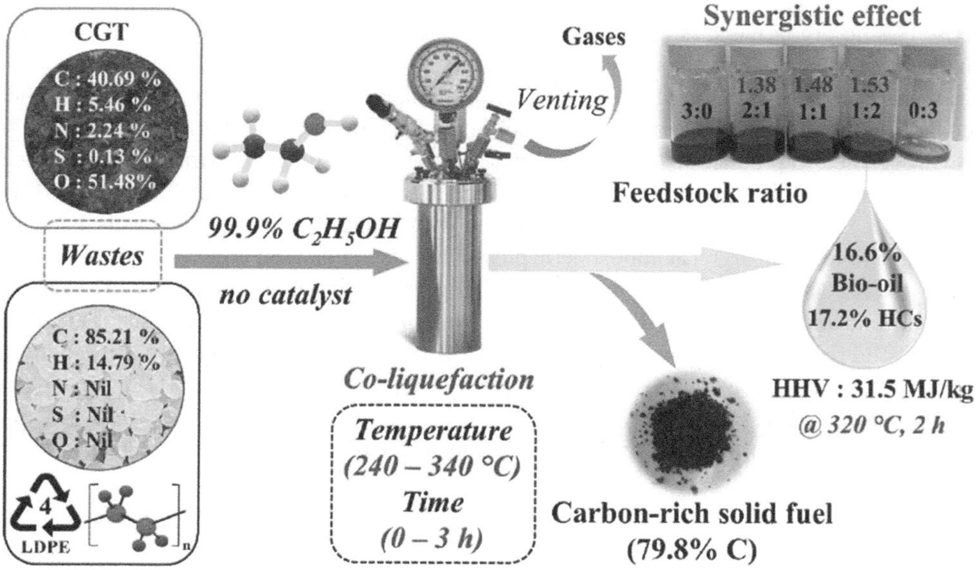

FIGURE 1.11 Supercritical ethanolysis for co-liquefaction of LCB (CGT) and plastic waste (LDPE). **(Adopted with permission from [101]).**

To overcome this challenge posed due to the heterogeneous catalyst, an alternative approach for co-liquefaction of CGT and low-density polyethylene in a 1:2 ratio using supercritical ethanolysis was attempted [101]. This process produced better quality hydrocarbon-rich oil and solid fuel at reaction temperatures below 340°C in less than 3 hours, as depicted in Figure 1.11. This process at temperatures below that of pyrolysis and gasification temperatures was reported to increase the yield of the bio-oil and improve the quality of the resulting bio-oil and the solid fraction. Co-liquefaction of 1:2 ratio CGT/LDPE feedstock at 320°C and 2 h had positive synergistic effects on oil yield, and the hydrocarbon content in the oil increased from 3.7 wt% to 17.2 wt% when the LDPE proportion was increased from 33.33% to 66.66%. Maximum higher heating values (HHV) of 31.5 MJ/kg for oil and 42.5 MJ/kg for solid products were reported. The co-processing of LCB and plastic wastes through supercritical ethanol co-liquefaction is a sustainable approach to improve fuel quality, facilitate plastic recycling and compensate for decreased biomass availability due to seasonal variations.

1.4 CHALLENGES AND OPPORTUNITIES IN LIGNIN-FIRST APPROACH TO LCB VALORISATION

Despite remarkable progress in lignin valorisation using the lignin-first approach or lignin-stabilisation strategies, very few studies have targeted grass lignocellulose due to challenges such as poor lignin monomer yields, use of toxic chemicals and harsh reaction conditions [99]. Moreover, the existing conversion strategies require high temperature and metal catalysts for hydrogenolysis of ether linkages, leading to low functionalised products, difficulty in separating the solid catalyst from cellulose and hemicellulose fraction, and are associated with safety issues. Hence, it is necessary to further explore the lignin-first approach, delving deeper into the impact of the reaction system (e.g., the type of external gases, use of water as co-solvent, choice of reactor type, the heating rate, catalyst recyclability, etc.) on lignin polymerisation. Such investigations are crucial for achieving significant yields of monoaromatic products, along with generating high-quality delignified pulps for producing value-added products. These advancements would pave the way for further progress in this field, using the entire lignocellulose for large-scale demonstration.

In general, biomass energy plays a pivotal role in the prospective energy landscape, and a comprehensive knowledge of biomass properties is indispensable. However, biomass materials have diverse properties and the design of biomass to energy conversion systems heavily depends on the specific biomass material, its morphology, moisture content and the extent of contaminants. One such promising, sustainable biomass valorisation technology that can incorporate most of these challenges is hydrothermal liquefaction (HTL). In HTL, liquid-phase catalysts enable additional degrees of flexibility for biomass conversion. The water medium (H_2O) actively participates in the reaction over the reduced catalysts and facilitates the hydrogenation of C-O bonds, forming a hydrogen-bonded water structure. However, large-scale applications are hampered due to the limited dissolution of biomass in pure water. Recent findings suggest the potential of utilising organic solvents in improving the selectivity for tailored products through metal-catalysed reactions. Moreover, hydrogenolysis can prevent condensation reactions of the reactive lignin intermediates, maximising the monomer yield. Thus, the use of water/organic solvent mixtures enables both isolation and transformation of lignin, leading to high yields of aromatic monomers. Yet, the HTL process encounters problems associated with lignin product repolymerisation at elevated reaction temperatures, even though catalytic activity improves at higher temperatures. Hence, mild reaction temperatures and a minimum heating period are essential to prevent side reactions and minimise energy consumption and heat recovery losses.

Yet another obstacle in this process lies in developing catalysts capable of selectively converting lignin into specific target products. On an industrial scale, the durability and potential for catalyst reuse play pivotal roles in ensuring the economic viability of HTL of lignin. Metal-support catalysts are susceptible to leaching of their active metal sites when exposed to water. However, this issue can be alleviated by enhancing the interactions between the metal and support. All the above aspects, including the composition of feedstocks and its lignin architecture (intra/inter-unit linkages, S/G/H ratio), need to be considered while targeting a sustainable biorefinery process. While the noncatalytic co-liquefaction of LCB with plastic waste shows promising results, producing better quality hydrocarbon oil, future work should be focussed on reducing the solid fractions by striking a balance between oil yield and fuel quality using a mix of different plastic wastes, solvent mixtures and higher temperatures. Improvements are still required, and research should be carried out to further enhance the overall biorefinery process. Future research activities are poised to explore the development of catalysts optimised for CGT processing, focusing on improving their longevity and recovery. Furthermore, deeper insights into the reaction kinetics should be sought to elucidate the liquefaction process mechanism and facilitate enhanced reactor designs.

1.5 CONCLUSION

The conventional approach to lignocellulosic biomass valorisation typically prioritises the breakdown and utilisation of cellulose and hemicellulose to produce biofuels, chemicals and materials. Lignin, which is more complex and recalcitrant, is often treated as a low-value by-product or burned for energy. In the lignin-first approach, lignin is depolymerised into valuable aromatic chemicals through solvolysis and catalytic hydrodeoxygenation before dealing with the cellulose and hemicellulose fractions. This approach produces high-value chemicals from lignin, improving the overall economic viability of the process and potentially enhancing the quality of the remaining cellulose for further conversion into biofuels or chemicals. By obtaining multiple value-added products and effectively utilising various fractions of lignocellulosic biomass, this approach improves the economic feasibility of lignocellulosic biomass valorisation.

REFERENCES

[1] V. K. Ponnusamy et al., "A review on lignin structure, pretreatments, fermentation reactions and biorefinery potential," *Bioresour. Technol.*, vol. 271, pp. 462–472, 2019, doi: 10.1016/j.biortech.2018.09.070

[2] R. Rinaldi et al., "Paving the way for lignin valorisation: recent advances in bioengineering, biorefining and catalysis," *Angew. Chemie Int. Ed.*, vol. 55, no. 29, pp. 8164–8215, July 2016, doi: 10.1002/anie.201510351

[3] R. G. Saratale et al., "Pretreatment of kenaf (Hibiscus cannabinus L.) biomass feedstock for polyhydroxybutyrate (PHB) production and characterization," *Bioresour. Technol.*, vol. 282, December, pp. 75–80, 2019, doi: 10.1016/j.biortech.2019.02.083

[4] H. V. Lee, S. B. A. Hamid, and S. K. Zain, "Conversion of lignocellulosic biomass to nanocellulose: structure and chemical process," *Sci. World J.*, vol. 2014, pp. 1–20, 2014, doi: 10.1155/2014/631013

[5] J. Zakzeski, P. C. A. Bruijnincx, A. L. Jongerius, and B. M. Weckhuysen, "The catalytic valorisation of lignin for the production of renewable chemicals," *Chem. Rev.*, vol. 110, no. 6, pp. 3552–3599, 2010, doi: 10.1021/cr900354u

[6] A. Barakat, H. de Vries, and X. Rouau, "Dry fractionation process as an important step in current and future lignocellulose biorefineries: A review," *Bioresour. Technol.*, vol. 134, pp. 362–373, April 2013, doi: 10.1016/j.biortech.2013.01.169

[7] J. R. Regalbuto, "Cellulosic biofuels—got gasoline?," *Science*, vol. 325, no. 5942, pp. 822–824, August 2009, doi: 10.1126/science.1174581

[8] M. Balat, "Production of bioethanol from lignocellulosic materials via the biochemical pathway: A review," *Energy Convers. Manag.*, vol. 52, no. 2, pp. 858–875, February 2011, doi: 10.1016/j.enconman.2010.08.013

[9] J. Pérez, J. Muñoz-Dorado, T. De La Rubia, and J. Martínez, "Biodegradation and biological treatments of cellulose, hemicellulose and lignin: An overview," *Int. Microbiol.*, vol. 5, no. 2, pp. 53–63, 2002, doi: 10.1007/s10123-002-0062-3

[10] S. Van den Bosch et al., "Reductive lignocellulose fractionation into soluble lignin-derived phenolic monomers and dimers and processable carbohydrate pulps," *Energy Environ. Sci.*, vol. 8, no. 6, pp. 1748–1763, 2015, doi: 10.1039/C5EE00204D

[11] A. J. Ragauskas et al., "Lignin valorisation: improving lignin processing in the biorefinery," *Science*, vol. 344, no. 6185, May 2014, doi: 10.1126/science.1246843

[12] N. Mahmood, Z. Yuan, J. Schmidt, and C. (Charles) Xu, "Depolymerization of lignins and their applications for the preparation of polyols and rigid polyurethane foams: A review," *Renew. Sustain. Energy Rev.*, vol. 60, pp. 317–329, July 2016, doi: 10.1016/j.rser.2016.01.037

[13] C. Xu, R. A. D. Arancon, J. Labidi, and R. Luque, "Lignin depolymerisation strategies: towards valuable chemicals and fuels," *Chem. Soc. Rev.*, vol. 43, no. 22, pp. 7485–7500, 2014, doi: 10.1039/C4CS00235K

[14] D. Kai, M. J. Tan, P. L. Chee, Y. K. Chua, Y. L. Yap, and X. J. Loh, "Towards lignin-based functional materials in a sustainable world," *Green Chem.*, vol. 18, no. 5, pp. 1175–1200, 2016, doi: 10.1039/C5GC02616D

[15] X. Wang, Y. Guo, J. Zhou, and G. Sun, "Structural changes of poplar wood lignin after supercritical pretreatment using carbon dioxide and ethanol–water as co-solvents," *RSC Adv.*, vol. 7, no. 14, pp. 8314–8322, 2017, doi: 10.1039/C6RA26122A

[16] C. Li, M. Zheng, A. Wang, and T. Zhang, "One-pot catalytic hydrocracking of raw woody biomass into chemicals over supported carbide catalysts: simultaneous conversion of cellulose, hemicellulose and lignin," *Energy Environ. Sci.*, vol. 5, no. 4, pp. 6383–6390, 2012, doi: 10.1039/C1EE02684D

[17] Q. Song et al., "Lignin depolymerization (LDP) in alcohol over nickel-based catalysts via a fragmentation–hydrogenolysis process," *Energy Environ. Sci.*, vol. 6, no. 3, p. 994, 2013, doi: 10.1039/c2ee23741e

[18] W. Boerjan, J. Ralph, and M. Baucher, "Lignin biosynthesis," *Annu. Rev. Plant Biol.*, vol. 54, no. 1, pp. 519–546, June 2003, doi: 10.1146/annurev.arplant.54.031902.134938

[19] P. Azadi, O. R. Inderwildi, R. Farnood, and D. A. King, "Liquid fuels, hydrogen and chemicals from lignin: A critical review," *Renew. Sustain. Energy Rev.*, vol. 21, pp. 506–523, May 2013, doi: 10.1016/j.rser.2012.12.022

[20] J.-K. Weng, X. Li, N. D. Bonawitz, and C. Chapple, "Emerging strategies of lignin engineering and degradation for cellulosic biofuel production," *Curr. Opin. Biotechnol.*, vol. 19, no. 2, pp. 166–172, April 2008, doi: 10.1016/j.copbio.2008.02.014

[21] H. M. Wang, B. Wang, J. L. Wen, T. Q. Yuan, and R. C. Sun, "Structural characteristics of lignin macromolecules from different eucalyptus species," *ACS Sustain. Chem. Eng.*, vol. 5, no. 12, pp. 11618–11627, 2017, doi: 10.1021/acssuschemeng.7b02970

[22] N.-E. El Mansouri and J. Salvadó, "Structural characterization of technical lignins for the production of adhesives: Application to lignosulfonate, kraft, soda-anthraquinone, organosolv and ethanol process lignins," *Ind. Crops Prod.*, vol. 24, no. 1, pp. 8–16, July 2006, doi: 10.1016/j.indcrop.2005.10.002

[23] C. Li, X. Zhao, A. Wang, G. W. Huber, and T. Zhang, "Catalytic transformation of lignin for the production of chemicals and fuels," *Chem. Rev.*, vol. 115, no. 21, pp. 11559–11624, 2015, doi: 10.1021/acs.chemrev.5b00155

[24] M. V. Galkin and J. S. M. Samec, "Lignin valorisation through catalytic lignocellulose fractionation: a fundamental platform for the future biorefinery," *ChemSusChem*, vol. 9, no. 13, pp. 1544–1558, July 2016, doi: 10.1002/cssc.201600237

[25] H. Chen, "Chemical composition and structure of natural lignocellulose," in *Biotechnology of Lignocellulose*, Dordrecht: Springer Netherlands, 2014, pp. 25–71. doi: 10.1007/978-94-007-6898-7_2

[26] C. Chio, M. Sain, and W. Qin, "Lignin utilization: A review of lignin depolymerization from various aspects," *Renew. Sustain. Energy Rev.*, vol. 107, pp. 232–249, June 2019, doi: 10.1016/j.rser.2019.03.008

[27] L. Shuai and B. Saha, "Towards high-yield lignin monomer production," *Green Chem.*, vol. 19, no. 16, pp. 3752–3758, 2017, doi: 10.1039/C7GC01676J

[28] H. Wang, M. Tucker, and Y. Ji, "Recent development in chemical depolymerization of lignin: a review," *J. Appl. Chem.*, vol. 2013, pp. 1–9, July 2013, doi: 10.1155/2013/838645

[29] R. Vanholme, B. Demedts, K. Morreel, J. Ralph, and W. Boerjan, "Lignin biosynthesis and structure," *Plant Physiol.*, vol. 153, no. 3, pp. 895–905, July 2010, doi: 10.1104/pp.110.155119

[30] N. Mahmood, Z. Yuan, J. Schmidt, and C. (Charles) Xu, "Hydrolytic depolymerization of hydrolysis lignin: Effects of catalysts and solvents," *Bioresour. Technol.*, vol. 190, pp. 416–419, August 2015, doi: 10.1016/j.biortech.2015.04.074

[31] X. Chen, H. Li, S. Sun, X. Cao, and R. Sun, "Effect of hydrothermal pretreatment on the structural changes of alkaline ethanol lignin from wheat straw," *Sci. Rep.*, vol. 6, pp. 1–9, November 2016, doi: 10.1038/srep39354

[32] A. K. Deepa and P. L. Dhepe, "Lignin depolymerization into aromatic monomers over solid acid catalysts," *ACS Catal.*, vol. 5, no. 1, pp. 365–379, 2015, doi: 10.1021/cs501371q

[33] R. Katahira et al., "Base-catalyzed depolymerization of biorefinery lignins," *ACS Sustain. Chem. Eng.*, vol. 4, no. 3, pp. 1474–1486, 2016, doi: 10.1021/acssuschemeng.5b01451

[34] J. S. Kruger et al., "Lignin depolymerization with nitrate-intercalated hydrotalcite catalysts," *ACS Catal.*, vol. 6, no. 2, pp. 1316–1328, 2016, doi: 10.1021/acscatal.5b02062

[35] T. Ročnik, B. Likozar, E. Jasiukaitytė-Grojzdek, and M. Grilc, "Catalytic lignin valorisation by depolymerisation, hydrogenation, demethylation and hydrodeoxygenation: Mechanism, chemical reaction kinetics and transport phenomena," *Chem. Eng. J.*, vol. 448, p. 137309, November 2022, doi: 10.1016/j.cej.2022.137309

[36] W. Schutyser, T. Renders, S. Van den Bosch, S.-F. Koelewijn, G. T. Beckham, and B. F. Sels, "Chemicals from lignin: an interplay of lignocellulose fractionation, depolymerisation, and upgrading," *Chem. Soc. Rev.*, vol. 47, no. 3, pp. 852–908, 2018, doi: 10.1039/C7CS00566K

[37] V. M. Roberts, V. Stein, T. Reiner, A. Lemonidou, X. Li, and J. A. Lercher, "Towards quantitative catalytic lignin depolymerization," *Chem. – A Eur. J.*, vol. 17, no. 21, pp. 5939–5948, May 2011, doi: 10.1002/chem.201002438

[38] C. Zhang and F. Wang, "Catalytic lignin depolymerization to aromatic chemicals," *Acc. Chem. Res.*, vol. 53, no. 2, pp. 470–484, 2020, doi: 10.1021/acs.accounts.9b00573

[39] S. Van den Bosch et al., "Catalytic strategies towards lignin-derived chemicals," *Top. Curr. Chem.*, vol. 376, no. 5, pp. 129–130, 2018, doi: 10.1007/s41061-018-0214-3

[40] J. M. Pepper and Y. W. Lee, "Lignin and related compounds. I. A comparative study of catalysts for lignin hydrogenolysis," *Can. J. Chem.*, vol. 47, no. 5, pp. 723–727, March 1969, doi: 10.1139/v69-118

[41] B. Feng, H. Kobayashi, H. Ohta, and A. Fukuoka, "Aqueous-phase hydrodeoxygenation of 4-propylphenol as a lignin model to n-propylbenzene over Re-Ni/ZrO2 catalysts," *J. Mol. Catal. A Chem.*, vol. 388–389, pp. 41–46, July 2014, doi: 10.1016/j.molcata.2013.09.025

[42] M. Á. González-Borja and D. E. Resasco, "Anisole and guaiacol hydrodeoxygenation over monolithic Pt-Sn catalysts," *Energy and Fuels*, vol. 25, no. 9, pp. 4155–4162, 2011, doi: 10.1021/ef200728r

[43] Y. C. Lin, C. L. Li, H. P. Wan, H. T. Lee, and C. F. Liu, "Catalytic hydrodeoxygenation of guaiacol on Rh-based and sulfided CoMo and NiMo catalysts," *Energy and Fuels*, vol. 25, no. 3, pp. 890–896, 2011, doi: 10.1021/ef101521z

[44] T. H. Parsell et al., "Cleavage and hydrodeoxygenation (HDO) of C–O bonds relevant to lignin conversion using Pd/Zn synergistic catalysis," *Chem. Sci.*, vol. 4, no. 2, pp. 806–813, 2013, doi: 10.1039/C2SC21657D

[45] J.-Y. Kim, J. Park, H. Hwang, J. K. Kim, I. K. Song, and J. W. Choi, "Catalytic depolymerization of lignin macromolecule to alkylated phenols over various metal catalysts in supercritical tert-butanol," *J. Anal. Appl. Pyrolysis*, vol. 113, pp. 99–106, May 2015, doi: 10.1016/j.jaap.2014.11.011

[46] J. Zhang et al., "A series of NiM (M = Ru, Rh, and Pd) bimetallic catalysts for effective lignin hydrogenolysis in water," *ACS Catal.*, vol. 4, no. 5, pp. 1574–1583, 2014, doi: 10.1021/cs401199f

[47] A. Rahimi, A. Azarpira, H. Kim, J. Ralph, and S. S. Stahl, "Chemoselective metal-free aerobic alcohol oxidation in lignin," *J. Am. Chem. Soc.*, vol. 135, no. 17, pp. 6415–6418, 2013, doi: 10.1021/ja401793n

[48] M. P. Pandey and C. S. Kim, "Lignin depolymerization and conversion: a review of thermochemical methods," *Chem. Eng. Technol.*, vol. 34, no. 1, pp. 29–41, January 2011, doi: 10.1002/ceat.201000270

[49] K. Lee, Y. Jing, Y. Wang, and N. Yan, "A unified view on catalytic conversion of biomass and waste plastics," *Nat. Rev. Chem.*, vol. 6, no. 9, pp. 635–652, 2022, doi: 10.1038/s41570-022-00411-8

[50] W. O. S. Doherty, P. Mousavioun, and C. M. Fellows, "Value-adding to cellulosic ethanol: Lignin polymers," *Ind. Crops Prod.*, vol. 33, no. 2, pp. 259–276, March 2011, doi: 10.1016/j.indcrop.2010.10.022

[51] K. Shimada, S. Hosoya, and T. Ikeda, "Condensation reactions of softwood and hardwood lignin model compounds under organic acid cooking conditions," *J. Wood Chem. Technol.*, vol. 17, no. 1–2, pp. 57–72, February 1997, doi: 10.1080/02773819708003118

[52] L. Shuai et al., "Formaldehyde stabilization facilitates lignin monomer production during biomass depolymerization," *Science*, vol. 354, no. 6310, pp. 329–333, October 2016, doi: 10.1126/science.aaf7810

[53] M. R. Sturgeon et al., "A mechanistic investigation of acid-catalyzed cleavage of aryl-ether linkages: Implications for lignin depolymerization in acidic environments," *ACS Sustain. Chem. Eng.*, vol. 2, no. 3, pp. 472–485, 2014, doi: 10.1021/sc400384w

[54] C. S. Lancefield, O. S. Ojo, F. Tran, and N. J. Westwood, "Isolation of functionalized phenolic monomers through selective oxidation and C-O bond cleavage of the β-O-4 linkages in lignin," *Angew. Chemie*, vol. 127, no. 1, pp. 260–264, January 2015, doi: 10.1002/ange.201409408

[55] Z. Cao, M. Dierks, M. T. Clough, I. B. Daltro de Castro, and R. Rinaldi, "A convergent approach for a deep converting lignin-first biorefinery rendering high-energy-density drop-in fuels," *Joule*, vol. 2, no. 6, pp. 1118–1133, 2018, doi: 10.1016/j.joule.2018.03.012

[56] T. J. Morgan, A. Youkhana, S. Q. Turn, R. Ogoshi, and M. Garcia-Pérez, "Review of biomass resources and conversion technologies for alternative jet fuel production in hawai'i and tropical regions," *Energy and Fuels*, vol. 33, no. 4, pp. 2699–2762, 2019, doi: 10.1021/acs.energyfuels.8b03001

[57] Y. Cao, C. Zhang, D. C. W. Tsang, J. Fan, J. H. Clark, and S. Zhang, "Hydrothermal liquefaction of lignin to aromatic chemicals: Impact of lignin structure," *Ind. Eng. Chem. Res.*, vol. 59, no. 39, pp. 16957–16969, 2020, doi: 10.1021/acs.iecr.0c01617

[58] Y. Y. Wang, M. Li, C. E. Wyman, C. M. Cai, and A. J. Ragauskas, "Fast fractionation of technical lignins by organic cosolvents," *ACS Sustain. Chem. Eng.*, vol. 6, no. 5, pp. 6064–6072, 2018, doi: 10.1021/acssuschemeng.7b04546

[59] L. Zhou, V. Budarin, J. Fan, R. Sloan, and D. Macquarrie, "Efficient method of lignin isolation using microwave-assisted acidolysis and characterization of the residual lignin," *ACS Sustain. Chem. Eng.*, vol. 5, no. 5, pp. 3768–3774, 2017, doi: 10.1021/acssuschemeng.6b02545

[60] Y. Cao et al., "Microwave-assisted depolymerization of various types of waste lignins over two-dimensional CuO/BCN catalysts," *Green Chem.*, vol. 22, no. 3, pp. 725–736, 2020, doi: 10.1039/C9GC03553B

[61] O. M. Terrett et al., "Molecular architecture of softwood revealed by solid-state NMR," *Nat. Commun.*, vol. 10, no. 1, pp. 1–11, 2019, doi: 10.1038/s41467-019-12979-9

[62] D. M. Alonso et al., "Increasing the revenue from lignocellulosic biomass: Maximizing feedstock utilization," *Sci. Adv.*, vol. 3, no. 5, May 2017, doi: 10.1126/sciadv.1603301

[63] Z. Zhang and G. W. Huber, "Catalytic oxidation of carbohydrates into organic acids and furan chemicals," *Chem. Soc. Rev.*, vol. 47, no. 4, pp. 1351–1390, 2018, doi: 10.1039/C7CS00213K

[64] L. Gan and X. Pan, "Phenol-enhanced depolymerization and activation of kraft lignin in alkaline medium," *Ind. Eng. Chem. Res.*, vol. 58, no. 19, pp. 7794–7800, 2019, doi: 10.1021/acs.iecr.9b01147

[65] T. Renders et al., "Influence of acidic (H3PO4) and alkaline (NaOH) additives on the catalytic reductive fractionation of lignocellulose," *ACS Catal.*, vol. 6, no. 3, pp. 2055–2066, 2016, doi: 10.1021/acscatal.5b02906

[66] C. G. Yoo, M. Li, X. Meng, Y. Pu, and A. J. Ragauskas, "Effects of organosolv and ammonia pretreatments on lignin properties and its inhibition for enzymatic hydrolysis," *Green Chem.*, vol. 19, no. 8, pp. 2006–2016, 2017, doi: 10.1039/C6GC03627A

[67] W. Schutyser et al., "Influence of bio-based solvents on the catalytic reductive fractionation of birch wood," *Green Chem.*, vol. 17, no. 11, pp. 5035–5045, 2015, doi: 10.1039/C5GC01442E

[68] Y. Li et al., "Mechanistic study of diaryl ether bond cleavage during palladium-catalyzed lignin hydrogenolysis," *ChemSusChem*, vol. 13, no. 17, pp. 4487–4494, September 2020, doi: 10.1002/cssc.202000753

[69] C. Cheng, J. Truong, J. A. Barrett, D. Shen, M. M. Abu-Omar, and P. C. Ford, "Hydrogenolysis of organosolv lignin in ethanol/isopropanol media without added transition-metal catalyst," *ACS Sustain. Chem. Eng.*, vol. 8, no. 2, pp. 1023–1030, 2020, doi: 10.1021/acssuschemeng.9b05820

[70] T. Renders, S. Van den Bosch, S.-F. Koelewijn, W. Schutyser, and B. F. Sels, "Lignin-first biomass fractionation: the advent of active stabilisation strategies," *Energy Environ. Sci.*, vol. 10, no. 7, pp. 1551–1557, 2017, doi: 10.1039/C7EE01298E

[71] A. Toledano, L. Serrano, and J. Labidi, "Improving base catalyzed lignin depolymerization by avoiding lignin repolymerization," *Fuel*, vol. 116, pp. 617–624, January 2014, doi: 10.1016/j.fuel.2013.08.071

[72] N. Giummarella, Y. Pu, A. J. Ragauskas, and M. Lawoko, "A critical review on the analysis of lignin carbohydrate bonds," *Green Chem.*, vol. 21, no. 7, pp. 1573–1595, 2019, doi: 10.1039/C8GC03606C

[73] L. Yao et al., "Insights of ethanol organosolv pretreatment on lignin properties of broussonetia papyrifera," *ACS Sustain. Chem. Eng.*, vol. 6, no. 11, pp. 14767–14773, 2018, doi: 10.1021/acssuschemeng.8b03290

[74] P. J. Deuss, M. Scott, F. Tran, N. J. Westwood, J. G. de Vries, and K. Barta, "Aromatic monomers by in situ conversion of reactive intermediates in the acid-catalyzed depolymerization of lignin," *J. Am. Chem. Soc.*, vol. 137, no. 23, pp. 7456–7467, June 2015, doi: 10.1021/jacs.5b03693

[75] F. Héroguel, X. T. Nguyen, and J. S. Luterbacher, "Catalyst support and solvent effects during lignin depolymerization and hydrodeoxygenation," *ACS Sustain. Chem. Eng.*, vol. 7, no. 20, pp. 16952–16958, October 2019, doi: 10.1021/acssuschemeng.9b03843

[76] M. M. Jensen et al., "Hydrothermal liquefaction of enzymatic hydrolysis lignin: biomass pretreatment severity affects lignin valorisation," *ACS Sustain. Chem. Eng.*, vol. 6, no. 5, pp. 5940–5949, May 2018, doi: 10.1021/acssuschemeng.7b04338

[77] X. Huang, T. I. Korányi, M. D. Boot, and E. J. M. Hensen, "Ethanol as capping agent and formaldehyde scavenger for efficient depolymerization of lignin to aromatics," *Green Chem.*, vol. 17, no. 11, pp. 4941–4950, 2015, doi: 10.1039/C5GC01120E

[78] A. Mittal et al., "Ammonia pretreatment of corn stover enables facile lignin extraction," *ACS Sustain. Chem. Eng.*, vol. 5, no. 3, pp. 2544–2561, March 2017, doi: 10.1021/acssuschemeng.6b02892

[79] L. da Costa Sousa et al., "Next-generation ammonia pretreatment enhances cellulosic biofuel production," *Energy Environ. Sci.*, vol. 9, no. 4, pp. 1215–1223, 2016, doi: 10.1039/C5EE03051J

[80] F. P. Bouxin, S. David Jackson, and M. C. Jarvis, "Isolation of high quality lignin as a by-product from ammonia percolation pretreatment of poplar wood," *Bioresour. Technol.*, vol. 162, pp. 236–242, June 2014, doi: 10.1016/j.biortech.2014.03.082

[81] L. da Costa Sousa et al., "Isolation and characterization of new lignin streams derived from extractive-ammonia (EA) pretreatment," *Green Chem.*, vol. 18, no. 15, pp. 4205–4215, 2016, doi: 10.1039/C6GC00298F

[82] L. Weigand, S. Mostame, A. Brandt-Talbot, T. Welton, and J. P. Hallett, "Effect of pretreatment severity on the cellulose and lignin isolated from Salix using ionoSolv pretreatment," *Faraday Discuss.*, vol. 202, pp. 331–349, 2017, doi: 10.1039/C7FD00059F

[83] K. H. Kim, B. A. Simmons, and S. Singh, "Catalytic transfer hydrogenolysis of ionic liquid processed biorefinery lignin to phenolic compounds," *Green Chem.*, vol. 19, no. 1, pp. 215–224, 2017, doi: 10.1039/C6GC02473D

[84] N. Sathitsuksanoh et al., "Lignin fate and characterization during ionic liquid biomass pretreatment for renewable chemicals and fuels production," *Green Chem.*, vol. 16, no. 3, pp. 1236–1247, 2014, doi: 10.1039/C3GC42295J

[85] A. Brandt, J. Gräsvik, J. P. Hallett, and T. Welton, "Deconstruction of lignocellulosic biomass with ionic liquids," *Green Chem.*, vol. 15, no. 3, p. 550, 2013, doi: 10.1039/c2gc36364j

[86] J. S. Luterbacher et al., "Nonenzymatic sugar production from biomass using biomass-derived γ-valerolactone," *Science*, vol. 343, no. 6168, pp. 277–280, January 2014, doi: 10.1126/science.1246748

[87] C. S. Lancefield, I. Panovic, P. J. Deuss, K. Barta, and N. J. Westwood, "Pre-treatment of lignocellulosic feedstocks using biorenewable alcohols: towards complete biomass valorisation," *Green Chem.*, vol. 19, no. 1, pp. 202–214, 2017, doi: 10.1039/C6GC02739C

[88] P. J. Deuss, C. S. Lancefield, A. Narani, J. G. de Vries, N. J. Westwood, and K. Barta, "Phenolic acetals from lignins of varying compositions via iron(iii) triflate catalysed depolymerisation," *Green Chem.*, vol. 19, no. 12, pp. 2774–2782, 2017, doi: 10.1039/C7GC00195A

[89] E. Feghali, G. Carrot, P. Thuéry, C. Genre, and T. Cantat, "Convergent reductive depolymerization of wood lignin to isolated phenol derivatives by metal-free catalytic hydrosilylation," *Energy Environ. Sci.*, vol. 8, no. 9, pp. 2734–2743, 2015, doi: 10.1039/C5EE01304F

[90] Y. M. Questell-Santiago, M. V. Galkin, K. Barta, and J. S. Luterbacher, "Stabilization strategies in biomass depolymerization using chemical functionalization," *Nat. Rev. Chem.*, vol. 4, no. 6, pp. 311–330, 2020, doi: 10.1038/s41570-020-0187-y

[91] A. Maiorana et al., "Structure property relationships of biobased n-alkyl bisferulate epoxy resins," *Green Chem.*, vol. 18, no. 18, pp. 4961–4973, 2016, doi: 10.1039/C6GC01308B

[92] A. Llevot, E. Grau, S. Carlotti, S. Grelier, and H. Cramail, "From lignin-derived aromatic compounds to novel biobased polymers," *Macromol. Rapid Commun.*, vol. 37, no. 1, pp. 9–28, January 2016, doi: 10.1002/marc.201500474

[93] P. Ferrini and R. Rinaldi, "Catalytic biorefining of plant biomass to non-pyrolytic lignin bio-oil and carbohydrates through hydrogen transfer reactions," *Angew. Chemie*, vol. 126, no. 33, pp. 8778–8783, August 2014, doi: 10.1002/ange.201403747

[94] M. V. Galkin and J. S. M. Samec, "Selective route to 2-propenyl aryls directly from wood by a tandem organosolv and palladium-catalysed transfer hydrogenolysis," *ChemSusChem*, vol. 7, no. 8, pp. 2154–2158, August 2014, doi: 10.1002/cssc.201402017

[95] T. Parsell et al., "A synergistic biorefinery based on catalytic conversion of lignin prior to cellulose starting from lignocellulosic biomass," *Green Chem.*, vol. 17, no. 3, pp. 1492–1499, 2015, doi: 10.1039/C4GC01911C

[96] E. M. Anderson et al., "Reductive catalytic fractionation of corn stover lignin," *ACS Sustain. Chem. Eng.*, vol. 4, no. 12, pp. 6940–6950, December 2016, doi: 10.1021/acssuschemeng.6b01858

[97] H. Luo et al., "Total utilization of miscanthus biomass, lignin and carbohydrates, using earth abundant nickel catalyst," *ACS Sustain. Chem. Eng.*, vol. 4, no. 4, pp. 2316–2322, April 2016, doi: 10.1021/acssuschemeng.5b01776

[98] S. Van den Bosch et al., "Integrating lignin valorisation and bio-ethanol production: on the role of Ni-Al 2 O 3 catalyst pellets during lignin-first fractionation," *Green Chem.*, vol. 19, no. 14, pp. 3313–3326, 2017, doi: 10.1039/C7GC01324H

[99] T. Tana et al., "Valorisation of native sugarcane bagasse lignin to bio-aromatic esters/monomers via a one pot oxidation–hydrogenation process," *Green Chem.*, vol. 21, no. 4, pp. 861–873, 2019, doi: 10.1039/C8GC03655A

[100] V. B. Subramani et al., "Cotton farming sustainability: Formation of trans-isoeugenol/bio-aromatics, 5-chloromethylfurfural, C13–C17 liquid hydrocarbons & fertilizer from cotton gin trash," *J. Clean. Prod.*, vol. 363, p. 132404, August 2022, doi: 10.1016/j.jclepro.2022.132404

[101] V. B. Subramani, L. Atanda, W. O. S. Doherty, D. Rackemann, and L. Moghaddam, "Co-liquefaction of cotton gin trash and low-density polyethylene wastes via supercritical ethanolysis for hydrocarbon-rich oil," *Energy Convers. Manag.*, vol. 290, p. 117216, August 2023, doi: 10.1016/j.enconman.2023.117216

2 Biomass Utilisation for Meeting Sustainable Development Goals

An Indian Context

Sreeda Pallipad

2.1 INTRODUCTION

Biomass as an energy source has been used since the day man came into this world, as food, fodder or heat source. Until fossil fuels were discovered and widely used, biomass, especially in the form of wood from forests, was used as the primary source of energy. Charcoal, derived from biomass (wood), was used in iron smelting before coal came into the picture (Rosillo-Calle et al., 2015). The importance of biomass as fuel was forgotten when other sources of energy were discovered. Yet, biomass remains a popular source for traditional purposes and in rural households.

For a nation like India, which is committed to providing clean, affordable and reliable energy source to its citizens and at the same time committed to fight climate change by focussing on alternative sources of energy, biomass has the same potential as solar or wind (Tyagi et al., 2019). The ill effects of stubble burning have long been recognised, but not until recently have they been recognised as a serious contributor to air pollution and smog events in the Northern states and extending the effects to the National Capital region. On December 20th, 2015, the National Green Tribunal banned the burning of crop residue in the states of Rajasthan, Uttar Pradesh, Haryana and Punjab. However, the incidents of such burning still continue to occur. With the National Policy for management of crop residue and the various policies by the Ministry of Power, the government is trying to provide an alternative to stubble burning and for the effective use of biomass, especially crop residue. This will also go a long way in reassuring the country's commitment to working towards achieving the Sustainable Development Goals. Energy is a prime requirement for the economic and social development of a country, and biomass is a carbon-neutral source of energy (Saidur et al., 2011). In the whole cycle of production of biomass, carbon is taken up from the atmosphere and then released while used as an energy source. It may be considered that there is no net addition of carbon due to the use of biomass (Sudha & Ravindranath, 1999).

The need for an exclusive governance infrastructure for the utilisation of renewable energy in India was realised early on, and in the year 1981, the Commission of Additional Source of Energy (CASE) was formed as part of the Department of Science and Technology (DST). In the same year, the Department of Non-Conventional Energy Sources (DNES) was formed, to which CASE was added. In the year 1982, the Ministry of Energy was expanded by including the Department of Petroleum and the Department of Non-Conventional Energy Sources (DNES). After 10 years, the Ministry of Non-Conventional Energy Sources was formed, and in October 2006, the Ministry was renamed to the current Ministry of Non-Renewable Energy (MNRE).

In India, 40% of the greenhouse gas emissions are contributed by the electricity sector, and despite the emphasis and growth of the renewable sector, coal dominates as the fuel source for the

DOI: 10.1201/9781003732907-2

production of electricity, which amounts to almost 74% of the total electricity produced in the nation. And as per the Central Electricity Authority, India shall continue to depend on coal for energy until 2030 and beyond. Thus, with a large amount of greenhouse gases (GHG) emissions, it is the need of the hour that necessary interventions be made in this sector to meet our goals of decarbonisation.

With the concerns over stubble burning increasing, governments at various levels have been taking initiative, framing policies to encourage farmers to adopt alternative methods of crop residue management. Biomass co-firing has been recognised as a method to utilise the biomass, reduce the burning of agricultural residue and also to reduce the carbon footprint of the power sector. This chapter intends to give an outline of the various policies which focus on management of crop residue and utilising biomass for co-firing in energy generation, the challenges faced during the utilisation of biomass and a discussion on the way forward to overcome the challenges.

2.2 BIOMASS

Researchers have classified biomass based on several criteria, including the source, the composition and the use of the biomass, according to their scope of study. The biomass could be purpose-grown energy crops or waste (Saidur et al., 2011). The waste biomass also includes agricultural residue, by-products from the wood processing industry, from municipal solid waste (Kumar, Kumar, Baredar, and Shukla, 2015; Saidur et al., 2011; Tumuluru et al., 2011). Further classification based on the kind of biomass includes (a) woody, (b) herbaceous, (c) aquatic, (d) animal and human waste, and (e) biomass mixtures (Tursi, 2019). In 2004, FAO recognised the need for the identification of biomass under a Unified Bioenergy Terminology to encourage sustainable management along with efficient, competitive and affordable delivery of modern energy services (FAO, 2004). FAO classified the biomass (especially for bioenergy purposes) as: wood fuels, agrofuels and municipal by-products.

2.2.1 The Indian Scenario

In India, biomass is a traditional form of energy source, and 32% of the primary energy requirement is met by biomass, and a population of more than 70% depends on biomass for its requirements. As per MNRE, the current availability of biomass is about 750 million metric tonnes per year, with a surplus of 230 million metric tonnes per year corresponding to about 28 GW of energy. Also, a potential of 14GW exists through the use of bagasse in sugar mills (MNRE, 2022). However, as of 31.05.2023, the installed generation capacity using biomass/cogeneration accounts for only about 2.5% (10248 MW), and waste to energy contributes to 0.1% (554 MW) of the total installed capacity.

In the year 2019–2021, rural households using solid fuel for cooking (including coal/lignite, charcoal, wood, straw/shrubs/grass, agricultural crop waste, and dung cakes) were 56.1% while in the urban areas it was 9.5 % (International Institute for Population Sciences (IIPS) and ICF, 2022). The rural households depend on the biomass in its available forms, even when access to LPG has improved and the uses are mainly restricted to cooking and heating purposes. The use across the nation is highly variable depending on the spatio-temporal distribution, climatic conditions and availability throughout the year (Hiloidhari et al., 2014). However, it is also to be noted that the traditional form of use of such biomass is often by inefficient means with impacts on both human health and the environment (Singh, 2017).

Apart from the forest-based (woody biomass) biomass which is being utilised currently, India has a huge potential for utilisation of agriculture residue-based biomass for energy generation. The utilisation of such residue can also curb the issues caused by residue burning practised in several parts of the country. An average of 114.5 million tonnes of rice, 102.5 million tonnes of wheat and 365.5

million tonnes of sugarcane was produced between 2016–2017 and 2020–2021 in the country MoAFW. (2023), and of all the crops, rice, wheat and sugarcane are more prone to crop residue burning (National Policy for Management of Crop Residues, NPMCR, 2014). Such crops are preferred for cultivation as they give higher economic returns to the farmers. In order to gain maximum income and to reduce downtime, the residue is disposed of by burning, taking the least effort and minimum time.

2.2.2 Stubble Burning

A major concern with biomass, especially with the agricultural residue in India today, is due to the stubble burning. This is a major problem that increases the pollution levels in the North and North-western states, especially during the winter months when the atmospheric conditions are also unfavourable. Stubble burning is a quick method for clearing land for the next sowing season and a low-cost economic solution to dispose of the agricultural residue (Porichha et al., 2021). Air pollution and loss of soil fertility are the two major environmental issues related to the widely prevalent method of burning biomass. The burning of crop stubble results in the emission of several harmful gases, including CO, NO_x and SO_2, apart from the particulate matter (Kumar, Kumar, and Joshi, 2015). It is estimated that the burning of one tonne of rice straw releases 3 kg particulate matter, 60 kg CO, 1460 kg CO_2, 199 kg ash and 2 kg SO_2 (GoI, 2023). These pollutants have a detrimental effect on human and animal health. Respiratory diseases, eye irritations and skin disorders are among the very few of the many conditions that are caused by the pollution caused by crop residue burning. The sensitive individuals are at a greater risk from the particulate matter which may trigger or aggravate diseases like asthma, etc. Burning of stubble also depletes the soil of its major nutrients, which are absorbed during the growth phase. Burning of the residue affects the organic content of the soil, the carbon and nitrogen equilibrium, and reduces the beneficial microflora of the soil. It is estimated that paddy straw contains approximately 5.5 kg N, 2.3 kg P_2O_5, 25 kg K_2O, 1.2 kg S, 50–70% of micro-nutrients absorbed by rice and 400 kg of carbon, which are lost during burning (GoI, 2023).

2.3 POLICIES PERTAINING TO BIOMASS UTILISATION

Several ministries in India have been involved in formulating and implementing the policies regarding the management and utilisation of biomass, especially crop residue, with the aim of discouraging the practice of stubble burning. Both in-situ and ex-situ methods for crop residue management have been encouraged. Ex-situ methods, such as use as raw material in other industries, transforming to other forms such as pellets for use in power generation, etc., also provide the farmers with income from the agricultural waste. The major policies formulated by different ministries are discussed briefly in Figure 2.1.

2.3.1 National Policy for Management of Crop Residues (NPMCR), 2014

The Ministry of Agriculture, in 2014, formulated the National Policy for Management of Crop Residues (NPMCR), regarding both the in-situ and ex-situ management of crop residue. The major objectives of the policy were to:

1. Control the burning of crop residue by promoting in-situ management
2. To diversify the use of crop residue for power generation and as other industrial raw materials
3. Spread awareness about the disadvantages of burning crop residue and capacity building
4. Formulate and implement suitable law/policy, etc., to prevent burning of crop residue.

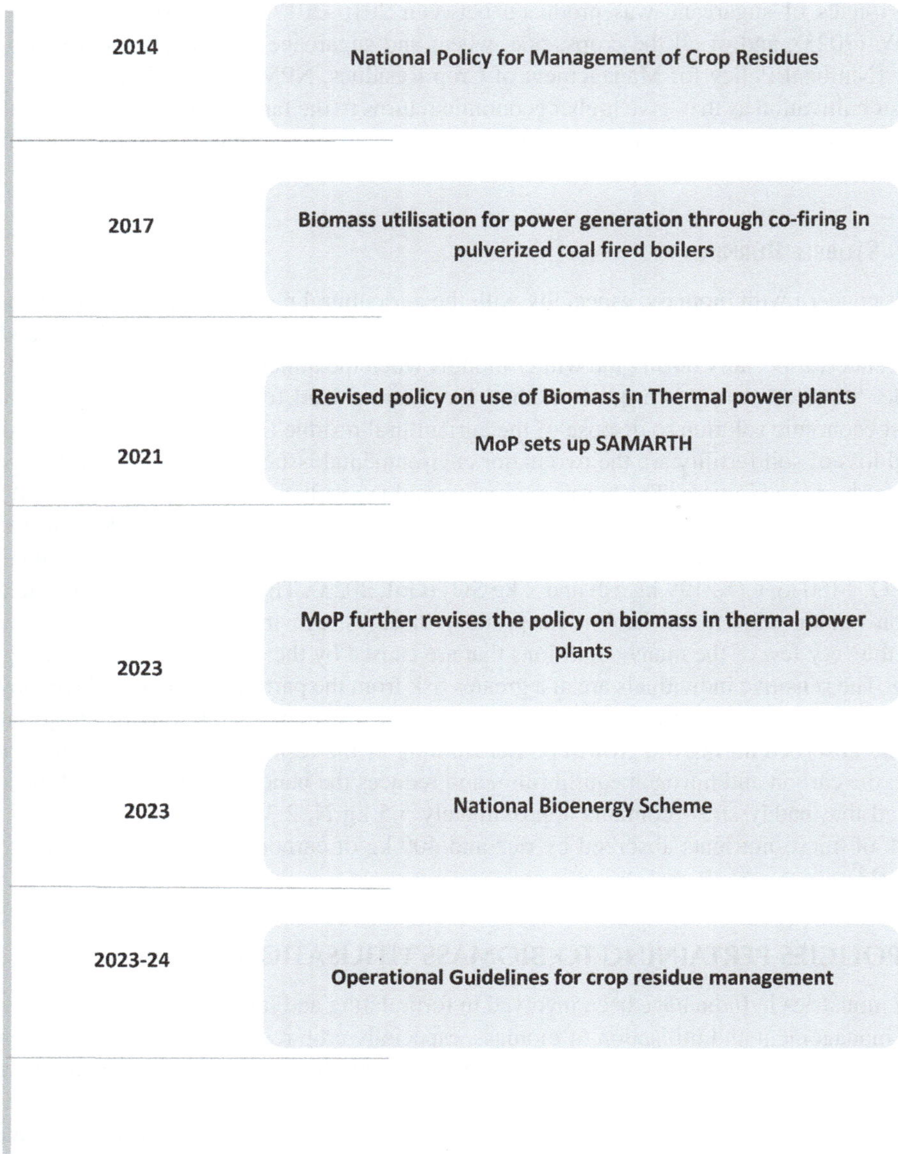

FIGURE 2.1 Policies pertaining to biomass utilisation – a timeline.

The policy states five interventions that need to be made to curb residue burning, which include:

1. Technological interventions: Use of technology for utilising crop residue for use as mulch material or to prevent soil erosion, etc., and further use as raw materials in the cultivation of mushrooms. Encouraging machinery usage by extending subsidy/financial incentives to the farmers for hiring/purchasing machinery.
2. Diversified uses of crop residue: promoting the use in other industries and for uses such as power generation or ethanol production in Public-Private Partnership (PPP) mode. Further extending the subsidy for the transport of crop residue.
3. Capacity building and awareness generation: Improving the awareness among the farmers to adopt alternative methods. Setting up a self-help group providing financial subsidy to the

unemployed youth to establish centres for hiring machinery. Demonstration of management technology by state agencies to create awareness and the dissemination of technologies.
4. Pilot studies on management of crop residue: Research and technology on the use of farm machinery and the use of interdepartmental resources for monitoring of crop residue management.
5. Laws and legislation to curb crop residue burning: Bringing law/rules/policy in the states to prevent/ban the practise of crop residue burning.

The policy document further elaborates on the financial resource mobilisation and states that, as it involves a multidisciplinary approach, the funds need to be mobilised from the various ministries involved, and agriculture being a state subject, the states may identify the need-based interventions for the locations and implement the same under the Annual Work Plan. The ministries at the centre shall provide financial support to the states. The State shall monitor the implementation, and, at regular intervals, the national level monitoring shall be done to take proactive steps.

In 2023–2024, the operational guidelines on crop residue management in the States of Punjab, Haryana, Uttar Pradesh, Madhya Pradesh and NCT of Delhi, with the objectives of

1. Protecting the environment from air pollution and preventing loss of nutrients and soil microorganisms caused by the burning of crop residue
2. Promoting management of crop residue by retention/incorporation into the soil and establishing an appropriate supply chain for further utilisation through the deployment of appropriate mechanisation inputs
3. Creating awareness among stakeholders through demonstration, capacity building activities and differentiated Information, Education and Communication strategies for effective utilisation and management of crop residue.

The management strategy included in the guidelines aims to achieve the objectives by making optimum use of the available machinery, establishing a supply chain of biomass through cluster-based approach and increasing the awareness among farmers through the participation of stakeholders. The fund sharing will be in the ratio of 60:40 among the Centre and the states, except NCT of Delhi which will have complete funding from the Centre. The guidelines require the states to submit an annual action plan with multiple components, including:

1. Financial assistance to farmers for the procurement of crop residue management machines on an individual ownership basis
2. Establishment of Custom Hiring Centres of Crop Residue Management Machines
3. Establishment of crop residue/paddy straw supply chain
4. Information, Education and Communication (IEC) for awareness on crop residue management

Thus, it aims to provide comprehensive guidelines for effective crop management by trying to focus on strengthening the infrastructure and supply chain.

2.3.2 Policy on Biomass Utilisation in Power Plants

In 2017, Dadri Thermal Power Plant, located in Ghaziabad, Uttar Pradesh (of National Thermal Power Corporation (NTPC)), successfully co-fired biomass in its existing pulverised coal (PC) fired boilers and reported very little reduction in efficiency. This was followed by the policy formulated by the Ministry of Power in 2017 entitled "Biomass utilisation for power generation through co-firing in pulverized coal fired boilers". The policy stated "*Stubble burning has been cited as the major cause of recent smog in north-west India*", thus recognising the major issue of addressing the problem of burning of stubble. The policy advised all Power plants/Utilities to promote the use of "biomass pellets", in all Fluidised bed and pulverised coal units (coal-based thermal power plants, with exception to ball and tube mill, which may be attributed to the safety concerns associated with such plants), power generation utilities regardless of public owned or private, shall use a 5–10% blend of biomass and of primarily agro-residue.

In 2021, the policy was revised to implement the mandatory use of biomass pellets (primarily of agro-residues) and defined the same for utilities with different mechanisms, including:

- a 5% blend on an annual basis with power generation utilities using bowl mill, which shall be increased to 7% in the next two years
- a 5% blend (torrefied only) on annual basis with power generation utilities using ball and race mill, which had to be complied within one year and which shall be increased to 7% in the next two years
- a 5% blend (torrefied with volatile content below 22%) on annual basis with power generation utilities using ball and tube mill which had to be compiled within one year
- a 5% blend on an annual basis with power generation utilities using Circulating Fluidised bed combustion, which shall be increased to 7% in the next two years. (MoP, 2022)

In 2023, the policy was further revised and a definite timeline was given for the implementation, including the mandatory use of 5% blend by the FY 2024–2025 and increase to 7% by FY 2025–2026, applicable for Bowl mill and ball and race mill. For utilities with ball and tube mills, the use has been mandated from 2024 to 2025. Further, considering the low availability of torrefied biomass pellets in the market, it was advised that the power plants which can use the non-torrefied (Bowl Mill) biomass pellets shall use the same, whereas the rest would use only torrefied biomass pellets.

2.3.3 National Mission on Use of Biomass in Coal-Based Thermal Power Plants

In May 2021, the Ministry of Power (MoP) announced the setting up of the "National Mission on use of biomass in coal based thermal power plant", in order to address the issue of burning of stubble and to reduce the carbon footprint of the thermal power plants. The mission also intends to support the nation's goal to move towards cleaner energy resources. The mission, which was later rechristened to SAMARTH (Sustainable Agrarian Mission on use of Agri Residue in Thermal Power Plant) mission, with six sub-groups covers a wide range of activities, to enable the utilisation of biomass in thermal power plants, including research on biomass and on improving the supply chain of biomass. Through the SAMARTH mission, MoP has also issued long-term model contract with a minimum contract period of seven years to establish biomass supply chain facilities. Awareness programmes and training programmes are also conducted under SAMARTH.

2.3.4 National Bioenergy Scheme

The Ministry of New and Renewable Energy notified the National Bioenergy Scheme in November 2022, for the year 2021–2022 to 2025–2026. The programme is to be implemented in two phases, and the first phase will have a budget outlay of Rs. 858 crores. The scheme with three sub-schemes:

1. Waste to Energy: A scheme to support the setting up of large-scale biogas, bio-CNG and power plants from urban, industrial and agricultural wastes
2. Biomass programme: A scheme to support the setting up of pellets and briquettes for use in power generation and cogeneration of power (non-bagasse)
3. Biogas programme: A scheme to support the setting up of family and middle-size biogas in rural areas.

TABLE 2.1
Central Financial Assistance to Set Up Plants Under Biomass Programme

Project Type	CFA
Briquette/Pellet Manufacturing plants	Rs. 9 Lakh per MTPH (metric tonne/hour) manufacturing capacity (maximum CFA of Rs 45 Lakhs per plant)
Biomass (Non-bagasse) cogeneration projects	Rs. 40 Lakhs/MW (on Installed Capacity) (maximum CFA of Rs. 5 Crores per project)

2.3.4.1 Scheme to Support Manufacturing of Briquettes and Pellets and Promotion of Biomass (Non-Bagasse) Based Cogeneration in Industries

Under the National Bioenergy Scheme, the sub-scheme "Biomass Programme" broadly aims to utilise the surplus agricultural residue, to prevent stubble burning and to encourage better environmental practices. It aims to support the setting up of briquette/pellet manufacturing plants and non-bagasse-based cogeneration plants in the country. The scheme is also aimed at providing additional income to farmers and thus discouraging them from burning agricultural residue. It provides Central Financial Assistance (CFA) to projects for the setting up of plants (as shown in Table 2.1) and service charges to implementing agencies. All kinds of biomass, including crop residue, plantation wood, agro-industry waste, wood industry waste, forest residue, etc., are included in the scheme but does not support activities involving bagasse and wastes such as municipal waste or processed industrial waste.

2.4 CHALLENGES ASSOCIATED WITH USE OF BIOMASS

The use of biomass for energy has several advantages associated. However, till today, the full potential of such resources has not been realised due to several challenges. The challenges for the use of biomass are several and at multiple levels, at the source of the biomass, at the processing and at the consumer end, including the entire supply chain management.

1. Availability of biomass: Biomass production from agriculture is a time-bound activity that produces the whole of the biomass within a short period during harvesting. Once the harvesting season is complete, the downtime for the next available season creates a wide gap affecting the continuous supply of the biomass. Further, the long-term storage of agricultural residue decreases its quality owing to decomposition, moisture absorption, etc., rendering it unusable for further processing and meeting the required standards as fuel.
2. Collection, storage and transportation: The biomass produced in the remote rural areas of the country needs to be collected post-harvesting and transported to sites/units where it can be processed for use in units for energy production. The "bulky" nature of the agricultural residues also increases the transportation costs. The storage of these bulky residues is also a major hindrance, at the harvested site or at the processing unit. For example, as per Kadam et al. (2000), 1% of the cultivation land is required for storing bales from paddy harvest. Further, these storage spaces should prevent any damage to the raw material, such as moisture absorption, microbial decomposition or prevent rodent damage to preserve the quality. Once the biomass residues have been transformed into pellets, they have to be stored to prevent any moisture absorption, and pellets with up to 14% moisture can be used for combustion along with coal Aggarwal (2022).
3. Processing of agro-residue: The processing of agro-residue is economically demanding and a major drawback. The procurement cost of raw agro-residue ranges between Rs. 2500 and 4500 per tonne, and the manufacturing cost of pellets is Rs. 4200–7200 and that of

briquettes is Rs. 3500–5600 (Aggarwal, 2023). Also, the cost of the machinery varies widely from Rs. 2.5 to 4 million for a briquette manufacturing to Rs 15 million for a pelletiser machine.

4. Inconsistent availability of biomass pellets: According to the Ministry of Power policy on biomass, nearly 0.25–0.3 million tonnes of biomass pellets are required for every GW capacity at 7% co-firing. The SMARTH provides a list of 54 vendors who have offered pellets (torrefied and non-torrefied), and there is a huge demand and supply gap. Also, the price of the pellets at which the coal plants procure is lower (Rs. 8–9 per kg in 2022) than the procurement by other industries such as textile, food processing, etc. (Rs. 12–13 per kg in 2022) (Aggarwal (2022)).
5. Time factor: The biomass pellets are procured by the coal plants through a tending process, which typically takes 6–8 months for the entire process until the award of the work order. Manufacturers are apprehensive of such long durations for the award of work order and the uncertainty associated with it (Aggarwal, 2023).

2.5 OVERCOMING CHALLENGES

Stubble burning has long been recognised as a menace, especially in the north, north-western states of the country. Several initiatives, policies, incentives, etc., to curb stubble burning have been put forward. However, these regions still continue to face the problem. Several initiatives at the local and national levels need to be taken to overcome these challenges, which include:

1. Aligning the policies and initiatives: Governing bodies at multiple levels have taken several initiatives and formulated policies to overcome stubble burning and encourage the utilisation of agricultural residue. However, each of these works as "islands", unable to meet their objectives. A much-needed intervention is to formulate a mechanism where the state and the central policies align with each other and strengthen the system. Further, at the central level, the policies under multiple ministries have to align and find common objectives (e.g., crop management and biomass utilisation).
2. Building a stronger supply chain system: The major roadblock in the biomass utilisation for use in the energy sector is the poor supply chain infrastructure. Beginning from source collection to processing and utilisation, the network of collection and storage is currently not sufficient to meet the policy objectives. Strengthening the basic infrastructure to enable the collection of biomass from farmers and transferring the same to processing plants is the first step, as farmers, especially in remote areas, are at a disadvantage. The development of low-cost storage systems for processed biomass is needed to ensure that the quality of the pellets does not deteriorate, and a year-round supply of pellets can be ensured. Also, simplifying the procurement system of processed biomass by the end users will further encourage investors in this field.
3. Ease in benefit transfer: The use of biomass utilisation was also conceptualised to generate extra income for the farmers. However, the monetary resources currently required to collect and transfer the biomass to processing units often outweigh the benefits that the farmers receive, and hence they revert to burning the residue. Thus, a robust system to ensure that the farmers get the right price will further ensure that the biomass is utilised effectively.
4. Building awareness among the farmers: Farmers often see stubble burning as a quick way to get rid of the agricultural residue and prepare the field for the next sowing season. As pointed out (Abdurrahman et al., 2020), 90% of the farmers are aware of the ill effects of burning. However, they may be ignorant of the various initiatives, supports and alternative methods for the management of agricultural residues. Also, farmers are demotivated when the alternative methods are not incentivised on time. Thus, by ensuring that the mechanisms have a strong ground-level reach and efficiency, the farmers should be made aware of the system in place.

2.6 CONCLUSION

The biomass utilisation has become necessary to combat the issues related to stubble burning and the consequent air pollution and other ill effects. Also, the use of biomass will aid in reducing the carbon footprint of units producing energy from fossil fuels. Even with several policies formulated and implemented, the incidents of stubble burning are still reported. Building a strong ground-level infrastructure where the farmers can access the machinery and the presence of a strong supply chain management system, which ensures that the farmers do not lose out on time or income, is a necessity of today. Only this would ensure that more and more farmers, who are well aware of the disadvantages of burning, are encouraged to opt for friendlier methods of crop residue management. Further push is required for research and development of technology in increasing the calorific value, decreasing moisture content, shelf-life, etc., making any biomass fit for use in the energy sector. Aligning the policies of the different ministries to achieve the common goal from the ground level up is also necessary. Although the menace of stubble burning and the availability of biomass are mostly concentrated in the northern region, the utilisation of biomass needs to be explored from other regions, and the scope of developing decentralised plants needs to be studied further. Thus, with a high potential for use of biomass, India has a long way to go to make biomass utilisation effective to achieve the larger goals.

REFERENCES

Abdurrahman, M. I., Chaki, S., & Saini, G. (2020). Stubble burning: Effects on health & environment, regulations and management practices. *Environmental Advances*, *2*(October), 100011. https://doi.org/10.1016/j.envadv.2020.100011

Aggarwal, A. (June 13, 2022). Biomass co-firing: Why India's target for coal power plants is challenging. *Down to Earth*. Accessed at https://www.downtoearth.org.in/news/energy/biomass-co-firing-why-india-s-target-for-coal-power-plants-is-challenging-83261 on 04.04.2023.

Aggarwal, A. (2023). *Status of Biomass Co-Firing in Coal Thermal Power Plants in Delhi NCR*. Centre for Science and Environment, 1–40.

FAO. (2004). *Unified Bioenergy Terminology*. https://openknowledge.fao.org/server/api/core/bitstreams/4817badd-b827-4411-a100-a22953692796/content

GoI. (2023). *Crop Residue Management Operational Guidelines*. Ministry of Agriculture & Farmers Welfare.

Hiloidhari, M., Das, D., & Baruah, D. C. (2014). Bioenergy potential from crop residue biomass in India. *Renewable and Sustainable Energy Reviews*, *32*, 504–512. https://doi.org/10.1016/j.rser.2014.01.025

International Institute for Population Sciences (IIPS) and ICF. (2022). *National Family Health Survey (NFHS-5) India 2019-21* (Volume I). https://dhsprogram.com/pubs/pdf/FR375/FR375.pdf

Kadam, K. L., Forrest, L. H., & Jacobson, W. A. (2000). Rice straw as a lignocellulosic resource: Collection, processing, transportation, and environmental aspects. *Biomass and Bioenergy*, *18*(5), 369–389. https://doi.org/10.1016/S0961-9534(00)00005-2

Kumar, A., Kumar, N., Baredar, P., & Shukla, A. (2015). A review on biomass energy resources, potential, conversion and policy in India. *Renewable and Sustainable Energy Reviews*, *45*, 530–539. https://doi.org/10.1016/j.rser.2015.02.007

Kumar, P., Kumar, S., & Joshi, L. (2015). Valuation of the health effects. In *Socioeconomic and Environmental Implications of Agricultural Residue Burning A Case study of Punjab, India* (pp. 35–67). https://doi.org/10.1007/978-81-322-2014-5_3

MNRE. (2022). Annual Report, Ministry of New and Renewable Energy Government of India https://cdnbbsr.s3waas.gov.in/s3716e1b8c6cd17b771da77391355749f3/uploads/2023/08/2023080211.pdf

MoAFW. (2023). Agricultural Statistics at a Glance, 2022. Govt. of India, 280. https://desagri.gov.in/wp-content/uploads/2023/05/Agricultural-Statistics-at-a-Glance-2022.pdf

MoP. (2022). Accessed at https://samarth.powermin.gov.in/content/policies/c5938aea-835b-4a68-a5bc-34f925b95e79.pdf on 03.04.2023.

NPMCR. (2014). National policy for management of crop residues (NPMCR). *Department of Agriculture and Cooperation, Ministry of Agriculture, Government of India*, 1–11. http://agricoop.nic.in/sites/default/files/NPMCR_1.pdf

Porichha, G. K., Hu, Y., Rao, K. T. V., & Xu, C. C. (2021). Crop residue management in India: Stubble burning vs. other utilizations including bioenergy. *Energies*, *14*(14), 1–17. https://doi.org/10.3390/en14144281

Rosillo-Calle, F., de Groot, P., Hemstock, S. L., & Woods, J. (2015). The biomass assessment handbook: Energy for a sustainable environment. In *The Biomass Assessment Handbook: Energy for a Sustainable Environment*. https://doi.org/10.4324/9781315723273

Saidur, R., Abdelaziz, E. A., Demirbas, A., Hossain, M. S., & Mekhilef, S. (2011). A review on biomass as a fuel for boilers. *Renewable and Sustainable Energy Reviews*, *15*(5), 2262–2289. https://doi.org/10.1016/j.rser.2011.02.015

Singh, J. (2017). Management of the agricultural biomass on decentralized basis for producing sustainable power in India. *Journal of Cleaner Production*, *142*, 3985–4000. https://doi.org/10.1016/j.jclepro.2016.10.056

Sudha, P., & Ravindranath, N. H. (1999). Land availability and biomass production potential in India. *Biomass and Bioenergy*, *16*(3), 207–221. https://doi.org/10.1016/S0961-9534(98)00083-X

Tumuluru, J. S., Sokhansanj, S., Wright, C. T., Boardman, R. D., & Yancey, N. A. (2011). A review on biomass classification and composition, co-firing issues and pretreatment methods. *American Society of Agricultural and Biological Engineers Annual International Meeting 2011, ASABE 2011*, *3*, 2053–2083. https://doi.org/10.13031/2013.37191

Tursi, A. (2019). A review on biomass: Importance, chemistry, classification, and conversion. *Biofuel Research Journal*, *6*(2), 962–979. https://doi.org/10.18331/BRJ2019.6.2.3

Tyagi, S. K. Kothari, R. & Tyagi, V. V. (2019). Recent advances in biofuels in India, *Biofuels*, 10:1, 1–2, https://doi.org/10.1080/17597269.2018.1532732

3 Hydrothermal Processing of Biomass

Arunodhaya Natarajan, Umaiyambika Neduvel Annal, and Renganathan Sahadevan

3.1 INTRODUCTION

Hydrothermal processing of biomass is an innovative and promising technology that harnesses the power of water and heat to convert organic materials into valuable energy products and resources. In a world striving for sustainable and renewable alternatives to fossil fuels, this process has garnered significant attention due to its ability to efficiently transform various types of biomass, including agricultural residues, forest waste, algae, biofuels, chemicals, and other valuable products. Hydrothermal processing involves subjecting biomass to high temperatures and pressures in a water-rich environment, typically within the supercritical or subcritical regime [1]. The unique properties of water under these conditions facilitate the breakdown of complex biomass structures, such as cellulose, hemicellulose, and lignin, into simpler components like sugars, organic acids, and other valuable intermediates. These products can then be further refined into biofuels like bioethanol and biodiesel, as well as platform chemicals used in various industrial processes. One of the key advantages of hydrothermal processing is its feedstock flexibility. Unlike traditional biofuel production methods that rely heavily on specific types of feedstocks, hydrothermal processing can utilize a wide range of biomass sources, including those that are considered low-value or waste materials [2]. This not only reduces the competition for food crops but also addresses waste management challenges by transforming agricultural and forestry residues into valuable resources. Furthermore, hydrothermal processing offers environmental benefits. The process can significantly reduce greenhouse gas emissions compared to fossil fuel counterparts, contributing to the mitigation of climate change. Additionally, the potential to recycle nutrients from biomass, such as phosphorus and nitrogen, can help minimize the environmental impact associated with traditional agriculture practices. As research and development in hydrothermal processing continue to advance, challenges related to process efficiency, cost-effectiveness, and scalability are being addressed. Engineers and scientists are working on optimizing reactor designs, understanding reaction kinetics, and exploring catalysts to enhance product yields and selectivity [3]. Moreover, the integration of hydrothermal processing with existing industrial infrastructure is being explored to enable a smoother transition toward a more sustainable energy landscape.

3.2 TYPES OF HYDROTHERMAL PROCESSING

There are three main types (Figure 3.1) of hydrothermal processing: hydrothermal carbonization (HTC), hydrothermal liquefaction (HTL), and hydrothermal gasification (HTG).

3.2.1 Hydrothermal Carbonization

Hydrothermal Carbonization (HTC), also known as hydrothermal carbonization or wet torrefaction, is a thermochemical process that converts biomass, organic waste, or other carbonaceous materials

DOI: 10.1201/9781003732907-3

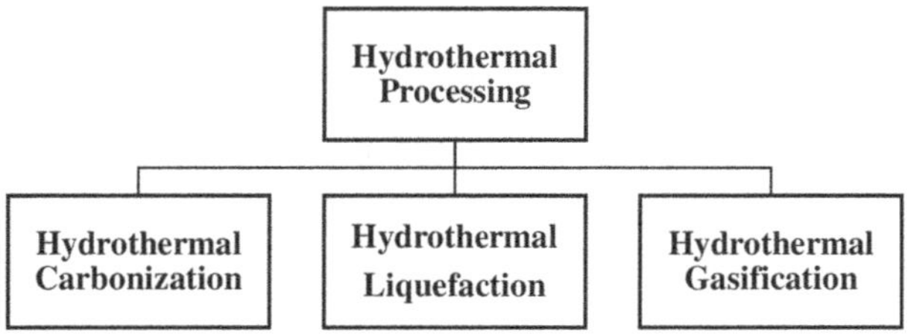

FIGURE 3.1 Types of hydrothermal processing.

into a carbon-rich solid material known as hydrochar or bio coal. This process occurs under high temperature and pressure conditions in the presence of water. The end product, hydrochar, resembles traditional coal in terms of energy content and physical properties but is produced from renewable sources.

The goal of HTC of biomass is to produce a charcoal-like carbonaceous fuel material (known as "hydrochar" or "HTC coal") with a high heating value. Typically, HTC of biomass processes the biomass at mild conditions with temperatures equal to 250°C or lower, pressure equal to 2 MPa or lower, and residence times of between 1 hour and several days [4]. Hydrothermal Carbonization method can convert between 50% and 80% of the feedstock into hydrochar. Hydrochar has a higher mass and energy density than raw biomass but is hydrophobic and fragile [5]. The byproducts of the HTC process water-soluble organic fractions include sugars, organic acids, and 5-Hydroxymethylfurfural (5-HMF), which have a variety of uses. Additionally, the so-called "high-temperature HTC process" allows for the HTC of biomass at higher temperatures between 300°C and 800°C, favoring the development of unique materials such as carbon nanotubes and graphitic carbon [4].

The mechanism of HTC is intricate and involves a number of simultaneous processes, including hydrolysis, dehydration, decarboxylation, condensation polymerization, and aromatization. To create oligomers and monomers, biomass components are hydrolyzed [6]. Dehydration and decarboxylation are the two main types of degradation events that occur with soluble extractables. Condensation polymerization processes continue in the interim, and hydrochar is subsequently produced. Other reaction pathways, such as transformation reactions, demethylation, pyrolytic reactions, and Fischer–Tropsch type reactions, have also been suggested. In addition to the aforementioned ones, dehydration and decarboxylation reactions dominate the HTC process, making it exothermic. In addition to being used as a solid fuel, hydrochars have found a number of promising applications due to the flexible operating conditions and end products, including absorbent for processing wastewater, carbon fixation, soil improvement, catalyst support, medicine, and gas sensors [7]. The physical and chemical characteristics (such as the shape, surface structure, and elemental makeup) of the produced hydrochar are influenced by a wide range of variables. The structure and composition of the biomass feedstock are two factors. For instance, carbonization of two different types of biomasses (bark mulch and sugar beet pulp) was examined under operating conditions (200°C, 3 h), and it was discovered that the aromaticity of bark hydrochars was higher than that of sugar beet hydrochar because the latter still contained lignin residues [8]. Lignin is thought to be rather robust during low-temperature HTC processing and hence persists in the hydrochars that are produced. Additionally, operational parameters such as temperature, pressure, and reaction time have a significant impact on the HTC process and hydrochar properties [9]. The response rate of HTC processes is significantly influenced by temperature. The speeds of paralleled reactions, like biomass hydrolysis and fragment polymerization, can be increased by raising the temperature. At increasing temperatures, the oxygen concentration of the generated hydrochar also decreases [10]. Reaction equilibrium changes to compressed solids and liquids as pressure is increased. Due to the increased density and pressure-capable penetration in the subcritical region, more compounds may be recovered from the biomass. The inclusion of catalysts has an impact on the characteristics of the hydrochar as well. Temperature and pressure changes during a process can be influenced by catalysts, which can result in higher carbonization efficiency and faster conversion rates. They also help regulate the final composition of the hydrochar and attributes, affecting its surface area, porosity, and chemical makeup. Catalysts can also change the characteristics of hydrochar, improving its suitability for a range of uses, including energy production, soil amendment, and carbon sequestration. Their influence on the process of producing hydrochar highlights how important it is to maximize the effectiveness and characteristics of this carbon-based substance.

For instance, Yu and his team found that the use of metal-based catalysts could have a considerable impact on the form of the carbon nanomaterials generated under low-temperature HTC conditions. While iron oxide nanoparticles might produce very fine, rope-like carbon nanostructures, the presence of iron ions could result in both hollow and large carbon microspheres [11].

3.2.2 Hydrothermal Carbonization Application

3.2.2.1 Pretreatment

For the majority of thermochemical processes, such as entrained flow gasification, pretreatment is required to eliminate the fibrous structure of biomass particles. HTC is a remarkable pretreatment method that improves the mechanical qualities of biomass, including its capacity to grind, as well as its heating value. The hydrochar is suited for boilers in heating systems because of several unique properties, such as a greater ignition temperature and a wide combustion range.

3.2.2.2 Human Waste/Sewage Sludge

According to statistics, Europe produces over 10 million tons of sewage sludge each year on a dry weight basis. Today, HTC is used in some plants to cleanse sewage sludge. The hydrochar created throughout the process demonstrates how the conversion increases the energy density, carbon content, and decreases its oxygen content of the biomass. Additionally, with proper temperature management, ash content is greatly reduced. Additionally, the liquid phase study reveals that it is biodegradable. By doing this, it is feasible to use the liquid phase energetically in the form of CH_4 [12].

3.2.2.3 Catalyst Production

Because of the high stability and customizable surface properties of hydrochar, catalyst synthesis from the HTC process has been mentioned multiple times for the creation of hydrogen (H_2). The experiment found that the HTC of algae with the goal of producing hydrogen created hydrochar with catalytic properties that increased phenol creation in the liquid phase while acid production was constrained.

3.2.2.4 Hydrothermal Liquefaction

Biomass is thermochemically transformed into liquid fuels by hydrothermal liquefaction when it is processed long enough in a hot, pressurized water environment to liquefy the solid biopolymeric structure primarily into liquid components. The temperature range for hydrothermal processing is 523–647 K, and the operating pressure range is 4–22 MPa. Using a liquid water processing medium, the procedure aims to obtain ionic reaction conditions while treating wet materials without drying them out. In biopolymers, the temperature is sufficient to begin pyrolytic mechanisms, and the pressure is adequate to maintain a liquid water processing phase. Based on how severe the working conditions are, hydrothermal processing is separated into three distinct procedures. It is referred to as hydrothermal carbonization when temperatures are lower than 520 K. The primary output is a hydrochar, which resembles low-rank coal in terms of its properties. It is possible to extract the lipids before hydrothermal carbonization in the case of microalgae, since the hydrochar is mostly made from the carbohydrate and protein fractions, while the lipid fraction is still whole. The process is known as hydrothermal liquefaction and results in the creation of a liquid fuel known as biocrude at intermediate temperature ranges between 520 and 647 K [13].

The HTL pathway at intermediate settings to manufacture a biocrude using continuous-flow reaction systems will be the main focus of the current review. Pyrolysis in hot liquid water is what is meant by hydrothermal liquefaction, also referred to as direct liquefaction. As a result, a catalyst is not necessary, but considerable research and development on catalytic processes in hydrothermal liquefaction has been done. The most frequently used "catalyst" has been the addition of alkali to the ionic medium in order to influence specific base-catalyzed condensation events, which can result in the synthesis of aromatic oils, over acid-catalyzed polymerization reactions, which result in the development of solid products. Although HTL uses pyrolytic processes, the biocrude it produces is very different from fast pyrolysis bio-oil. Through condensation reactions of the light fragments produced by the biomass, among other reaction routes, it becomes progressively deoxygenated, resulting in a more hydrophobic phase with less dissolved water [14]. Analyses of HTL products made from woody biomass demonstrate that the biocrude product is obviously not an analog of petroleum. It is a complicated mixture of substances that include oxygen, which differ widely in molecular weight. Acids, alcohols, cyclic ketones, phenols, methoxy-phenols (guaiacols from softwood lignin), and more condensed structures like naphthols and benzofurans were among the biocrude products. The amount of distillate product was determined by the amount of biocrude that was chromatographed. Because of inadequate resolution, peak overlap, or the absence of the complex, oxygenated isomers in the common mass spectrometry libraries, a significant portion of the chromatographed oil was not included in the quantified recognized peaks. Therefore, novel analytical methods are being investigated to comprehend, organize the biocrude peaks, and identify the composition of the biocrude accurately.

3.2.2.5 Biomass as an HTL Feedstock

The types of biomass taken as feedstock are given in Figure 3.2.

3.2.2.5.1 Carbohydrates

The polysaccharides cellulose, hemicellulose, and starch are the ones that make up the majority of the carbohydrates in biomass. Carbohydrates quickly hydrolyze in hydrothermal conditions to produce glucose and other saccharides, which are subsequently further broken down. Various carbohydrates hydrolyze at various speeds. While cellulose has a mostly crystalline structure, hemicelluloses and starch are hydrolyzed significantly more quickly than cellulose. Several writers have thoroughly researched the breakdown of carbohydrates in subcritical and supercritical water.

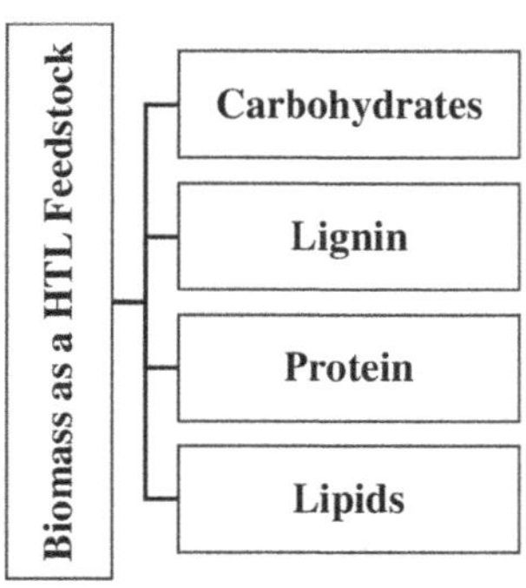

FIGURE 3.2 Types of biomasses taken as feedstock.

3.2.2.5.2 Cellulose

The glucose units that make up cellulose are connected by b-(1-4)-glycosidic linkages, as opposed to starch. Its straight chain structure allows for the development of potent intra- and intermolecular hydrogen bonding [4]. Consequently, cellulose has a high level of crystallinity and structural integrity. Its crystallinity prevents it from dissolving in water and protects it from being attacked by enzymes. However, cellulose is quickly dissolved in subcritical conditions and hydrolyzed to its component parts via hydrolysis. A suggested mechanism of action for the conversion of the microcrystalline cellulose employs 25 MPa subcritical and supercritical water.

Excellent research was done on the kinetics of the hydrolysis of several biopolymers. The authors calculated the rate constants for the cellulose, starch, and protein deterioration. The three separate biopolymers' hydrolysis rates varied quite a bit. These findings suggest that quick heating is necessary to prevent the depolymerization and degradation of some biopolymers before the proper reaction temperature is attained. It was shown that between 240 and 310°C, the rate of cellulose hydrolysis in water at 25 MPa increased tenfold, and that cellulose hydrolysis was much slower than starch hydrolysis. One hundred percent of the cellulose was converted in about 2 minutes at 280°C. The rate of glucose breakdown accelerated with temperature and surpassed the rate of glucose release between 250 and 270°C.

3.2.2.5.3 Hemicelluloses

Hemicelluloses amount to 20–40% of plant biomass. It is a heteropolymer made up of different monosaccharides, such as glucose, galactose, xylose, and mannose. In contrast to the wood hemicelluloses, which are primarily made of mannan, glucan, and galactan, grass hemicellulose is primarily made of xylan. Hemicellulose has a substantially lower degree of crystallinity than cellulose because of its abundance of side-groups and less regular structure. At temperatures above roughly 180°C, hemicelluloses are easily dissolved in water and hydrolyzed, with both acid and base catalyzing the process. Mok and Antal discovered that at 230°C, 2 min, and 34.5 MPa, nearly all of the hemicellulose in diverse wood and herbaceous biomass sources was hydrolyzed [15].

3.2.2.5.4 Starch

Another important component of biomass is starch, a polysaccharide made of glucose. A-(1-6) and b-(1-4) linkages join monomers. There are two main types of starch: amylopectin, which has a more branching structure than amylose, and amylose which has linear structure. Starch can be hydrolyzed more easily than cellulose. This article summarizes a few studies on starch hydrolysis at subcritical circumstances. Sweet potato starch was broken down in a batch reactor at hydrothermal conditions (180–240°C) without the use of a catalyst. Within a minute, the reaction temperature was attained. Fast heating is essential. When investigating these settings for hydrolysis and deterioration, it was noticed that starch was entirely solubilized after only 10 minutes at 180°C, although glucose production was insignificant [15].

3.2.2.5.5 Lignin

Lignin is a key component of plant materials, along with cellulose and hemicellulose. The p-hydroxy phenylpropanoid units are bound together by C-C or C-O-C bonds to form an aromatic heteropolymer. Trans-p-coumaryl alcohol, coniferyl alcohol, and sinapyl alcohol are the three fundamental building blocks.

Lignin is not easily broken down by enzymes or chemicals. However, other phenols and methoxy phenols are created during hydrothermal degradation by the hydrolysis of ether bonds. These products may potentially deteriorate much worse, though the benzene ring remains stable in these conditions as the hydrolysis primarily affects the methoxy groups. Alkaline pH catalyzes the hydrolysis of lignin. Liquification caused by the heat of lignin results in large quantities of solid residue; hence, the amount of the substrate's lignin content needs to be properly balanced [16].

3.2.2.5.6 Lipids

Fats and oils are primarily aliphatic nonpolar molecules. Triacyl glycerides (TAGs), which are triesters of fatty acids and glycerol, are what they are known as chemically. When the temperature is ambient, fats are insoluble in water. Water is more miscible under subcritical circumstances because its dielectric constant is much lower. In hot compressed water, TAGs are easily hydrolyzed; hence, catalysts are typically not needed. The generated free fatty acids, however, are comparatively stable in subcritical water [17]. At 330–340°C and 13.1 MPa, fast hydrolysis of soybean oil in subcritical water is successfully accomplished. 90–100% yields of free fatty acids were reached in just a few minutes. The phase behavior at subcritical settings could be studied using a reactor with transparent windows [18].

3.2.2.5.7 Proteins

Major components of biomass, proteins are particularly prevalent in animal and microbial biomass. One or more peptide chains, which are themselves polymers of amino acids, make up a protein. The peptide bond and amide bond between the carboxyl and amine groups of the amino acids are the structural bonds that hold amino acids together. Understanding protein degradation is crucial, since a sizeable portion of the nitrogen in proteins will be absorbed into the bio-oil during a hydrothermal liquefaction process, altering odor, combustion, and several other qualities. Only slow hydrolysis happens below 230°C because the peptide links in proteins are more durable than the glycosidic bonds in cellulose and starch. Since amino acids degrade rather quickly under hydrothermal circumstances compared to other biomass monomers, the amino acid yield during hydrothermal liquefaction is often much lower than attained during normal low-temperature acid hydrolysis.

3.2.3 Process Variations and Pathways

Different hydrothermal processing pathways are explored in this section. It covers subcritical and supercritical water gasification, hydrothermal liquefaction, and hydrothermal carbonization. Each pathway is discussed in terms of its specific reaction conditions, feedstock suitability, and major product outputs.

3.2.3.1 Pathway of HTL Process

The pathway of the hydrothermal liquefaction reaction is complex and not fully understood. However, it is generally believed that the following steps occur:

1. **Pretreatment**: The biomass is subjected to a pretreatment step, such as grinding or milling, to increase its surface area and make it more accessible to the reaction conditions.
2. **Hydrolysis**: The biomass is heated in water at high temperatures and pressures, which causes the breakdown of the complex molecules in the biomass into smaller, more soluble molecules.
3. **Reactions**: The soluble molecules then undergo a series of reactions, including dehydration, decarboxylation, and polymerization. These reactions lead to the formation of a variety of products, including bio-oil, bio-gas, and solid residue.
4. **Separation**: The products of the reaction are then separated from each other. The bio-oil is the most valuable product and can be used as a fuel or feedstock for other applications.

The specific reactions that occur during hydrothermal liquefaction depend on the composition of the biomass, the reaction conditions, and the presence of any catalysts. However, the overall process is a promising way to convert biomass into a variety of valuable products.

The main reactions that occur during hydrothermal liquefaction are as follows:

- **Hydrolysis**: The polysaccharides in the biomass are hydrolyzed into sugars.
- **Dehydration**: The sugars are dehydrated to form furans and aldehydes.
- **Decarboxylation**: The carboxylic acids in the biomass are decarboxylated to form alcohols and alkanes.
- **Polymerization**: The smaller molecules can react to form larger molecules, such as bio-oil.
- **Gasification**: Some of the smaller molecules can also be gasified to form syngas.

The steps in the hydrothermal carbonization (HTC) pathway are as follows:

1. **Pretreatment**: The biomass is subjected to a pretreatment step, such as grinding or milling, to increase its surface area and make it more accessible to the reaction conditions.
2. **Hydrolysis**: The biomass is heated in water under pressure at a temperature of 180–300°C, which causes the breakdown of the complex molecules in the biomass into smaller, more soluble molecules.
3. **Reactions**: The soluble molecules then undergo a series of reactions, including dehydration, decarboxylation, polymerization, and aromatization. These reactions lead to the formation of a variety of products, including hydrochar, bio-oil, and gas.
4. **Separation**: The products of the reaction are then separated from each other. The hydrochar is the most valuable product and can be used as a fuel or adsorbent.

The main reactions that occur during hydrothermal carbonization are as follows:

- **Hydrolysis**: The polysaccharides in the biomass are hydrolyzed into sugars.
- **Dehydration**: The sugars are dehydrated to form furans and aldehydes.
- **Decarboxylation**: The carboxylic acids in the biomass are decarboxylated to form alcohols and alkanes.
- **Polymerization**: The smaller molecules can react to form larger molecules, such as hydrochar.
- **Aromatization**: The smaller molecules can also be aromatized to form aromatic compounds, such as benzene and toluene.

3.2.4 Hydrothermal Gasification

Biocrude is comparable to petroleum crude and can be refined to the full range of fuel products made from petroleum distillates. The process is known as hydrothermal gasification (Figure 3.3) and produces synthetic fuel gas when temperatures rise over 647 K, where gasification reactions begin to predominate. The reduced amount of organic carbon present in the water phase after gasification results in high carbon efficiency, which is one advantage of hydrothermal gasification over liquefaction. To produce a product with a higher energy density by removing oxygen is the ultimate goal in each scenario.

3.2.4.1 High-Temperature Gasification

In this instance, biomass is gasified at high temperatures of 500–800°C in supercritical water without catalysts or with nonmetal catalysts, producing hydrogen-rich gas. The generation of H_2 is considerably accelerated by free radical reactions and water-gas shift reactions, which could eliminate

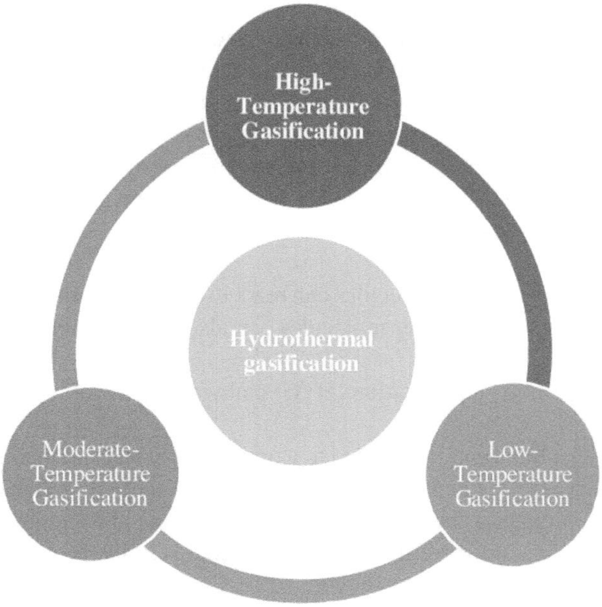

FIGURE 3.3 Different methodologies under the hydrothermal gasification process.

the need for catalysts. In contrast to pressure, temperature has a clear impact on the yield and composition of gases. Activated carbon and basic alkali salts are two frequent catalysts that may change the course of chemical reactions [19].

3.2.4.2 Moderate-Temperature Gasification

In this process, biomass is heated to temperatures between near-critical (about 300°C) and 500°C in subcritical or supercritical water with the help of catalysts in order to produce methane-rich gas. In most cases, catalysts are necessary to speed up the reaction and provide satisfactory product selectivity. In order to minimize the polymerization reactions and prevent the development of tars and chars, catalysts should be able to gasify the reactive intermediates (mostly phenols and furfurals) produced from the hydrolysis and dehydration of biomass feedstock quickly [19].

3.2.4.3 Low-Temperature Gasification

In this instance, biomass is processed to produce a gas combination in subcritical water at temperatures substantially lower than the water critical point (374°C). Since hydrogen may be created at far lower temperatures than in usual studies, this work is extremely important. For instance, the Pt/$Al20_3$ catalyst was used to create hydrogen from the gasification of glucose at low temperatures of about 200–300°C and pressures of 3–6 MPa [20].

3.2.5 Subcritical and Supercritical Gasification Process

Subcritical water gasification operates at temperatures and pressures below the critical point of water (critical point of water is approximately 374°C and 22.1 MPa). Typically, temperatures range from 100°C to 374°C, and pressures are below 22.1 MPa [21]. Subcritical water (also known as hot water or pressurized hot water) behaves as both a solvent and a reactant in this process. It has unique properties that enhance the breakdown of organic materials.

In subcritical water gasification, organic feedstock is mixed with hot water, and the mixture is pressurized. Under these conditions, the water acts as a reactant, participating in hydrolysis,

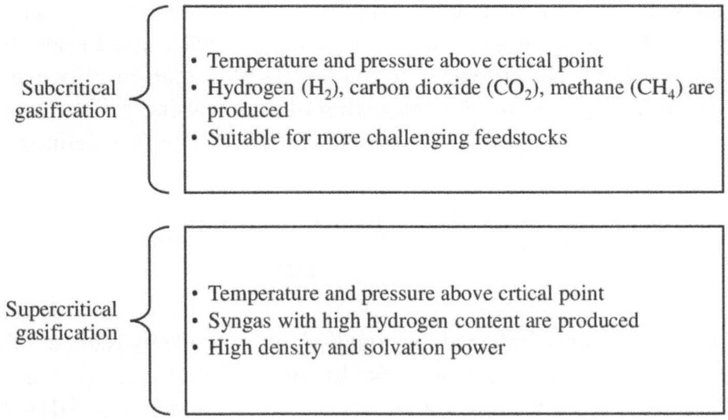

FIGURE 3.4 Comparison of subcritical and supercritical gasification.

oxidation, and other reactions to convert the organic compounds into gases, primarily hydrogen (H_2), carbon dioxide (CO_2), and methane (CH_4).

Subcritical water gasification has applications in waste-to-energy conversion, wastewater treatment, and the production of renewable hydrogen and synthesis gas (syngas) for various industrial processes [22].

3.2.5.1 Supercritical Water Gasification

Operating Conditions: Supercritical water gasification takes place at temperatures and pressures above the critical point of water. The typical operating conditions are temperatures above 374°C and pressures above 22.1 MPa. Supercritical water exhibits remarkable properties, including high density and solvation power, which enable efficient gasification of various organic materials. In supercritical water gasification, organic feedstock is mixed with supercritical water. The high temperature and pressure conditions lead to rapid and efficient gasification reactions, resulting in the production of gases like hydrogen, carbon monoxide, methane, and other hydrocarbons [23].

Supercritical water gasification is used in advanced applications, such as the conversion of organic waste into clean syngas, which can be used for electricity generation or as a feedstock for chemical processes. It is also explored as a means of efficiently converting biomass and plastics into valuable products. Both subcritical and supercritical water gasification processes utilize the unique properties of water at different temperature and pressure conditions to efficiently convert organic materials into useful gases and products. While subcritical water gasification is suitable for a range of applications, supercritical water gasification offers more advanced and high-temperature processing capabilities, making it suitable (Figure 3.4) for more challenging feedstocks and applications [24].

3.3 CATALYSTS AND ENHANCEMENTS

The utilization of catalysts to enhance reaction rates, selectivity, and overall process efficiency is a crucial aspect of hydrothermal processing. The interaction between catalysts and biomass constituents is examined to provide insights into the underlying catalytic mechanisms.

The impact of the catalyst can be comprehended on hydrothermal decomposition, and the products can be intricate. Catalysts influence multiple aspects of the process, leading to variations in both the quantity and quality of the resulting biocrude. A significant objective of catalysis is to maximize biocrude yield while minimizing the production of other byproducts like char, gases, and water-soluble compounds. Maximizing biocrude production is fundamental as it directly influences the process' economic viability.

These two tendencies, namely the enhancement of yield and the improvement of higher heating value (HHV), do not always occur simultaneously, creating challenges in determining the overall advantages of catalysis [25]. Nevertheless, it can be asserted that an effective catalytic process should have the capability to increase the proportion of biomass energy that is converted into biocrude. This can be quantified through the "Energy Ratio" (ER), which is defined as

$$Energy\ Ratio = \frac{Yield_{Biocrude}HHV_{Biocrude}}{HHV_{Feed}}$$

Conversely, catalysis also aims to enhance the "quality" of biocrude by reducing the content of oxygen, nitrogen, and other heteroatoms, thereby facilitating easier upgrading. This improvement can be quantified through the higher heating value (HHV), where an increased HHV directly correlates with a reduced heteroatom content.

While activity and selectivity are commonly used as primary metrics for evaluating catalyst performance, the stability of catalytic materials represents a critical third dimension. Stability, in this context, refers to the capacity of the catalyst to maintain its activity and selectivity over extended periods of time during continuous operation. Initial activity and selectivity are undoubtedly crucial starting points, but they must be complemented by long-term viability [26]. Assessing stability can be challenging due to the necessity for extended testing durations and the difficulty in replicating real-world, large-scale reaction conditions in laboratory-scale experiments. However, the catalyst's lifetime is of paramount importance, especially when considering further development and scaling up for industrial applications. Industrial catalytic processes typically achieve productivities that are 10^3 to 10^4 times the weight of the catalyst used before disposal. This underscores the necessity for catalysts to possess an adequate operational lifetime and/or the capability for straightforward regeneration to meet specific durability criteria [27, 28].

The biocrude produced by HTL is of greater quality than that produced by conventional pyrolysis because it has a higher heating value (HHV) and a lower content of oxygenated chemicals. During HTL, processes that result in CO_2, CO, and H_2O are used to partially remove the oxygen that originally existed in biomass macromolecules. Even if the quality is higher, the oxygen concentration is still high, producing a biocrude that is highly viscous, acidic, unstable, and has a poor heating value (HHV). Evidently, the operating circumstances and utilized biomass have a significant impact on its qualities [29, 30]. After upgrading processes like hydrotreating and catalytic cracking, the gasoline can be utilized as a substitute for transportation fuel.

3.3.1 Homogeneous Catalysts

Homogeneous catalysts are those that readily dissolve in the aqueous phase. This category encompasses substances like alkali metal carbonates and hydroxides (e.g., Na, K, Ca), as well as various acids such as formic acid, acetic acid, hydrochloric acid, sulfuric acid, and their aqueous solutions, which are widely recognized as effective homogeneous catalysts [31].

Homogeneous catalysts offer advantages such as affordability, accessibility, and their ability to effectively inhibit char formation. However, they also come with certain drawbacks. For instance, their recovery after the reaction can pose challenges due to their high solubility in the aqueous phase [32]. Furthermore, alkali hydroxides may exhibit corrosive properties toward the reactor system, although this issue can be mitigated by selecting appropriate anti-corrosive materials for the hydrothermal liquefaction (HTL) reactor [33].

3.3.1.1 Base Catalysts

Over the years, alkali catalysts have demonstrated significant promise in enhancing both the yield and quality of biocrude [34, 35]. Alkali catalysts facilitate biomass conversion by promoting extensive depolymerization, deoxygenation, desulfurization, and denitrogenation processes. This leads to a reduction in the concentration of heteroatoms present in the resulting biocrude [36]. Additionally, alkali catalysts play a vital role in catalyzing water gas shift reactions and effectively preventing the formation of char during the hydrothermal liquefaction (HTL) process. It is worth noting that addressing char formation can be particularly challenging when dealing with high-lignin biomass, which tends to yield substantial amounts of char, possibly due to its polyaromatic ring structure [37].

To investigate the role and reactivity of catalysts with various biomass components such as carbohydrates, proteins, and lipids, a diverse range of biomass sources has been selected, including lignocellulosic materials [38], manures [39], microalgae, sewage sludge [40, 41], and model compounds [42].

Numerous studies have highlighted the favorable impact of alkali catalysts on lignocellulosic and carbohydrate-rich biomass. However, investigations involving biomass with higher proportions of proteins (>50%) and lipids (7 to 30%) have yielded varying conclusions, largely dependent on the carbohydrate content [43]. In nitrogen-containing feedstocks, the peptide linkages in proteins exhibit greater stability compared to carbohydrates, rendering alkaline hydrolysis less effective in cleaving peptide linkages when compared to glycosidic bonds [44].

In the case of high-protein feedstocks like sewage sludge and microalgae, the catalytic influence of alkalis is contingent on factors such as the presence of carbohydrates and ash content [45]. In some instances, the inherent ash content, comprising alkaline inorganic compounds like carbonates and oxides, within the feedstock can lead to the deactivation of the catalyst as it dissolves in the aqueous phase [46].

Aquatic biomasses, particularly microalgae and macroalgae, have long been considered promising candidates for biofuel production. Generally, for the liquefaction of microalgae, it is preferable to employ low to moderate temperatures (280–350°C) [47]. Numerous studies have demonstrated that the use of alkali catalysts can have detrimental effects on the conversion of microalgae, resulting in reduced biocrude yields and quality. The primary reason for the inefficacy of alkali catalysts in this context is their incompatibility with cleaving peptide linkage bonds found in protein-rich biomass [48].

3.3.1.2 Acid Catalyst

Acid catalysts, including formic acid (HCOOH), acetic acid (CH_3COOH), hydrochloric acid (HCl), and sulfuric acid (H_2SO_4), have found application in numerous instances of biomass conversion [49–52]. However, it's worth noting that acid catalysts are generally considered less promising than alkali salts, particularly when it comes to lignocellulosic biomass. For instance, in a specific case involving desert shrub biomass, treatment in an acetic acid medium resulted in a moderate increase in biocrude yield, from 26% to 30%, with no discernible impact on calorific values [53].

3.3.2 Heterogeneous Catalyst

Heterogeneous catalysts play a crucial role in the hydrothermal liquefaction (HTL) process of biomass. Catalysts enhance the efficiency and selectivity of the HTL process. Following are some common types of heterogeneous catalysts used in HTL of biomass (Table 3.1).

3.3.2.1 Transition Metal Catalysts

Transition metals serve as highly active catalysts in various chemical reactions, encompassing oxidation, hydrogenation, hydrocracking, gasification, and more [61]. In the context of hydrothermal liquefaction (HTL) of lignocellulosic biomass, several key transition metals have been explored, including Nickel, Copper, Cobalt, Iron, Manganese, Zinc, Vanadium, Palladium, and Ruthenium. These metals can promote the decomposition of biomass components and facilitate hydrogenation reactions. Incorporating these metals into the HTL process leads to a notable increase in liquid

TABLE 3.1
Different Homogeneous Catalysts Used in Earlier Works

Homogeneous Catalyst	Biomass Type	Important Observations	Reference
KOH, K_2CO_3	Castor residue	Bio-oil yield was improved	[54]
Ethanol and acetic acid	Lipids	Hydrolysis, transesterification, and esterification process yields were improved	[55]
NaOH	Blackcurrant pomace	Bio-Oil yield was more, and biochar yield was less	[56]
K_2CO_3, Ca $(OH)_2$, and Ba $(OH)_2$	Pulp and paper sludge	Large number of organic compounds were shifted to water phase	[57]
KOH and K_2CO_3	Water hyacinth	High bio-oil yield at low residence time	[58]
Fe SO_4	Birchwood sawdust	Improved HHV and yield of bio-oil	[59]
K_2HPO_4	Cellulose	Less nitrogen content and more HHV	[60]

TABLE 3.2
Transition Metal Oxides Used as Heterogeneous Catalyst

Transition Metal	Catalyst wt %	Important Observations	Reference
Ni	10	Hydrogenation activity had been increased	[53]
Ni/Si-Al	2	Enhancement in the yield of biocrude	[54]
CO_3O_4	5	Fischer–Tropsch mechanism induced high yield of biocrude	[55]
Fe	10	Improvement in biocrude yield with best decomposition of phenolic compounds	[56]
FeS	5	Less biochar formation and more biocrude yield	[57]
Zn	20	Better quality of biocrude yield due to the activity of Zn in reaction pathway mechanisms	[58]
Zn	10	Better catalyst for gas and aqueous phase	[59]

product yields, primarily attributable to their catalytic involvement in isomerization, retro-aldol condensation, and hydrogenation reactions. Additionally, many of these metals can engage in redox reactions with water under HTL conditions, generating active H_2. Among these catalysts, iron-based catalysts appear to exhibit the highest activity in converting biomass into liquid products [62].

Nickel-based catalysts, such as Ni-Mo and Ni-Co catalysts, are known for their ability to enhance hydrogenation reactions during HTL, leading to higher biocrude oil yields (Table 3.2).

3.3.2.2 Alkaline Metal Oxide Catalysts

Heterogeneous catalysts based on alkaline earth metals find extensive use owing to their alkaline properties, which closely resemble those of commonly employed homogeneous alkaline catalysts. These catalysts primarily facilitate the decomposition of carbohydrates. Alkaline earth metal-based catalysts investigated in the hydrothermal decomposition encompass calcium oxide, colemanite ($Ca_2B_6O_{11}\cdot5H_2O$), magnesium-based oxides (MgO and MgMOx where M can represent Mn, Ni, Fe, Cr, Zn, and Al) and hydrotalcite ($Mg_6Al_2(CO_3)$ $(OH)_{16}\cdot4H_2O$). With the exception of $MgMnO_2$, all

TABLE 3.3
Alkaline Metal Oxides Used as Heterogeneous Catalyst

Alkaline Metal Oxides	Catalyst wt %	Important Observations	Reference
CaO	10	Less improvement in biocrude	[70]
Colemanite	10	High yield in biocrude	[71]
MgO	5	Biocrude improvement is less because of catalyst deactivation caused by the reactivity of the basic catalyst sites with CO_2 and H_2O	[72]
$MgMnO_2$	10	High biocrude yield due to changes in lignin depolymerization reactions	[73]
Hydrotalcite	5	High yield of biocrude	[72]
SiO_2	5	High HHV of 37 MJ/Kg	[74]
CaO	5	The yield was 1.4 times more than the process in the absence of catalyst	[75]

studies indicate that the incorporation of alkaline earth metals results in an increased yield of biocrude, albeit with a decline in its quality. When compared to biocrude produced without catalysts, the catalyzed biocrudes exhibit a lower higher heating value (HHV) due to an elevated concentration of oxygenated compounds (Table 3.3).

3.3.2.3 Zeolites

Zeolites are crystalline aluminosilicate materials with well-defined pores and acidic sites. They can act as catalysts and adsorbents in HTL processes [76]. Zeolites are characterized by their highly porous aluminosilicate crystal structures, forming an intricate three-dimensional network [77]. These porous structures are instrumental in trapping molecules as small as 1 nanometer, giving rise to the creation of channels or cages. The chemical composition plays a pivotal role in shaping this three-dimensional framework, and zeolites undergo chemical alterations during their production process. These modifications encompass charge balancing, the incorporation of organic groups, and the substitution of oxygen. Zeolites are used to enhance the cracking and deoxygenation of biomass components, which can increase the yield of biocrude oil and reduce oxygen content.

Zeolite catalysts exhibit an intermediate behavior in terms of their impact on deoxygenation reactions in comparison to their role in enhancing the quality of biocrude. The mean relative values of higher heating value (HHV) and biocrude yields are considered as classification-related indicators of the catalytic conversion process. The introduction of alkaline earth metals tends to increase the water-insoluble (WI) content, which adversely affects the quality of biocrude. Conversely, when zeolites are employed, there is a decrease in biocrude yield but with an accompanying increase in HHV, indicating a trade-off between quantity and quality.

Technologies for the production of renewable fuels have been extensively utilized for an extended period. Commercially available options like biodiesel and bioethanol are derived from processes such as vegetable oil transesterification and sugar fermentation, respectively (Table 3.4).

3.3.2.4 Mixed-Metal Catalysts

Mixed-metal catalysts combine two or more metals to create synergistic effects that enhance the catalytic activity and selectivity. Catalysts like Ru-Ni, Ru-Cu, or Co-Mo are examples of mixed-metal catalysts employed in HTL to improve the deoxygenation and hydrogenation of biomass (Table 3.5).

TABLE 3.4
Zeolites Used as Heterogeneous Catalyst

Zeolites	Catalyst wt %	Important Observations	Reference
HZSM-5	20	Oxygenated compounds were broken into hydrocarbons	[78]
ZSM-5	30	Quality of biocrude was improved due to esterification and decarburization	[79]
Y zeolite	29.3	High de-oxygenation of algal biomass	[80]
ZSM-5 (from bagasse)		Prefer hydrogenation and methanation reaction	[81]
Zeolite Na-Y	10	High thermal stability	[82]

TABLE 3.5
Mixed Metals Used as Heterogeneous Catalyst

Mixed Metal	Catalyst wt %	Important Observations	Reference
Zno-ZrO_2	10.05	Hydrothermal degradation occurred at comparatively low temperature and pressure	[83]
TiO_2 +ZnO	10–30	Better thermal stability	[84]
CeZrOx	-	Biocrude oil was more than biochar formation	[85]
CaO, CeO_2, La_2O_3, MnO_2, and Al_2O_3	1:10	Decreased nitrogen content and increased oxygen content	[86]
$MgMnO_2$	-	Value-added phenolic monomers are processed more easily with the aid of a catalyst	[87]
ZrO_2	10	The yield and quality of biocrude were negatively impacted by ZrO_2	[88]
$ZnCl_2$, $CuCl_2$, and $NiCl_2$	-	Low water content and high HHV	[89]

3.3.2.5 Supported Catalysts

Catalysts can be supported on various materials, such as alumina (Al_2O_3), silica (SiO_2), or activated carbon. These supported catalysts provide a stable and high-surface-area platform for catalytic reactions. Supported metal catalysts, like Ni/Al_2O_3 or Ru/SiO_2, are commonly employed to improve the overall efficiency of biomass liquefaction (Table 3.6).

3.3.2.6 Sulfided Catalysts

Sulfided catalysts, often derived from metal sulfides, can be effective in promoting the removal of sulfur and nitrogen impurities during HTL. Molybdenum sulfide (MoS_2) and cobalt-molybdenum sulfide (CoMoS) catalysts are used to reduce sulfur and nitrogen content in the biocrude.

3.3.2.7 Acidic Catalysts

Acidic catalysts, such as solid acid catalysts and zeolites with Bronsted or Lewis acid sites, can facilitate acid-catalyzed reactions, including dehydration and isomerization, in HTL processes.

The choice of catalyst in HTL depends on the specific biomass feedstock and the desired product composition. Catalysts can enhance the overall yield of valuable products, reduce reaction times, and improve the quality of biocrude oil produced during the HTL process.

TABLE 3.6
Supported Heterogeneous Catalyst Used in Earlier Works

Supported Catalysts	Biomass Type	Important Observations	Reference
Co/CNTs	Algal biomass	Low O/C ratio	[90]
NiO/SAPO-34, NiO/ZSM-5,NiO/USY, NiO/γ-Al_2O_3, NiO/SiO_2	Spirulina	Higher % of low light components were obtained	[91]
CaO supported on CeO_2, Al_2O_3, and ZrO_2	Sargassum	High bio-oil yield and high HHV	[92]
5%Ga/NiFe-LDO/AC	Macroalgae	Bio-oil yield was better	[93]
Ni-Al/MCM-41	Lignin	Decomposition of lignin was perfect	[94]

3.4 APPLICATIONS AND PRODUCTS

The diverse range of products obtained from hydrothermal processing is discussed in this section. Emphasis is placed on bioenergy carriers like bio-oil and syngas, as well as high-value bioproducts such as specialty chemicals, platform molecules, and carbonaceous materials. The potential applications of these products in the bioenergy and chemical industries are explored.

The valuable products obtained from bio-oil are given in Figure 3.5. Each of the products has a commercial value.

The use of syngas is growing as a way to produce clean and renewable energy. Syngas can be produced from a variety of feedstocks, including biomass, coal, and natural gas. This makes it a versatile and sustainable fuel source. The important uses of syngas are as follows:

Fuel: Syngas can be used as a fuel for vehicles, power plants, and other applications. It can be burned directly in engines or gas turbines, or it can be converted into other fuels, such as methanol or dimethyl ether.

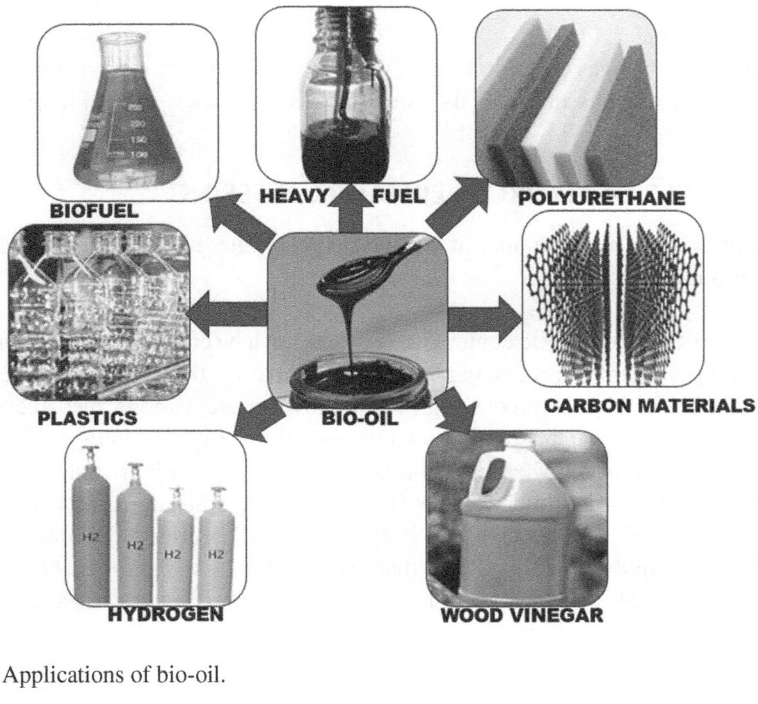

FIGURE 3.5 Applications of bio-oil.

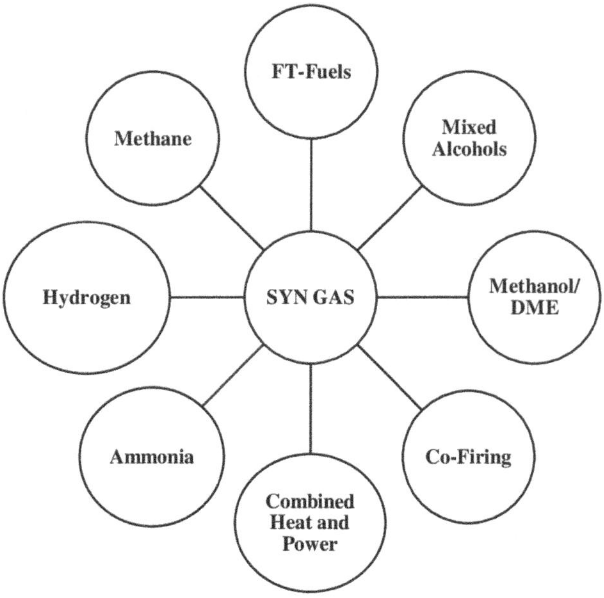

FIGURE 3.6 Applications of syngas.

Chemicals: Syngas can be used to produce a variety of chemicals, including ammonia, methanol, and synthetic fuels. Ammonia is used as a fertilizer and in the production of plastics and explosives. Methanol is used as a solvent and in the production of biodiesel and other fuels. Synthetic fuels can be used to replace gasoline and diesel fuel.

Hydrogen production: Syngas can be used to produce hydrogen, which is a clean and renewable fuel. Hydrogen can be used in fuel cells to generate electricity or in other applications.

Iron ore reduction: Syngas can be used to reduce iron ore to iron, which is a key step in the production of steel.

Other applications: Syngas can also be used in other applications, such as the production of glass, ceramics, and pharmaceuticals. The uses of syngas are listed in Figure 3.6.

3.5 ADVANTAGES OF HYDROTHERMAL PROCESSING

Hydrothermal processing has a number of advantages over other biomass conversion technologies. These advantages include:

- More than 99% of the organic content in feedstocks can be converted to fuels or inert products using hydrothermal processing.Wet wastes make up the majority of the feedstocks suitable for hydrothermal processing. The following are various feedstocks that have undergone hydrothermal processing testing, including: Lignocellulosic Biomass: Wood (e.g., pine sawdust, beech wood, oak wood), lignin, simple carbohydrates (cellulose, glucose, starch), and agricultural residues (rice husk, barley straw, sugarcane bagasse, sorghum bagasse) [3], [7], [12], [39], [42], [48], [64]); Aquatic Biomass: Microalgae (Spirulina, low-lipid microalgae) and macroalgae (*Fucus vesiculosus*, *Gracilaria corticata*). [14], [22], [24], [91], [93]; Waste Streams: Food processing wastes, wastewater solids, and pulp/paper-mill sludge. [1], [16], [57].

- Hydrothermal processing solves three problems at once while processing wet wastes:
  - It disposes of the wet waste material in a clean manner.
  - It generates renewable fuel.
  - It generates clean, clear, sterile water.
- No other wet or dry process can match hydrothermal processing's technical and financial efficiency. While wet processes like anaerobic digestion typically only convert half or less of the feedstock to usable energy, dry processes like pyrolysis waste a significant amount of energy simply drying the feedstock.
- Hydrothermal processing produces oil and gas of a higher caliber than any other method. Compared to pyrolysis oil, biocrude oil is significantly less oxygenated and contains significantly less water, making it easier and less expensive to refine. It is comparable to a light, sweet crude, since it has a low sulfur content. The gas is pure and has very no water, silanes, nitrogen, or sulfur in it. It may be quickly stripped of carbon dioxide and burned directly or inserted into a natural gas pipeline.
- The world's energy supply might be significantly increased by the renewable fuels produced by hydrothermal processing, which could account for up to 25% or more of the current fossil fuel supply for transportation. This goes well beyond what wind and solar power now contribute.

3.6 ISSUES WITH HYDROTHERMAL TECHNOLOGIES

3.6.1 Economic Considerations

Economic factors matter when implementing new technologies as well as when evaluating new technologies themselves. The profitability of a technology and the expenditures involved in optimizing it are both influenced by a plant's economic viability.

Process competitiveness and viability can be compared to well-known conventional technologies based on an economic analysis. Cost evaluations are available for the biomass-based hydrothermal HTL or HTG process, even though various techno-economic assessments have been conducted for thermochemical-based conversion processes, such as fast pyrolysis and traditional gasification. Under the auspices of the National Advanced Biofuels Consortium (NABC), the Pacific Northwest National Laboratory (PNNL) carried out bench-scale HTL and upgrading studies with woody biomass. For two cases, a state-of-technology (SOT) case with experimental results from the HTL process and a goal case assuming plausible future advancements for mature HTL technology, the techno-economic study included the development of a large-scale commercial HTL and upgrading platform for bio-oil production. The findings demonstrated that the goal case's production costs were lower because higher product yields and less wastewater treatment were predicted as a result of decreased organics loss to the aqueous phase.

The cost outcomes from the SOT scenario demonstrated that the cost of producing bio-oil based on the current HTL technique is not comparable with gasoline made from petroleum. Although the goal case findings for the production of bio-oil from woody biomass by HTL appear promising, there are financial concerns due to a lack of process expertise and conception. Feedstock costs, product yield, and the cost of upgrading equipment are the key variables affecting the cost of producing bio-oil. In future research investigations, it will be required to identify key parameters. Additionally, hydrothermal liquefaction (HTL) and upgrading procedures for lipid-extracted microalgae (LEA) liquid fuels have been the subject of techno-economic studies. Compared to traditional gasoline and diesel, the results showed that the HTL process is promising for the generation of liquid fuel. The price of the feedstock was unpredictable, yet this had a significant impact on production costs. Product yield and equipment upgrade costs were two other important variables affecting manufacturing costs.

3.6.2 Gaps in Knowledge

The design and configuration of the reactor are key factors in the operation of the process and have an impact on reaction kinetics. Enhancing heat integration, managing potentially poor heat transfer caused by contact between incoming reactor effluent and reactor feed due to the feed's high viscosity, and lowering the costs of the reactor system itself when operating at high pressures are the main challenges in reactor design. To address these issues, it is necessary to conduct an experimental investigation of the necessary heat transfer coefficients at various points throughout the process. Additionally, the material choice for the HTL reactor design must be assessed while taking into account the abrasive reaction conditions and any corrosive consequences. To increase the acceptable liquid hourly space velocity (LHSV) in the HTL reactor system, extensive research is required.

High solids content must be handled by the pump. At the pressure and temperature of the reactor, it is yet unknown whether it is possible to separate the bio-oil from the water. This is crucial since effective bio-oil separation from an aqueous phase will boost bio-oil production. We will learn how to optimize reactor design by looking at typical reaction rates and byproducts from biomass processing using hydrothermal technology. Temperature, feed material content, biomass type, and residence duration are only a few of the variables that affect HTL bio-oil output. It is necessary to characterize each of the hydrothermal processing's outputs in great detail, including the bio-oil, aqueous, gaseous, and solid products. Understanding bio-oil stability and quality will help you better understand ongoing process reactions and upgrading requirements, but it will take some work. Where solutions for moving bio-oil offsite (where an upgrading facility is not located alongside an HTL plant) are identified, this effort will also be crucial.

Understanding the types of reactive species influencing product quality and yield requires using characterization techniques that make use of tools like GC/MS, NMR, and HPLC for product analysis. However, due to the low resolution and restricted selectivity of analytical techniques like chromatography, it is impossible to forecast high molecular weight substances with any degree of accuracy. To comprehend process development for commercial applications, a study on the continuous flow system is required. Catalyst performance and yield are highly influenced by the catalyst. Research on catalyst upkeep, stability, conceivable regeneration, and further lifetimes is lacking.

A method may be practically practicable if it is economically viable. Cost analyses alone, however, are insufficient to enable the development of large-scale processes. Testing with several feedstocks is necessary for commercialization in order to understand the process. Additionally, understanding energy flow and consumption, which have a direct impact on cost, is made easier by combining a technoeconomic study with energy tools. Internal heat and power recovery in an integrated system would lower the need for and expense of external energy, lowering technology costs. To adequately assess the technology, sensitivity analysis and Monte Carlo simulations for risk assessments are required.

Although there are a number of difficulties with hydrothermal technologies, including environmental issues, the depletion of fossil fuel reserves, etc., research is still being done on specialized fuels for the transportation industry as well as raw materials for the chemical industry. In light of this, hydrothermal technologies show great promise, and research and development work to remove the obstacles to the technology's market integration is ongoing.

3.7 CURRENT STATUS OF HYDROTHERMAL TECHNOLOGIES

There have been many recent advancements in the hydrothermal processing of biomass. The most important one is the development of new catalysts to improve the efficiency and selectivity of the process. One of the most promising recent advancements is the development of new catalysts.

Catalysts can help to accelerate the reaction rate and improve the selectivity of the process, leading to higher yields of valuable products. Enzymes are biological catalysts that can be used to accelerate the hydrothermal conversion of biomass. Enzymes are specific to their substrate, meaning that they can only catalyze the conversion of one or a few specific compounds. This specificity can be an advantage in the hydrothermal conversion of biomass, as it can help to improve the selectivity of the process.

Cellulases are a group of enzymes that break down cellulose, a major component of biomass. Cellulases can be used to produce sugars from cellulose, which can then be used to produce bioethanol or other fuels and chemicals. Hemicellulases are a group of enzymes that break down hemicellulose, another major component of biomass. Hemicellulases can be used to produce sugars from hemicellulose, which can then be used to produce bioethanol or other fuels and chemicals. Ligninases are a group of enzymes that break down lignin, a third major component of biomass. Ligninases can be used to produce bio-oil from biomass, which can then be used as a fuel or feedstock for other applications.

Enzymes can be used in the hydrothermal conversion of biomass in a number of ways. They can be used to pretreat the biomass in order to break down the complex molecules into smaller, more easily converted molecules. They can also be used to catalyze the conversion of the biomass into specific products, such as bioethanol or bio-oil. For example, the use of copper catalysts has been shown to improve the yield of bio-oil from hydrothermal processing of biomass. Another recent advancement is the use of microwaves or ultrasound to enhance the reaction rate. Microwaves and ultrasound can both be used to increase the temperature of the reaction mixture, which can lead to faster reaction rates. Additionally, microwaves can also help to break down the biomass into smaller molecules, which can also improve the efficiency of the process. The use of continuous processing systems is another recent advancement that is helping to improve the economics of hydrothermal processing. Continuous processing systems can produce bio-oil more efficiently than batch processing systems, which can lead to lower cost (Figure 3.7).

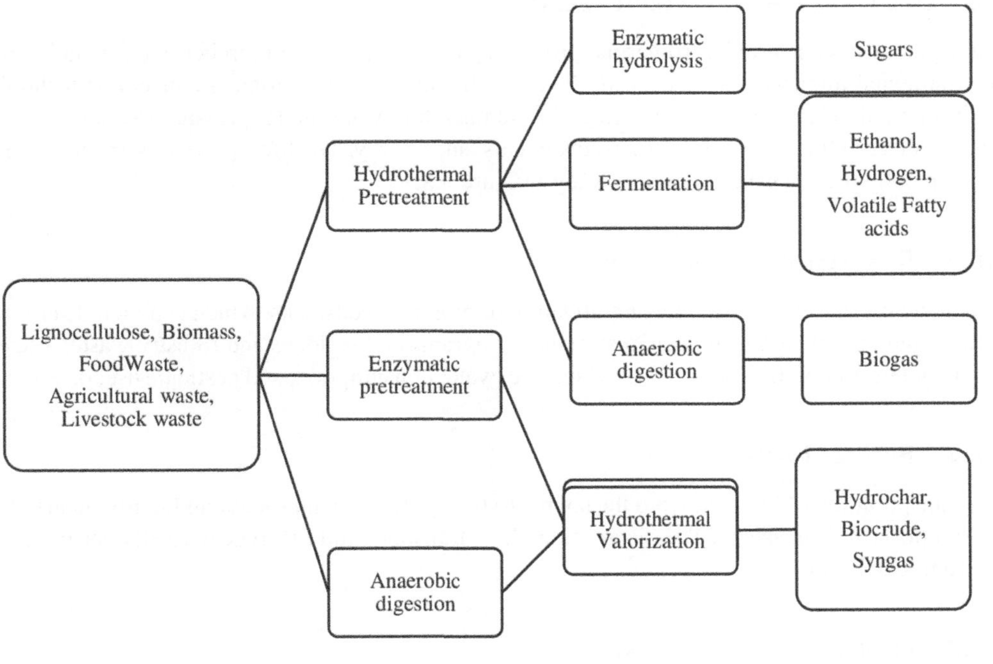

FIGURE 3.7 Recent works in hydrothermal processing.

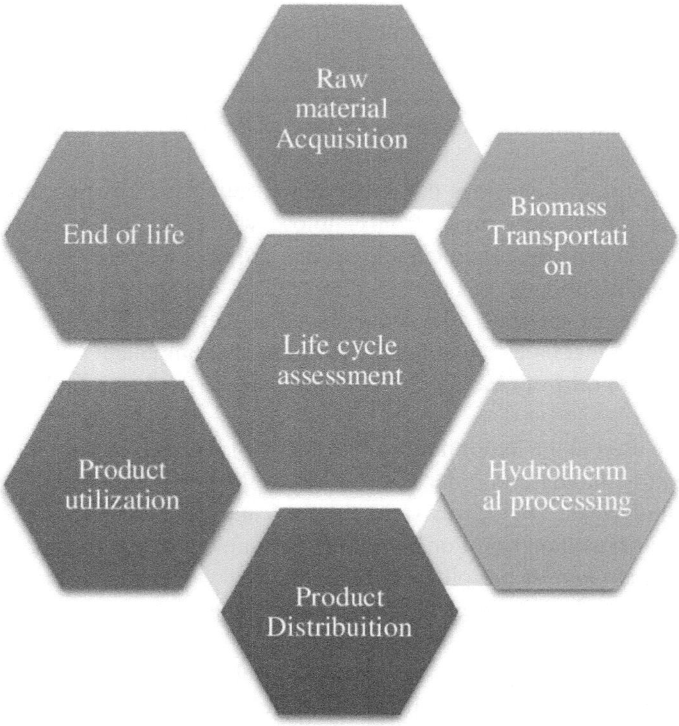

FIGURE 3.8 Stages in life cycle assessment of hydrothermal processing.

3.8 LIFE CYCLE ASSESSMENT (LCA)

Life Cycle Assessment of hydrothermal processing of biomass is a comprehensive evaluation of the environmental impacts associated with the entire life cycle of this biomass conversion technology. Hydrothermal processing involves using high-temperature water under pressure to convert biomass into various products, such as biofuels, chemicals, and energy. The life cycle assessment of the process typically includes the stages as given in Figure 3.8.

3.8.1 Raw Material Acquisition

This stage involves the extraction or cultivation of biomass feedstocks, which can include crops like corn, sugarcane, or lignocellulosic materials like agricultural residues and forestry waste. The environmental impact of this stage includes land use, water consumption, and pesticide use, if applicable.

3.8.2 Biomass Transportation

The transportation of biomass from the source to the hydrothermal processing facility incurs energy consumption and emissions. The distance, mode of transport, and fuel type used all contribute to the environmental impact.

3.8.3 Hydrothermal Processing

This is the core of the LCA, focusing on the energy and water consumption during the hydrothermal treatment, as well as the emissions of greenhouse gases and other pollutants. It's essential to consider the energy source used for heating the water and the efficiency of the process.

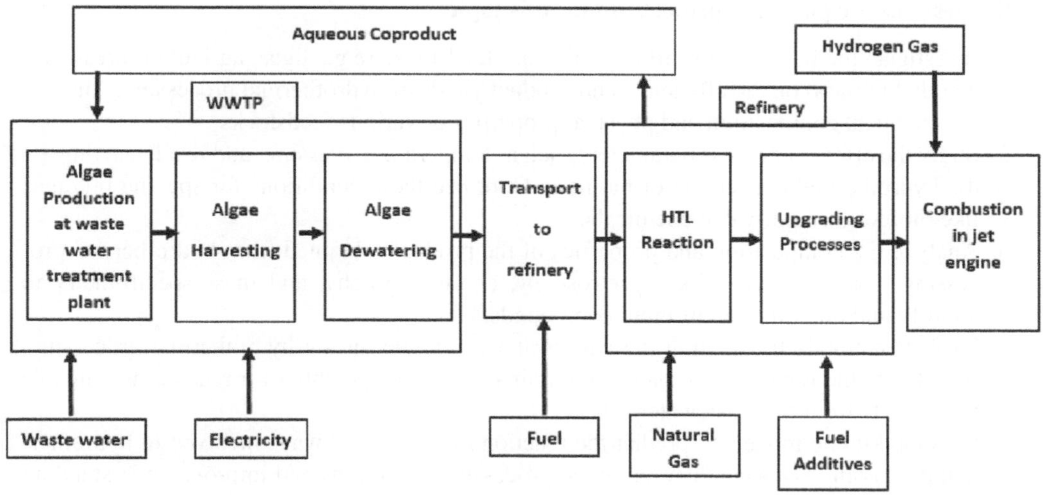

FIGURE 3.9 Life cycle assessment of hydrothermal processing of biomass.

3.8.4 Product Distribution

The transportation of hydrothermal processing products to end-users or markets is another stage to evaluate. This stage includes the energy and emissions associated with transportation.

3.8.5 Product Utilization

Depending on the products obtained from hydrothermal processing (e.g., biofuels, chemicals, or heat and electricity), their utilization can have various environmental impacts. For instance, if biofuels are produced, their combustion in engines would generate emissions, which need to be accounted for.

3.8.6 End-of-Life

This stage considers the disposal or recycling of equipment and materials used in the hydrothermal processing facility at the end of their life cycle.

3.8.7 Byproducts and Waste Management

Any waste generated during the hydrothermal processing must be managed appropriately, and the environmental impact of waste disposal or treatment should be included (Figure 3.9).

3.9 CONCLUSION

Hydrothermal processing of biomass is a promising method that shows great potential in converting different types of biomass into valuable products like biofuels and biochemicals. This approach comprises techniques such as hydrothermal liquefaction, gasification, and carbonization, each with its unique advantages and pathways. The use of catalysts is crucial in improving the efficiency of these processes. While hydrothermal processing offers benefits in terms of feedstock flexibility and potential environmental advantages, it is important to consider the energy requirements for the management of catalysts and wastewater. To ensure sustainability, it is essential to conduct a life cycle assessment (LCA) to evaluate and optimize the impact of the entire process. In all, hydrothermal processing holds promise as an important contributor toward a more sustainable and circular utilization of biomass.

Suggestions for pursuing projects under this topic

1. Investigate the influence of different biomass feedstocks (e.g., algae, agricultural residues, woody biomass) on the efficiency and product yields of hydrothermal processing. Analyze the chemical composition and physical properties of various feedstocks.
2. Study the effect of process parameters such as temperature, pressure, and residence time on the hydrothermal conversion of biomass. Optimize these conditions for specific products like biofuels or high-value chemicals.
3. Analyze the composition and properties of the products obtained from hydrothermal processing. Use techniques like spectroscopy, chromatography, and mass spectrometry to identify and quantify the different compounds.
4. Explore methods for treating the wastewater generated during hydrothermal processing. Investigate the removal of organic contaminants and the potential for recovering valuable byproducts from the wastewater.
5. Develop kinetic models to predict the reaction rates and pathways involved in hydrothermal processing. This can help optimize processing conditions and improve understanding of the underlying chemistry.

REFERENCES

[1] Pavlovic, I., Knez, Z. and Skerget, M., 2013. Hydrothermal reactions of agricultural and food processing wastes in sub-and supercritical water: a review of fundamentals, mechanisms, and state of research. *Journal of Agricultural and Food Chemistry*, 61(34), pp. 8003–8025.
[2] Libra, J.A., Ro, K.S., Kammann, C., Funke, A., Berge, N.D., Neubauer, Y., Titirici, M.M., Fühner, C., Bens, O., Kern, J. and Emmerich, K.H., 2011. Hydrothermal carbonization of biomass residuals: a comparative review of the chemistry, processes and applications of wet and dry pyrolysis. *Biofuels*, 2(1), pp. 71–106.
[3] Reza, M.T., 2011. *Hydrothermal carbonization of lignocellulosic biomass*. University of Nevada, Reno.
[4] Hu, B., Wang, K., Wu, L., Yu, S.H., Antonietti, M. and Titirici, M.M., 2010. Engineering carbon materials from the hydrothermal carbonization process of biomass. *Advanced Materials*, 22(7), pp. 813–828.
[5] Funke, A. and Ziegler, F., (2010). Hydrothermal carbonization of biomass: A summary and discussion of chemical mechanisms for process engineering. *Biofuels Bioproducts & Biorefining*, 4(2), pp. 160–177.
[6] Mok, W.S.L., Antal Jr, M.J., Szabo, P., Varhegyi, G. and Zelei, B., 1992. Formation of charcoal from biomass in a sealed reactor. *Industrial & Engineering Chemistry Research*, 31(4), pp. 1162–1166.
[7] Minowa, T., Fang, Z., Ogi, T. and Várhegyi, G., 1998. Decomposition of cellulose and glucose in hot-compressed water under catalyst-free conditions. *Journal of Chemical Engineering of Japan*, 31(1), pp. 131–134.
[8] Peterson, A.A., Vogel, F., Lachance, R.P., Froling, M., Antal Jr, M.J. and Tester, J.W., 2008. Thermochemical biofuel production in hydrothermal media: a review of sub-and supercritical water technologies. *Energy & Environmental Science*, 1(1), pp. 32–65.
[9] Jiang, W., Zhang, X., Sun, Z., Fang, Y., Li, F., Chen, K. and Huang, C., 2011. Preparation and mechanism of magnetic carbonaceous polysaccharide microspheres by low-temperature hydrothermal method. *Journal of Magnetism and Magnetic Materials*, 323(22), pp. 2741–2747.
[10] Cao, X., Ro, K.S., Libra, J.A., Kammann, C.I., Lima, I., Berge, N., Li, L., Li, Y., Chen, N., Yang, J. and Deng, B., 2013. Effects of biomass types and carbonization conditions on the chemical characteristics of hydrochars. *Journal of Agricultural and Food Chemistry*, 61(39), pp. 9401–9411.
[11] Yu, S.H., Cui, X.J., Li, L.L., Li, K., Yu, B., Antonietti, M. and Colfen, H., 2004. From starch to metal/carbon hybrid nanostructures: hydrothermal metal-catalyzed carbonization. *Advanced Materials*, 16(18), pp. 1636–1640.
[12] Cui, X., Antonietti, M. and Yu, S.H., 2006. Structural effects of iron oxide nanoparticles and iron ions on the hydrothermal carbonization of starch and rice carbohydrates. *Small*, 2(6), pp. 756–759.
[13] Toor, S.S., Rosendahl, L.A., Hoffmann, J., Pedersen, T.H., Nielsen, R.P. and Søgaard, E.G., 2014. Hydrothermal liquefaction of biomass. In *Application of Hydrothermal Reactions to Biomass Conversion* (ed. Fangming Jin), pp. 189–217. Springer, New York.

[14] Biller, P., 2018. Hydrothermal liquefaction of aquatic Feedstocks. In *Direct thermochemical liquefaction for energy applications* (pp. 101–125). Woodhead Publishing.

[15] Gollakota, A.R.K., Kishore, N. and Gu, S., 2018. A review on hydrothermal liquefaction of biomass. *Renewable and Sustainable Energy Reviews*, 81, pp. 1378–1392.

[16] Marrone, P.A., Elliott, D.C., Billing, J.M., Hallen, R.T., Hart, T.R., Kadota, P., Moeller, J.C., Randel, M.A. and Schmidt, A.J., 2018. Bench-scale evaluation of hydrothermal processing technology for conversion of wastewater solids to fuels: Marrone et al. *Water Environment Research*, 90(4), pp. 329–342.

[17] Yang, J., Hong, C., Xing, Y., Zheng, Z., Li, Z., Zhao, X. and Qi, C., 2021. Research progress and hot spots of hydrothermal liquefaction for bio-oil production based on bibliometric analysis. *Environmental Science and Pollution Research*, 28, pp. 7621–7635.

[18] Basar, I.A., Liu, H., Carrere, H., Trably, E. and Eskicioglu, C., 2021. A review on key design and operational parameters to optimize and develop hydrothermal liquefaction of biomass for biorefinery applications. *Green Chemistry*, 23(4), pp. 1404–1446.

[19] Ghaziaskar, A., McRae, G.A., Mackintosh, A. and Basu, O.D., 2019. Catalyzed hydrothermal carbonization with process liquid recycling. *Energy & Fuels*, 33(2), pp. 1167–1174.

[20] Jia, J., Wang, R., Chen, H., Liu, H., Xue, Q., Yin, Q. and Zhao, Z., 2022. Interaction mechanism between cellulose and hemicellulose during the hydrothermal carbonization of lignocellulosic biomass. *Energy Science & Engineering*, 10(7), pp. 2076–2087.

[21] Yang, F., Zhang, S., Cheng, K. and Antonietti, M., 2019. A hydrothermal process to turn waste biomass into artificial fulvic and humic acids for soil remediation. *Science of the Total Environment*, 686, pp. 1140–1151.

[22] Yu, G., Zhang, Y., Schideman, L., Funk, T. and Wang, Z., 2011. Distributions of carbon and nitrogen in the products from hydrothermal liquefaction of low-lipid microalgae. *Energy & Environmental Science*, 4(11), pp. 4587–4595.

[23] Barreiro, D.L., Zamalloa, C., Boon, N., Vyverman, W., Ronsse, F., Brilman, W. and Prins, W., 2013. Influence of strain-specific parameters on hydrothermal liquefaction of microalgae. *Bioresource Technology*, 146, pp. 463–471.

[24] Jazie, A.A., Haydary, J., Abed, S.A. and Al-Dawody, M.F., 2022. Hydrothermal liquefaction of Fucus vesiculosus algae catalyzed by Hβ zeolite catalyst for Biocrude oil production. *Algal Research*, 61, p. 102596.

[25] Sekar, M., Mathimani, T., Alagumalai, A., Chi, N.T.L., Duc, P.A., Bhatia, S.K., Brindhadevi, K. and Pugazhendhi, A., 2021. A review on the pyrolysis of algal biomass for biochar and bio-oil–Bottlenecks and scope. *Fuel*, 283, p. 119190.

[26] Aragon-Briceno, C.I., Pozarlik, A.K., Bramer, E.A., Niedzwiecki, L., Pawlak-Kruczek, H. and Brem, G., 2021. Hydrothermal carbonization of wet biomass from nitrogen and phosphorus approach: A review. *Renewable Energy*, 171, pp. 401–415.

[27] Mathimani, T. and Mallick, N., 2019. A review on the hydrothermal processing of microalgal biomass to bio-oil-Knowledge gaps and recent advances. *Journal of Cleaner Production*, 217, pp. 69–84.

[28] Hao, B., Xu, D., Jiang, G., Sabri, T.A., Jing, Z. and Guo, Y., 2021. Chemical reactions in the hydrothermal liquefaction of biomass and in the catalytic hydrogenation upgrading of biocrude. *Green Chemistry*, 23(4), pp. 1562–1583.

[29] Hussin, F., Hazani, N.N., Khalil, M. and Aroua, M.K., 2023, Environmental life cycle assessment of biomass conversion using hydrothermal technology: A review. *Fuel Processing Technology*, 246, p. 107747.

[30] Libra, J.A., Ro, K.S., Kammann, C., Funke, A., Berge, N.D., Neubauer, Y., Titirici, M.M., Fühner, C., Bens, O., Kern, J. and Emmerich, K.H., 2011. Hydrothermal carbonization of biomass residuals: a comparative review of the chemistry, processes and applications of wet and dry pyrolysis. *Biofuels*, 2(1), pp. 71–106.

[31] Sun, P., Heng, M., Sun, S. and Chen, J., 2010. Direct liquefaction of paulownia in hot compressed water: Influence of catalysts. *Energy*, 35(12), pp. 5421–5429.

[32] Bridgwater, A.V., 2003. Renewable fuels and chemicals by thermal processing of biomass. *Chemical Engineering Journal*, 91(2–3), pp. 87–102.

[33] Goyal, H.B., Seal, D. and Saxena, R.C., 2008. Bio-fuels from thermochemical conversion of renewable resources: a review. *Renewable and Sustainable Energy Reviews*, 12(2), pp. 504–517.

[34] Demirbaş, A., 2001. Biomass resource facilities and biomass conversion processing for fuels and chemicals. *Energy Conversion and Management*, 42(11), pp. 1357–1378.

[35] Ong, H.C., Chen, W.H., Farooq, A., Gan, Y.Y., Lee, K.T. and Ashokkumar, V., 2019. Catalytic thermochemical conversion of biomass for biofuel production: A comprehensive review. *Renewable and Sustainable Energy Reviews*, 113, p. 109266.
[36] Nanda, S. A., Kozinski, J. and Dalai, A. K., 2016. Lignocellulosic biomass: a review of conversion technologies and fuel products. *Current Biochemical Engineering*, 3(1), pp. 24–36.
[37] Yin, S. and Tan, Z., 2012. Hydrothermal liquefaction of cellulose to bio-oil under acidic, neutral and alkaline conditions. *Applied Energy*, 92, pp. 234–239.
[38] Strassberger, Z., Tanase, S. and Rothenberg, G., 2014. The pros and cons of lignin valorisation in an integrated biorefinery. *RSC Advances*, 4(48), pp. 25310–25318.
[39] Schuler, J., Hornung, U., Kruse, A., Dahmen, N. and Sauer, J., 2017. Hydrothermal liquefaction of lignin. *Journal of Biomaterials and Nanobiotechnology*, 8(1), p. 96.
[40] Zhu, Z., Toor, S.S., Rosendahl, L., Yu, D. and Chen, G., 2015. Influence of alkali catalyst on product yield and properties via hydrothermal liquefaction of barley straw. *Energy*, 80, pp. 284–292.
[41] Nazari, L., Yuan, Z., Souzanchi, S., Ray, M.B. and Xu, C.C., 2015. Hydrothermal liquefaction of woody biomass in hot-compressed water: catalyst screening and comprehensive characterization of bio-crude oils. *Fuel*, 162, pp. 74–83.
[42] Long, J., Li, Y., Zhang, X., Tang, L., Song, C. and Wang, F., 2016. Comparative investigation on hydrothermal and alkali catalytic liquefaction of bagasse: Process efficiency and product properties. *Fuel*, 186, pp. 685–693.
[43] Marinković, D.M., Stanković, M.V., Veličković, A.V., Avramović, J.M., Miladinović, M.R., Stamenković, O.O., Veljković, V.B. and Jovanović, D.M., 2016. Calcium oxide as a promising heterogeneous catalyst for biodiesel production: Current state and perspectives. *Renewable and Sustainable Energy Reviews*, 56, pp. 1387–1408.
[44] Widayat, W., Darmawan, T., Rosyid, R.A. and Hadiyanto, H., 2017, July. Biodiesel production by using CaO catalyst and ultrasonic assisted. *Journal of Physics: Conference Series*, 877(1), p. 012037. IOP Publishing.
[45] Aysu, T. and Bengü, A.S., 2014. Bio-Oil Production from Cirsium yildizianum through pyrolysis in a fixed-bed reactor. *Journal of Applied Solution Chemistry and Modeling*, 3(3), p. 135.
[46] Durak, H., 2016. Pyrolysis of Xanthium strumarium in a fixed bed reactor: Effects of boron catalysts and pyrolysis parameters on product yields and character. *Energy Sources, Part A: Recovery, Utilization, and Environmental Effects*, 38(10), pp. 1400–1409.
[47] Elkhalifa, E.A. and Friedrich, H.B., 2018. Magnesium oxide as a catalyst for the dehydrogenation of n-octane. *Arabian Journal of Chemistry*, 11(7), pp. 1154–1159.
[48] Shi, W., Li, S., Jin, H., Zhao, Y. and Yu, W., 2013. The hydrothermal liquefaction of rice husk to bio-crude using metallic oxide catalysts. *Energy Sources, Part A: Recovery, Utilization, and Environmental Effects*, 35(22), pp. 2149–2155.
[49] Sikander, U., Sufian, S. and Salam, M.A., 2017. A review of hydrotalcite based catalysts for hydrogen production systems. *International Journal of Hydrogen Energy*, 42(31), pp. 19851–19868.
[50] Wilson, K., Lee, A.F. and Dacquin, J.P., 2012. Heterogeneous catalysts for converting renewable feedstocks to fuels and chemicals. In *Catalysis for alternative energy generation* (pp. 263–304). Springer, New York, NY.
[51] Aramendía, M.A., Borau, V., Jiménez, C., Marinas, J.M., Ruiz, J.R. and Urbano, F.J., 2003. Influence of the preparation method on the structural and surface properties of various magnesium oxides and their catalytic activity in the Meerwein–Ponndorf–Verley reaction. *Applied Catalysis A: General*, 244(2), pp. 207–215.
[52] Elkhalifa, E.A. and Friedrich, H.B., 2018. Magnesium oxide as a catalyst for the dehydrogenation of n-octane. *Arabian Journal of Chemistry*, 11(7), pp. 1154–1159.
[53] De Caprariis, B., Bracciale, M.P., Bavasso, I., Chen, G., Damizia, M., Genova, V., Marra, F., Paglia, L., Pulci, G., Scarsella, M. and Tai, L., 2020. Unsupported Ni metal catalyst in hydrothermal liquefaction of oak wood: Effect of catalyst surface modification. *Science of the Total Environment*, 709, p. 136215.
[54] Ma, Q., Chen, D., Wei, L., Shen, Q., Ji, Z., Chen, Y., Zou, X., Xu, C. and Zhou, J., 2018. Bio-oil production from hydrogenation liquefaction of rice straw over metal (Ni, Co, Cu)-modified CeO2 catalysts. *Energy Sources, Part A: Recovery, Utilization, and Environmental Effects*, 40(2), pp. 200–206.
[55] Watanabe, M., Bayer, F. and Kruse, A., 2006. Oil formation from glucose with formic acid and cobalt catalyst in hot-compressed water. *Carbohydrate Research*, 341(18), pp. 2891–2900.
[56] Sun, P., Heng, M., Sun, S. and Chen, J., 2010. Direct liquefaction of paulownia in hot compressed water: Influence of catalysts. *Energy*, 35(12), pp. 5421–5429.

[57] Zhang, L., Champagne, P. and Xu, C.C., 2011. Bio-crude production from secondary pulp/paper-mill sludge and waste newspaper via co-liquefaction in hot-compressed water. *Energy*, 36(4), pp. 2142–2150.
[58] Yim, S.C., Quitain, A.T., Yusup, S., Sasaki, M., Uemura, Y. and Kida, T., 2017. Metal oxide-catalyzed hydrothermal liquefaction of Malaysian oil palm biomass to bio-oil under supercritical condition. *The Journal of Supercritical Fluids*, 120, pp. 384–394.
[59] Tekin, K., Karagöz, S. and Bektaş, S., 2012. Hydrothermal liquefaction of beech wood using a natural calcium borate mineral. *The Journal of Supercritical Fluids*, 72, pp. 134–139.
[60] Nazari, L., Yuan, Z., Souzanchi, S., Ray, M.B. and Xu, C.C., 2015. Hydrothermal liquefaction of woody biomass in hot-compressed water: catalyst screening and comprehensive characterization of bio-crude oils. *Fuel*, 162, pp. 74–83.
[61] Long, J., Li, Y., Zhang, X., Tang, L., Song, C. and Wang, F., 2016. Comparative investigation on hydrothermal and alkali catalytic liquefaction of bagasse: Process efficiency and product properties. *Fuel*, 186, pp. 685–693.
[62] Shi, W., Li, S., Jin, H., Zhao, Y. and Yu, W., 2013. The hydrothermal liquefaction of rice husk to bio-crude using metallic oxide catalysts. *Energy Sources, Part A: Recovery, Utilization, and Environmental Effects*, 35(22), pp. 2149–2155.
[63] de Caprariis, B., Bracciale, M.P., Bavasso, I., Chen, G., Damizia, M., Genova, V., Marra, F., Paglia, L., Pulci, G., Scarsella, M., Tai, L. and De Filippis, P., 2020. Unsupported Ni metal catalyst in hydrothermal liquefaction of oak wood: effect of catalyst surface modification. *Science of The Total Environment*, 709, p. 136215.
[64] Bi, Z., Zhang, J., Peterson, E., Zhu, Z., Xia, C., Liang, Y. and Wiltowski, T., 2017. Biocrude from pretreated sorghum bagasse through catalytic hydrothermal liquefaction, *Fuel*, 188, pp. 112–120. https://doi.org/10.1016/j.fuel.2016.10.039
[65] Watanabe, M., Bayer, F. and Kruse, A., 2006. Oil formation from glucose with formic acid and cobalt catalyst in hot-compressed water. *Carbohydrate Research*, 341, pp. 2891–2900. https://doi.org/10.1016/j.carres.2006.10.011
[66] Sun, P., Heng, M., Sun, S. and Chen, J., 2010. Direct liquefaction of paulownia in hot compressed water: influence of catalysts. *Energy*, 35, pp. 5421–5429. https://doi.org/10.1016/j.energy.2010.07.005
[67] Zhang, L., Champagne, P. and Charles, C., 2011. Bio-crude production from secondary pulp/paper-mill sludge and waste newspaper via co-liquefaction in hot-compressed water. *Energy*, 36, pp. 2142–2150. https://doi.org/10.1016/j.energy.2010.05.029
[68] Cheng, S., Wei, L. and Rabnawaz, M., 2018. Catalytic liquefaction of pine sawdust and in-situ hydrogenation of bio-crude over bifunctional Co-Zn/HZSM-5 catalysts. *Fuel*, 223, pp. 252–260. https://doi.org/10.1016/j.fuel.2018.03.043
[69] Cheng, S., Wei, L., Alsowij, M., Corbin, F., Boakye, E., Gu, Z. and Raynie, D., 2017. Catalytic hydrothermal liquefaction (HTL) of biomass for bio-crude production using Ni/HZSM-5 catalysts. *AIMS Environmental Science*, 4(3), pp. 417–430.
[70] Yim, S.C., Quitain, A.T., Yusup, S., Sasaki, M., Uemura, Y. and Kida, T., 2017. Metal oxide-catalyzed hydrothermal liquefaction of Malaysian oil palm biomass to bio-oil under supercritical condition. *The Journal of Supercritical Fluids*, 120, pp. 384–394. https://doi.org/10.1016/j.supflu.2016.05.044
[71] Tekin, K., Karagöz, S., and Bektas, S., 2012. Hydrothermal liquefaction of beech wood using a natural calcium borate mineral. *The Journal of Supercritical Fluids*, 72, pp. 134–139.
[72] Nazari, L., Yuan, Z., Souzanchi, S., Ray, M.B. and Xu, C.C., 2015. Hydrothermal liquefaction of woody biomass in hot-compressed water: catalyst screening and comprehensive characterization of bio-crude oils, *Fuel*, 162, pp. 74–83. https://doi.org/10.1016/j.fuel.2015.08.055
[73] Long, J., Li, Y., Zhang, X., Tang, L., Song, C. and Wang, F., 2016. Comparative investigation on hydrothermal and alkali catalytic liquefaction of bagasse: process efficiency and product properties. *Fuel*, 186, pp. 685–693. https://doi.org/10.1016/j.fuel.2016.09.016
[74] Yan, X., Ma, J., Wang, W., Zhao, Y. and Zhou, J., 2018. The effect of different catalysts and process parameters on the chemical content of bio-oils from hydrothermal liquefaction of sugarcane bagasse. *BioResources*, 13(1), pp. 997–1018.
[75] Long, J., Li, Y., Zhang, X., Tang, L., Song, C. and Wang, F., 2016. Comparative investigation on hydrothermal and alkali catalytic liquefaction of bagasse: Process efficiency and product properties. *Fuel*, 186, pp. 685–693.
[76] Qiao, K., Shi, X., Zhou, F., Chen, H., Fu, J., Ma, H.X. and Huang, H., 2017. Catalytic fast pyrolysis of cellulose in a microreactor system using hierarchical zsm-5 zeolites treated with various alkalis. *Applied Catalysis A: General*, 547, pp. 274–282.

[77] Hasan, N.H. and Jazie, A.A., 2021, June. Hydrothermal liquefaction conversion of lignocelluloses and waste biomass using zeolite catalyst. In *IOP Conference Series: Earth and Environmental Science* (Vol. 790, No. 1, p. 012036). IOP Publishing.
[78] Park, H.J., Heo, H.S., Jeon, J.K., Kim, J., Ryoo, R., Jeong, K.E. and Park, Y.K., 2010. Highly valuable chemicals production from catalytic upgrading of radiata pine sawdust-derived pyrolytic vapors over mesoporous MFI zeolites. *Applied Catalysis B: Environmental*, 95, pp. 365–373.
[79] Chen, H., Shi, X., Zhou, F., Ma, H.X., Qiao, K., Lu, X.Y., Fu, J. and Huang, H., 2018. Catalytic fast pyrolysis of cellulose to aromatics over hierarchical nanocrystalline ZSM-5 zeolites prepared using sucrose as a template. *Catalysis Communications*, 110, pp. 102–105.
[80] Li, J., Li, X.Y., Zhou, G.Q., Wang, W., Wang, C.W., Komarneni, S. and Wang, Y.J., 2014. Catalytic fast pyrolysis of biomass with mesoporous ZSM-5 zeolites prepared by desilication with NaOH solutions. *Applied Catalysis A: General*, 470, pp. 115–122.
[81] Puertolas, B., Veses, A., Callen, M.S., Mitchell, S., Garcia, T. and Perez-Ramirez, J., 2015. Porosity-acidity interplay in hierarchical ZSM-5 zeolites for pyrolysis oil valorization to aromatics. *Chemsuschem*, 8, pp. 3283–3293.
[82] Lee, H.I., Park, H.J., Park, Y.K., Hur, J.Y., Jeon, J.K. and Kim, J.M., 2008. Synthesis of highly stable mesoporous aluminosilicates from commercially available zeolites and their application to the pyrolysis of woody biomass. *Catalysis Today*, 132, pp. 68–74.
[83] Wang, X., Meng, L., Wu, F., Jiang, Y., Wang, L. and Mu, X., 2012. Efficient conversion of microcrystalline cellulose to 1, 2-alkanediols over supported Ni catalysts. *Green Chemistry*, 14(3), pp. 758–765.
[84] Vlazan, P., Ursu, D.H., Irina-Moisescu, C., Miron, I., Sfirloaga, P. and Rusu, E., 2015. Structural and electrical properties of TiO2/ZnO core–shell nanoparticles synthesized by hydrothermal method. *Materials Characterization*, 101, pp. 153–158.
[85] Maag, A.R., Paulsen, A.D., Amundsen, T.J., Yelvington, P.E., Tompsett, G.A. and Timko, M.T., 2018. Catalytic hydrothermal liquefaction of food waste using CeZrOx. *Energies*, 11(3), p. 564.
[86] Sánchez-Bayo, A., Rodríguez, R., Morales, V., Nasirian, N., Bautista, L.F. and Vicente, G., 2019. Hydrothermal liquefaction of microalga using metal oxide catalyst. *Processes*, 8(1), p. 15.
[87] Long, J., Li, Y., Zhang, X., Tang, L., Song, C. and Wang, F., 2016. Comparative investigation on hydrothermal and alkali catalytic liquefaction of bagasse: Process efficiency and product properties. *Fuel*, 186, pp. 685–693.
[88] Wang, S., Zeng, F., Liu, Y., Meng, Y., Wang, W., Liu, C., Zhang, J., Du, H. and Li, J., 2022. Preparation and application of ZrO2–SiO2 complex oxide for efficient biocrude generation by hydrothermal liquefaction of Spirulina. *Fuel*, 317, p. 123325.
[89] Reddy, G.R., 2022. Experimental investigation on CI engine using bio-oil obtained from hydrothermal of bio-mass waste as fuel, Undergraduate Thesis, Hindustan Institute of Technology and Science.
[90] Chen, Y., Mu, R., Yang, M., Fang, L., Wu, Y., Wu, K., Liu, Y. and Gong, J., 2017. Catalytic hydrothermal liquefaction for bio-oil production over CNTs supported metal catalysts. *Chemical Engineering Science*, 161, pp. 299–307.
[91] Wang, H., Tian, W., Zeng, F., Du, H., Zhang, J. and Li, X., 2020. Catalytic hydrothermal liquefaction of Spirulina over bifunctional catalyst to produce high-quality biofuel. *Fuel*, 282, p. 118807.
[92] Sharma, N., Jaiswal, K.K., Kumar, V., Vlaskin, M.S., Nanda, M., Rautela, I., Tomar, M.S. and Ahmad, W., 2021. Effect of catalyst and temperature on the quality and productivity of HTL bio-oil from microalgae: A review. *Renewable Energy*, 174, pp. 810–822.
[93] Li, Y., Zhu, C., Jiang, J., Yang, Z., Feng, W., Li, L., Guo, Y. and Hu, J., 2021. Catalytic hydrothermal liquefaction of Gracilaria corticata macroalgae: Effects of process parameter on bio-oil up-gradation. *Bioresource Technology*, 319, p. 124163.
[94] Feng, L., Li, X., Wang, Z. and Liu, B., 2021. Catalytic hydrothermal liquefaction of lignin for production of aromatic hydrocarbon over metal supported mesoporous catalyst. *Bioresource Technology*, 323, p. 124569.

4 Biomass Processing Technologies

Uma Krishnakumar and Feba C. Varghese

4.1 INTRODUCTION

In this era of increasing concern for the environment and an increasing need for energy sources, biomass opens up a door where engineering techniques, along with new technologies, can face the emerging needs of society. Biomass is a major source of energy and can also be used for many other purposes. Biomass plays a major role in the sustainable development of society. Biomass can be of different kinds, and specific techniques have to be used for each kind of biomass to extract the maximum potential of the substance. Bioenergy consists of solid, liquid, or gaseous fuels that can be obtained from the available conventional technologies.

Liquid fuels find common usage in vehicles but are also suitable for stationary engines. Solid fuels, on the other hand, are burned directly to produce heat, power, or combined heat and power (CHP). Gaseous fuels offer versatility and can be employed across a wide spectrum of applications. These various conversions require specific technologies. The extent to which the technology is being used efficiently depends on the quality of the feedstock. People have been practicing biomass conversion techniques for decades. But it has recently gained importance or a place on an industrial scale. Industrial-level biomass conversion technologies have brought many advancements in conversion technologies. Conversion technologies are developing day by day to meet economic standards as well as to make the processes more efficient. This chapter deals with some of the major advancements that have happened in the field of biomass conversion [1, 2].

4.2 BIOMASS

Biomass is generally referred to as organic matter derived from various plant or animal sources. Simply, biomass is composed of carbon, hydrogen, and oxygen arranged in a particular manner using different kinds of bonds and is one of the major sources of renewable energy available [3]. Some of the major sources of biomass are:

1. **Wood and forestry residues**: Biomass derived from woody and forest residues is a valuable and renewable source of energy that plays a significant role in sustainable energy production and environmental conservation. Woody biomass includes materials like branches, bark, sawdust, and wood chips, while forest residues encompass leftover materials from logging and timber processing activities. Woody and forest residues offer a sustainable solution to energy needs while promoting environmental conservation and economic growth. By harnessing the energy potential of these residues, societies can move toward a greener, more sustainable energy future, ensuring a balance between human needs and the preservation of our natural resources [4].
2. **Agricultural crops and residues**: Biomass derived from agriculture plays a vital role in sustainable energy production and environmental conservation. Agricultural biomass refers to organic materials sourced from farming activities that can be used for energy generation. These resources include crop residues (such as corn stalks, wheat straw, and rice husks),

DOI: 10.1201/9781003732907-4

animal manure, dedicated energy crops (like switchgrass and miscanthus), and organic waste from farms. One of the significant advantages of agricultural biomass is its renewability. These resources can be replenished seasonally, making them a constant and reliable source of energy. By utilizing agricultural biomass, farmers can diversify their income streams and reduce waste by converting crop residues and organic farm waste into valuable energy resources [5].

3. **Municipal solid waste**: Biomass derived from municipal waste, also known as municipal solid waste (MSW), presents a compelling solution to two significant challenges: waste management and renewable energy generation. MSW biomass includes organic materials from households, commercial establishments, and institutions, such as food waste, yard waste, paper, cardboard, and certain plastics. This has become a great approach toward sustainability and the generation of renewable energy [6, 7, 8].
4. **Livestock manure**: This includes animal waste from farms, which can be used as a feedstock for anaerobic digestion. It is a valuable source of energy, often utilized for the sustainable production of energy. Livestock manure is a great source of methane, and hence utilization of this to produce biogas is very convenient and highly efficient. The usage of livestock manure as a biomass helps in efficient waste management and sustainable development [9].
5. **Aquatic plants and algae**: Aquatic plants and algae are valuable sources of biomass that hold tremendous potential for sustainable energy production and environmental conservation. These organisms, found in oceans, lakes, rivers, and ponds, are rich in organic matter and can be harnessed for various purposes. These can be utilized for the production of bioenergy, nutrient recycling, etc. and hence can play a vital role in sustainable development. Efficient utilization of aquatic plants and algae can help in economic development [10, 11].
6. **Wastewater treatment plants**: Wastewater treatment plants, by harnessing the energy potential of organic waste through anaerobic digestion, not only address environmental concerns associated with wastewater management but also contribute significantly to the production of renewable energy. This innovative approach transforms what is traditionally considered a waste product into a valuable resource, promoting a greener and more sustainable energy future. This can be utilized for energy generation like heat energy, electricity, etc. and also be utilized for the production of biogas [12, 13].
7. **Industrial waste**: This includes waste products from various industries, such as food processing waste, pulp and paper mill waste, and waste from timber, etc. Industrial waste, often considered a problem, can be transformed into a valuable resource through the concept of utilizing it as biomass. Industrial waste refers to the byproducts generated from various manufacturing processes and industries, including food processing, paper production, and chemical manufacturing. When managed properly, industrial waste can serve as a sustainable source of biomass for several applications. Industrial biomass can be utilized for energy production, waste management, etc. [14].

4.2.1 Types of Biomass Based on Composition

Biomass can be broadly categorized into different types based on their chemical composition and sources. There are many kinds of biomass based on composition. Some of the types of biomass based on composition are:

Lignocellulosic biomass (LCB): It constitutes the majority of biomass on Earth, including agricultural residues, forestry residues, energy crops, and food waste. LCB's main constituents are cellulose, hemicellulose, and lignin. Cellulose, a linear polysaccharide composed of glucose units, provides structural support. Hemicellulose, a ramified polysaccharide made of various sugars, and lignin, a phenylpropanoid polymer, are also abundant.

Starch-based biomass: Starch-based biomass, like corn, sugarcane, and potatoes, contains high levels of starch, which can be converted into sugars and further processed into biofuels and other products.

Oil-rich biomass: Seeds such as sunflower, soybean, and a large number of aquatic plants act as biomass sources for the production of biodiesel and bio-oils.

Protein-rich biomass: Certain bacteria and fungi can be cultivated to produce proteins suitable for animal feed and other applications. Algae can also be a source of protein-rich biomass for animal feed [15, 16].

4.2.2 Significance of Biomass

In the present scenario, non-renewable energy has become one of the unavoidable parts of the global economy. These non-renewable resources cause large amounts of pollution and are an integral part of global warming. This can also, in turn, become a cause for the vaping out of our existence, and as a precaution, biomass plays an important role in the sustainable development of countries. Biomass is one of the most important renewable energy sources, contributing about 9–13% of total global energy. Energy from biomass is mostly utilized by developing countries. Many parts of Africa and Asia are traditional users of biomass as a major source of energy for cooking and heating.

While biomass opens up a huge opportunity for sustainability, it also causes some kinds of concerns, as well as comes with some sorts of disadvantages. Some of the disadvantages are that biomass has low bulk density, leading to storage, handling, and transportation challenges as compared to that of crude oil, coal, etc., which makes biomass less efficient. Some of the major concerns are forest and agriculture, the main source of biomass, and this can sometimes lead to deforestation and land usage issues, i.e., it can cause a significant reduction in the area used for growing edible crops. Water usage and soil quality are also factors of concern in biomass production and utilization [17].

4.2.3 Biomass as Energy Source

Biomass is a versatile and sustainable energy source, offering a range of applications, including:

Electricity Generation: Biomass can be transformed into electricity through various conversion techniques. Combustion, pyrolysis, gasification, and anaerobic digestion are major methods for this purpose. In combustion, materials like wood chips, agricultural residues, or municipal solid waste are burned directly to produce heat, which powers a turbine connected to a generator, generating electricity. Gasification involves heating biomass with limited oxygen to create syngas, which can be used directly in gas engines or turbines for electricity production. Pyrolysis heats biomass without oxygen to produce bio-oil, biochar, and syngas. Bio-oil can be used for steam generation, biochar improves soil, and syngas can be burned for electricity. Anaerobic digestion breaks down organic materials, like agricultural waste and manure, to produce biogas (mainly methane), which can be used in gas engines or turbines for electricity generation. Biomass fuel cells employ electrochemical reactions to directly convert biomass into electricity, without combustion, offering high efficiency and environmental benefits. They can use various feedstocks, including organic waste and plant materials.

Cooking and Heating in Households: Biomass is a valuable fuel for cooking and heating in households, particularly in rural areas [18, 19]. Biogas plays a crucial role in this context.

Heat and Steam Production: Biomass is employed as a fuel to generate heat and steam for industrial processes, space heating, and hot water production. The energy produced using methods like pyrolysis and gasification is primarily in the form of heat or biofuel. Heat can be utilized for steam production, and biofuels can be burned to produce heat and steam.

Biofuel Production: Biofuels come in various forms, such as bioethanol, biodiesel, and biogas. They are produced through thermochemical and biochemical processes. Pyrolysis is commonly used for bioethanol and biodiesel production, while biogas is produced through gasification, fermentation, or anaerobic digestion. Biofuels contribute to sustainable development and significantly reduce global warming, serving as a promising alternative to non-renewable energy sources.

Bioproducts: Biomass can be used to produce a variety of bioproducts, such as bioplastics, biofertilizers, and biochemicals [20, 21].

4.3 BIOMASS CONVERSION TECHNOLOGIES

Some of the standard methods for deriving energy from biomass are:

- Combustion: Biomass is burned to produce heat, which can generate steam and produce electricity, or used directly for space heating, hot water, or industrial processes.
- Gasification: Biomass is converted into a gas by heating it with a controlled amount of oxygen and steam. The resulting gas can be used for electricity generation or as a fuel for industrial processes and transportation.
- Pyrolysis: Biomass is heated in the absence of oxygen, which causes it to break down into liquid, gas, and solid products. The resulting liquid product, called bio-oil, can be used as a fuel or further processed into transportation fuels.
- Anaerobic digestion: Biomass is broken down by microorganisms in the absence of oxygen, producing biogas (a mixture of methane and carbon dioxide) as a bioproduct. The biogas can be used to generate electricity, heat, or fuel for transportation.
- Fermentation: Biomass is broken down by microorganisms (such as yeast) into ethanol, which can be used as a transportation fuel or further processed into other chemicals. [22]

Chemical conversion of biomass is mainly divided into three parts:

- Direct combustion process
- Thermochemical process
- Biochemical process

4.3.1 Direct Combustion Process

Direct combustion is generally used for the production of heat energy. The obtained heat energy can be further utilized for boiling water, which leads to steam production, which can be utilized for rotating turbines, leading to electricity production. Hence, direct combustion can be used for obtaining both heat energy and electrical energy. Wood is the largest source of biomass available for direct discharge. This is a method in which biomass is burned in the open air or in the presence of excess air. Generally, direct combustion is carried out inside a furnace or a boiler with a temperature ranging between 800 and 1000°C. In this method of biomass conversion, the cost of production is comparatively very low. The efficiency of the process depends on the moisture content in biomass. The lower the amount of biomass, easier the process will be. Even though the process leads to the production of a high amount of energy, it is not eco-friendly due to the emission of a large number of greenhouse gases [23] (Figure 4.1).

4.3.2 Thermochemical Process

The thermochemical process converts biomass into energy under controlled conditions like specific temperatures, pressure, etc. Rather than combustion, this method is more suitable for obtaining

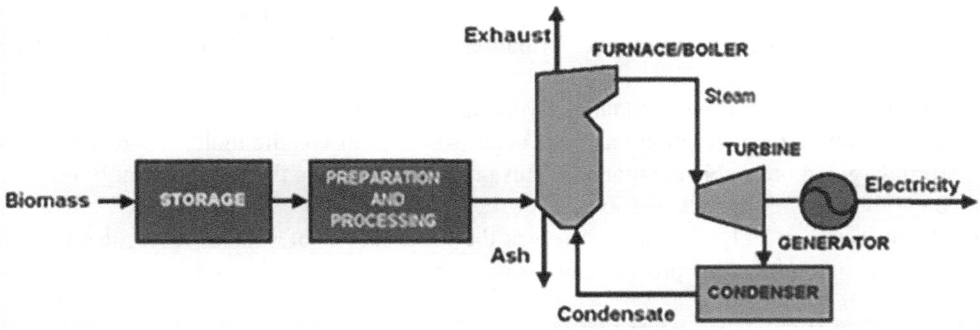

FIGURE 4.1 Direct combustion as a source of electricity production.

From *Biomass for Electricity Generation*, by Whole Building Design Guide (WBDG), U.S. General Services Administration. Reproduced from WBDG.

products. Biodiesel is an example of a product formed by thermochemical processes. This process mainly involves the reaction of biomass in a controlled environment. There are different kinds of thermochemical processes. Some of them are: Pyrolysis, Gasification, and Combustion [24].

4.3.2.1 Pyrolysis

Pyrolysis is a process characterized by subjecting organic matter, including biomass, to high temperatures in the presence of nitrogen or in the absence of oxygen. The term "pyrolysis" originates from the Greek elements "pyro," signifying fire, heat, or fever, and "lysis," meaning separation. This method involves the thermal breakdown of organic substances, leading to changes in their chemical composition. It's an endothermic reaction, meaning it absorbs heat energy, resulting in the formation of vapors and residual carbon material. These vapors consist of components derived from cellulose, hemicellulose, and lignin polymers and can be condensed into a flowing organic liquid known as bio-oil. Simultaneously, the carbon residues that remain are referred to as biochar.

Pyrolysis is a widely applied technique in the chemical industry. It is utilized to manufacture essential compounds like ethylene, various carbon-based products, and other chemicals. This process can efficiently convert feedstocks such as petroleum, coal, and even wood, or transform coal into coke. Additionally, it plays a pivotal role in converting natural gas, mainly methane, into hydrogen gas and solid carbon char, a practice that has been recently introduced on an industrial scale. Furthermore, pyrolysis holds great promise for aspirational applications. These include the conversion of biomass into syngas and biochar, the transformation of waste plastics into usable oil, and the conversion of waste materials into safe and environmentally friendly substances [25]. The chemical reactions taking part in pyrolysis depend upon the nature and components of biomass used and the type of process utilized for the conversion process. Some of the chemical reactions are:

- Dehydration: Biomass contains water, which is removed as steam during fast pyrolysis. This reaction involves breaking the bonds in cellulose, hemicellulose, and other organic compounds, releasing water vapor.
  e.g.: Cellulose → Char + Water vapor
- Decarboxylation: Decarboxylation involves the removal of carboxyl groups (-COOH) from organic compounds. This process releases carbon dioxide gas.
  e.g.: Acetic Acid → Char + Carbon Dioxide
- Depolymerization: Large, complex polymers in biomass, such as cellulose and lignin, are broken down into smaller, simpler molecules, including sugars and phenolic compounds.
  e.g.: Cellulose → Char + Sugars

- Cracking: Cracking reactions involve the breaking of carbon-carbon bonds in organic compounds. This process leads to the formation of smaller hydrocarbons and other volatile compounds.
  e.g.: Large Hydrocarbon → Smaller Hydrocarbons + Char
- Condensation: Condensation reactions occur when small volatile molecules recombine to form larger, more stable compounds. This process can lead to the formation of bio-oil.
  e.g.: Volatile Compounds → Bio-oil
- Polymerization: Polymerization involves the combination of smaller molecules to form larger compounds. This process can result in the formation of biochar, which is a stable form of carbon.
  e.g.: Small Hydrocarbons → Biochar
- Gasification: Gasification reactions involve the conversion of organic materials into gases like methane, carbon monoxide, and hydrogen. These gases can be utilized as energy sources.
  e.g.: Organic Compounds → Methane + Carbon Monoxide [26]

Processes involved in pyrolysis are the same for all types of pyrolysis except the temperature at which the process takes place, the time taken for the process, etc. The common processes involved in pyrolysis are:

1. Biomass preparation: Biomass feedstock, such as wood, agricultural residues, or organic waste, is collected, dried, and sometimes ground into small particles. Proper preparation ensures uniform heating and efficient pyrolysis.
2. Loading into bioreactor: The prepared biomass is loaded into a pyrolysis reactor, which can be a fixed-bed, fluidized-bed, rotary kiln, or other configurations. The reactor is designed to withstand high temperatures and is operated in an oxygen-free or oxygen-limited environment to prevent combustion.
3. Heating and drying: The biomass is gradually heated to the pyrolysis temperature (typically between 300 and 800°C). During this stage, moisture within the biomass evaporates, and the biomass dries.
4. Pyrolysis reactions: At the pyrolysis temperature, biomass undergoes thermal decomposition through a series of complex chemical reactions:
   a. Dehydration: Removal of water molecules from the biomass.
   b. Decarboxylation: Removal of carboxyl groups (-COOH) from organic compounds, releasing carbon dioxide.
   c. Depolymerization: Breakdown of large, complex polymers into smaller molecules, such as sugars and phenolic compounds.
   d. Cracking and Fragmentation: Breaking of carbon-carbon bonds within the organic molecules, forming smaller hydrocarbons, aldehydes, ketones, and other volatile compounds.
   e. Gasification: Conversion of biomass components into gases such as methane (CH4), carbon monoxide (CO), hydrogen (H2), and various hydrocarbons.
5. Vapor cooling and condensation: The hot gases and vapors produced during pyrolysis are rapidly cooled, causing them to condense into bio-oil. The cooling process needs to be quick to prevent further reactions. Solid residue formation (Biochar): The remaining solid material, known as biochar, is carbon-rich and stable. It contains the non-volatile components of biomass and can be used as a soil conditioner, sequestering carbon in the soil and improving its fertility.
6. Gas collection: Gaseous byproducts, such as syngas (synthesis gas), are also produced during pyrolysis. These gases can be collected and utilized for various purposes, including energy generation.

7. Gas cleaning: The gases produced during pyrolysis may contain impurities such as tars and particulates. These gases can be cleaned and purified through filtration and scrubbing processes to remove contaminants.
8. Product separation and cleaning: The collected bio-oil, gases, and biochar are separated and processed further if necessary. Bio-oil may undergo additional treatments to remove impurities and water content, improving its quality [27].

On the basis of temperature, residence time, heating rate, and products made, pyrolysis can be divided into:

- Fast pyrolysis
- Ultra-fast/flash pyrolysis
- Intermediate pyrolysis
- Conventional/slow pyrolysis
- Torrefaction
- Microwave-assisted pyrolysis

4.3.2.1.1 Fast Pyrolysis

Fast pyrolysis is a biomass conversion process characterized by its rapid and high-temperature decomposition of biomass to produce liquid bio-oil, solid biochar, and gaseous products. This process is distinct from slow pyrolysis due to its high heating rates, short residence times, and operating temperatures between 400°C and 600°C. The product distribution in fast pyrolysis typically consists of approximately 60–70% bio-oil, 12–15% biochar, and 13–25% gases on a dry basis.

It offers the potential for converting a variety of biomass feedstocks into valuable bio-oil, which can be used as a precursor for biofuels and various chemical products. Careful control of process parameters, feedstock characteristics, and moisture content is necessary to optimize bio-oil yields [28].

4.3.2.1.2 Ultra-Fast/Flash Pyrolysis

In the case of fast pyrolysis, biomass is finely ground and subjected to rapid heating under very high temperatures (300–700°C). The particle size of the feedstock must be less than 1 mm. Fast pyrolysis is mainly used for the production of bio-oil. Studies say that fast pyrolysis is one of the most efficient methods in the production of oil since it converts more than 75% of the biomass into bio-oil. During this process, biomass decomposes to produce vapors, aerosols, and charcoal-like char. The vapors and aerosols produced condense to produce bio-oil [29].

4.3.2.1.3 Intermediate Pyrolysis

Intermediate pyrolysis lies between fast and slow pyrolysis and usually has an optimum range of temperature 300–500°C. This method of pyrolysis is well known for its flexibility in the process. It has a good distribution of products and can be utilized in the coproduction of biochar, bio-oil, and gas. In intermediate pyrolysis, chemical reactions are controlled and can be optimized according to the type of feed used and the kind of product required. In this process, large-sized biomass, i.e., not finely ground feedstocks, can be used. Nowadays, this method is being noticed worldwide due to its ability to treat waste biomass. The liquid product obtained during this process is less viscous, and it is also observed that the production yield in this is low as compared to that of fast pyrolysis (nearly 55%) [30].

4.3.2.1.4 Conventional/Slow Pyrolysis

It is the oldest technique in the process of pyrolysis. This method is mainly used for the production of solid fuels like charcoal. Here, biomass is being subjected to comparatively low heat for a longer period of time. The process is carried out in batches. During the process, the moisture content present gets evaporated, and the carbon content in the substance gets concentrated. Due to this reason, the process is also known as carbonization. The carbonization process includes the dry phase, thermal decomposition, and cooling. The biomass used here can be of any size, i.e., from fine particles to a whole big log [31].

4.3.2.1.5 Torrefaction

It is a kind of slow pyrolysis carried out at temperatures of 225–300°C. The main aim of the process is to increase biomass energy density and fuel properties. It can also be categorized as incomplete carbonization. This process occurs in the absence of oxygen. The residence time during torrefaction varies depending on the particle size and can range from several seconds to an hour. The decomposition process during torrefaction can be categorized into several stages, including drying, depolymerization and recondensation, limited devolatilization and carbonization, and extensive devolatilization and carbonization. Different types of biomass exhibit varying torrefaction kinetics due to their inherent differences in composition and structure. The torrefied biomass is hydrophobic and soft, which can easily be crushed, ground, and pulverized. It enhances the energy content, transportability, and storability of biomass feedstocks, making them more suitable for use in various energy generation processes [32].

4.3.2.1.6 Microwave-Assisted Pyrolysis

Microwave-assisted pyrolysis is a unique and innovative heating technique that differs significantly from conventional heating methods. In conventional heating, such as through convection, conduction, and radiation, heat is transferred to a material's surface and gradually penetrates into the interior due to temperature gradients. In contrast, microwave energy is directly delivered into materials through an electromagnetic field. This electromagnetic field enters the material and generates thermal energy throughout the penetration depth using a process called dielectric heating. This occurs through interaction with polarizable dipoles present in the material, effectively heating the material from the inside. Some of the advantages are uniform heating, rapid heating rate, cleaner products, volumetric and selective heating, ease of operation, reduced hazardous product formation, lower pollutant emissions, high heating efficiency, ability to handle wet biomass, and suitability for large biomass particles. Some of the major disadvantages are its reliance on electricity, which can be expensive compared to the heat generated by the combustion of pyrolysis gases in conventional pyrolysis; it often requires pre-treatment and the use of catalysts before the heating process to optimize results [33]. Based on the type of pyrolysis performed, there are different kinds of end products or forms of converted biomass. Some of them and the various fields of application of those products are:

- **Biochar**: Biochar is a carbon-rich, porous material produced from biomass pyrolysis. It is stable, resistant to decomposition, and rich in nutrients. Uses are: Soil Amendment: Biochar enhances soil fertility, structure, and water retention. It acts as a long-term carbon sink, improving soil health and agricultural productivity. Carbon Sequestration: Biochar helps sequester carbon dioxide, mitigating climate change by storing carbon in the soil for centuries. Water Filtration: Biochar is used in water treatment to remove contaminants and improve water quality. Livestock Bedding: Biochar can be used as animal bedding, providing a comfortable and absorbent surface [34].
- **Bio-oil**: Bio-oil is a dark, viscous liquid obtained from the condensation of vapors produced during pyrolysis. It contains a complex mixture of organic compounds.
  Uses: Biofuel Production: Bio-oil can be upgraded to produce transportation fuels, such as biodiesel and gasoline substitutes. Heat and Power Generation: Bio-oil can be burned to generate heat and electricity, serving as a renewable energy source. Chemical Feedstock:

Bio-oil contains valuable chemicals that can be extracted and used in the production of chemicals and materials [35].

- **Syngas**: Syngas (synthesis gas) is a mixture of hydrogen (H_2), carbon monoxide (CO), and other hydrocarbons. It is a versatile energy carrier. Uses: Power Generation: Syngas can be used in gas turbines or engines to generate electricity and heat. Production of Fuels: Syngas can be converted into liquid fuels like synthetic diesel or methanol. Chemical Synthesis: Syngas is a precursor for various chemicals, including ammonia and methanol, used in the chemical industry [36].
- **Carbon nanotube**: Carbon nanotubes are cylindrical structures composed of carbon atoms. They have exceptional mechanical, electrical, and thermal properties.
  Uses: Nanotechnology: Carbon nanotubes find applications in nanoelectronics, nanocomposites, and drug delivery systems. Materials Enhancement: They are used to strengthen and enhance the properties of materials, including polymers and composites [37].
- **Activated carbon**: Activated carbon is a highly porous form of carbon with a large surface area. It has a strong adsorption capacity. Uses: Water Purification: Activated carbon is used in water treatment and filtration systems to remove impurities, odors, and contaminants. Air Purification: It is employed in air filters to capture pollutants and improve indoor air quality. Chemical and Pharmaceutical Industries: Activated carbon is used in the purification and separation of chemicals and pharmaceuticals [38].

4.3.2.2 Gasification

Biomass gasification is a process employed to convert various biomass materials into gaseous fuel, which can be further harnessed for heat or electricity generation. This procedure involves the incomplete combustion of biomass, resulting in the creation of combustible gases that primarily consist of carbon monoxide, hydrogen, and small amounts of methane. Typically, the gasification process occurs within a temperature range of 800–1300°C. The effectiveness and efficiency of the gasifier are influenced by factors such as the size, moisture content, shape, heating value, carbon content, and ash content of the biomass feedstock being used. When dealing with feedstocks possessing low volatile content, partial oxidation gasification is employed, whereas those with high volatile content undergo an indirect gasification process. The size and distribution of the particles used as feedstock are crucial in determining the gasifier's efficiency, which is why uniform-sized feedstock is preferred. Additionally, the operation of the gasifier is sensitive to the moisture content within the biomass. High moisture content in biomass can diminish the efficiency of the gasification process.

The gasifier consists of several components, as shown in Figure 4.2, including:

1. Fuel Feeding System: This system delivers the biomass feedstock to the gasifier. It may include a hopper, screw feeder, or other mechanisms to ensure a continuous and controlled feed of biomass.
2. Gasifier Reactor: The gasifier reactor is where the actual gasification process takes place. It consists of a fixed bed of biomass, a combustion zone, and a reduction zone. The reactor may be designed in an updraft, downdraft, or cross-draft configuration.
3. Oxidation Medium: The oxidation medium, which can be air, oxygen, or a mixture of both, is introduced into the gasifier to facilitate the gasification process. The flow and distribution of the oxidation medium are crucial for efficient gasification.
4. Syngas Outlet: The syngas, which is a mixture of carbon monoxide, hydrogen, and other gases, is collected and extracted from the gasifier through a syngas outlet. This outlet may be connected to a gas cleaning system for further processing.
5. Ash Removal System: As the biomass is gasified, ash and other solid residues are left behind. An ash removal system, such as a grate or a mechanical ash extractor, is used to remove these residues from the gasifier.

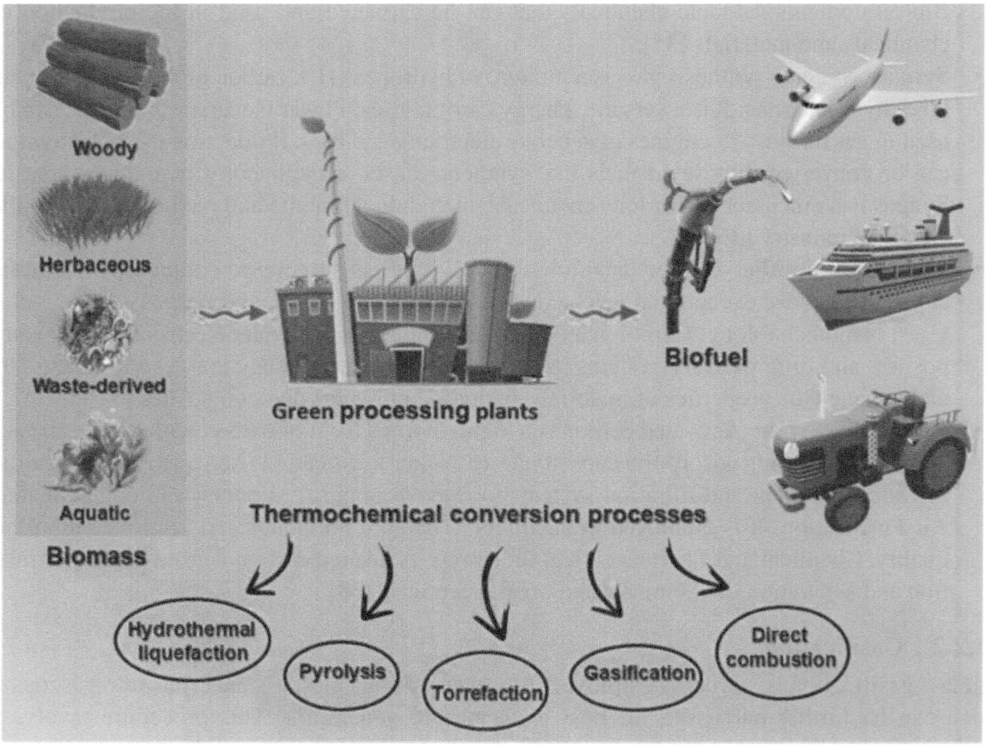

FIGURE 4.2 Types of thermochemical processes.

Jha, S., Nanda, S., Acharya, B., & Dalai, A. K. (2022). A review of thermochemical conversion of waste biomass to biofuels. Energies, 15(17), 6352. https://doi.org/10.3390/en15176352.

6. Gas Cleaning System: The syngas produced in the gasifier may contain impurities, such as tars, particulate matter, and sulfur compounds. A gas cleaning system, which can include filters, scrubbers, and catalytic converters, is used to remove these impurities and ensure the quality of the syngas.
7. Heat Recovery System: The heat generated during the gasification process can be recovered and used for various purposes, such as steam generation or space heating. A heat recovery system, such as a boiler or a heat exchanger, is used to capture and utilize this heat.
8. Control and Monitoring System: A biomass gasifier requires a sophisticated control and monitoring system to ensure safe and efficient operation. This system may include sensors, actuators, and a central control unit to regulate the various parameters of the gasification process [39].

Biomass conversion using gasification involves the following steps/operating principles:

- Biomass preparation: The biomass is first prepared by removing any unwanted materials, such as rocks or soil, and then sized and dried to the appropriate moisture content for gasification.
- Gasification: The biomass is then fed into a gasifier, which is a high-temperature reactor that converts the biomass into syngas. In the gasifier, the biomass is partially oxidized in a low-oxygen environment to produce a mixture of gases including carbon monoxide, hydrogen, and methane. This process is typically accomplished by introducing the biomass into a hot bed of charcoal or other high-carbon material, which provides the necessary heat and reduces the need for external fuel.

- Gas clean-up: The syngas produced by gasification typically contains impurities such as tar, particulate matter, and sulfur compounds that can damage downstream equipment or affect the quality of the end product. The syngas are therefore passed through a series of gas clean-up systems to remove these impurities.
- Shifting: Transitioning to the next step, the syngas is directed to a "shift reactor." In this reactor, a shift reaction takes place, wherein carbon monoxide is transformed into a higher proportion of hydrogen and carbon dioxide by combining it with steam. As a result, the syngas primarily comprises hydrogen and carbon dioxide.
- Purification: After the syngas has undergone the shift reaction, it is divided into separate streams of hydrogen and carbon dioxide. The purified hydrogen is then prepared for utilization, while the carbon dioxide is captured and transported for sequestration.
- Energy conversion: The cleaned syngas can be used for a variety of energy conversion processes, such as generating electricity or producing heat for industrial processes. The syngas can also be further processed to produce a variety of chemicals, such as methanol or synthetic natural gas [40].

The use of gasification is considered advantageous as it is environmentally friendly and produces very low amounts of greenhouse gases. At the same time, it is costly as it involves many processes, such as clean-up of the gas produced, etc. Based on the method by which gasification is carried out, it can be divided into: Fixed bed gasification, Fluidized bed gasification, Plasma gasification, and Hydrothermal gasification.

4.3.2.2.1 Fixed Bed Gasification

Fixed bed gasification is a biomass gasification technique where biomass is positioned on a stationary grate or bed within a gasifier. It is then subjected to elevated temperatures with a restricted air supply, leading to partial combustion for the production of heat or electricity. This gasification method is commonly employed for small to medium-scale energy applications due to its relatively straightforward operation. However, a notable drawback of fixed bed gasification is the challenge of maintaining consistent temperatures and ensuring efficient mixing within the reaction zone. Fixed bed gasifiers can be categorized into two types: updraft and downdraft, a classification based on the direction of the feedstock and airstream [41]. In updraft gasifiers, feedstock is fed from the top portion, and the airstream enters from the bottom part. In downdraft gasifiers, the feedstock and airstream are fed from the top portion. Some of the major advantages of updraft gasifiers are: Simplicity: Updraft gasifiers are simpler in design, making them easier to construct and maintain. Versatility: They can handle various biomass types and sizes, allowing for flexibility in feedstock selection. Suitable for Small-Scale Applications: Updraft gasifiers are often used in small-scale applications due to their simplicity and cost-effectiveness. Less Sensitive to Fuel Moisture: Updraft gasifiers are less sensitive to high fuel moisture content compared to downdraft gasifiers.

Major applications are:

- Small-Scale Power Generation: Updraft gasifiers are suitable for decentralized power generation in rural areas or small communities.
- Residential Heating: They are used for residential heating and cooking in some regions, providing an eco-friendly alternative to traditional fuels.
- Experimental and Educational Purposes: Updraft gasifiers are often used in educational institutions and experimental setups due to their simplicity [42].

The downdraft gasifiers are mainly used for the gasification of highly volatile feedstocks. Here, ash is collected in the bottom and helps in the production of high-quality gas. The updraft gasifiers are generally used for feedstocks with diverse inner compositions. Some of the advantages of downdraft gasifiers are: Higher Efficiency: Downdraft gasifiers generally offer higher efficiency due to better combustion control and reduced heat loss. Lower Tar Content: Downdraft gasifiers produce syngas with lower tar content, making the gas cleaner and suitable for use in internal combustion engines. Better Emission Control: Downdraft gasifiers have better emission control capabilities, leading to lower emissions of pollutants such as nitrogen oxides (NOx) and particulate matter. Higher Energy Content: The syngas produced in downdraft gasifiers has a higher energy content due to the more efficient combustion process.

Applications of downdraft fixed bed gasifiers are:

- Industrial Power Generation: Downdraft gasifiers are commonly used in industrial applications for large-scale power generation due to their efficiency and cleaner syngas production.
- Biomass to Biofuel Production: Downdraft gasifiers are employed in the production of biofuels from biomass feedstocks, providing a cleaner alternative to fossil fuels.
- Waste-to-Energy Plants: Downdraft gasifiers are used in waste-to-energy facilities, converting various types of waste into syngas for electricity generation [43].

The residue produced during the fixed bed has to be handled carefully. One of the main residues during the process is ash, and this has to be handled and disposed of properly. The steps followed for efficient handling of the ash are

1. Collection and Storage: Ash is collected from the gasifier and stored in designated containers. Proper storage prevents contamination and ensures easy transportation for disposal.
2. Cooling: Ash is often hot when removed from the gasifier. Allowing it to cool down before handling minimizes the risk of burns or other injuries.
3. Protection: Workers handling ash should wear appropriate protective gear, including gloves and masks, to avoid direct contact and inhalation of harmful particles.

Different methods of reuse of the ash produced are:

1. Reuse in Agriculture: In some cases, ash can be rich in nutrients and minerals. It can be used as a soil conditioner in agriculture, providing essential elements to plants and improving soil structure
2. Landfill Disposal: If the ash is not suitable for agricultural use, it is disposed of in landfills designed to handle hazardous waste. It is essential to follow local regulations and guidelines for proper disposal.
3. Treatment and Encapsulation: Ash can be treated to reduce its hazardous properties and then encapsulated in a stable form, making it safer for disposal. This method ensures that harmful substances do not leach into the environment.
4. Recycling: Depending on its composition, ash might contain materials that are recyclable, such as metals. Separating recyclable components before disposal reduces the overall waste volume and promotes sustainability.
5. Incineration: In some cases, ash can be incinerated in controlled environments, ensuring complete combustion of organic matter and reducing the volume of waste for disposal [44].

4.3.2.2.2 Fluidized Bed Gasification

Fluidized bed gasification is a more intricate process when compared to fixed bed gasification. It offers advantages such as consistent temperature distribution, enhanced carbon conversion, and reduced tar production. Fluidized bed gasifiers can be further categorized into various types, including bubbling fluidized bed, circulating fluidized bed, and entrained flow fluidized bed gasifiers. In bubbling fluidized bed gasification, biomass is placed on a distributor plate and is fluidized using a gasifying medium, typically air or nitrogen gas. The bed temperature typically ranges from 700°C to 900°C. For lower-quality biomass, a circulating bed is employed, with fluidization velocity exceeding the terminal settling velocity. On the other hand, entrained flow gasifiers are utilized under high pressure and elevated temperatures [45]. Some of the features that distinguish a fixed bed from a fluidized bed are:

Feature	Fixed Bed	Fluidized Bed
Operating Principle	In a fixed bed, the fuel is placed on a stationary bed, and air or oxygen is supplied from underneath. Combustion occurs as the air flows through the stationary bed of fuel.	In a fluidized bed, the fuel particles are suspended in a stream of air or gas, creating a fluidized state. This fluidized bed enhances the mixing and interaction between the fuel and the oxidizing agent.
Efficiency	Fixed bed systems generally have lower efficiency compared to fluidized bed systems. The stationary bed restricts the contact between the fuel and the oxidizing agent, leading to incomplete combustion and lower thermal efficiency.	Fluidized bed systems offer higher efficiency due to improved mixing and better heat transfer between the fuel particles and the fluidizing medium. This results in more complete combustion and higher thermal efficiency.
Fuel Flexibility	Fixed bed systems are often limited in terms of fuel flexibility. They work well with specific types of fuels and may not efficiently combust a wide range of biomass or waste materials.	Fluidized bed systems are highly flexible and can efficiently combust a variety of fuels, including biomass, coal, and waste materials. The fluidized state allows for better adaptation to different fuel types.
Control of Emissions	Controlling emissions, especially nitrogen oxides (NOx) and particulate matter, can be challenging in fixed bed systems due to incomplete combustion. Advanced technologies are required to meet stringent emission standards.	Fluidized bed systems inherently have better emission control capabilities. The higher combustion efficiency and lower operating temperatures help in reducing NOx and particulate emissions. Additionally, sulfur dioxide (SO_2) emissions can be controlled effectively.
Applications	Fixed bed systems are often used in small-scale applications such as residential heating, cooking stoves, and some industrial heating processes. They are also used in certain biomass power plants, particularly those with lower capacities.	Fluidized bed systems find widespread applications in large-scale power plants, industrial boilers, and waste-to-energy facilities. Their efficiency and fuel flexibility make them suitable for handling a wide range of fuels and high-capacity operations.

Source: [46].

4.3.2.2.3 Plasma Gasification

Plasma gasification is a thermal technique employed to transform organic materials into syngas through the utilization of plasma. This process involves the use of a plasma torch, which is energized by an electric arc to ionize the gas and catalyze the conversion of organic matter into syngas. In smaller plasma torches, inert gases like argon are commonly employed, while in larger torches, nitrogen gas is used. The electrodes used in these torches can vary, ranging from materials like copper or tungsten to hafnium or zirconium, among other alloys. The temperature of the plasma reaction plays a pivotal role in determining both the structure of the plasma and the composition of the resulting gas. Plasma gasification has found applications in several industries and waste management scenarios, including:

1. Municipal Solid Waste Management: Plasma gasification is used to treat municipal solid waste, reducing the volume of waste in landfills and generating syngas that can be used for electricity generation or as a feedstock for various chemical processes.

2. Hazardous Waste Treatment: Plasma gasification can handle hazardous waste materials, including medical waste, chemical waste, and industrial hazardous waste. It helps in destroying harmful components and reducing waste volume.
3. Biomass Gasification: Plasma gasification technology is applied to biomass, converting organic materials into syngas. This syngas can then be used for power generation or converted into biofuels.
4. Wastewater Sludge Treatment: Plasma gasification is used in treating sewage sludge, reducing the volume and producing energy-rich syngas.
5. Tire and Plastic Waste Management: Plasma gasification can process tires and plastic waste, converting them into syngas and carbon black. This is an environmentally friendly way of handling these non-biodegradable materials.
6. Coal Gasification: Plasma gasification is employed in coal gasification processes, enabling cleaner energy production by converting coal into syngas while capturing pollutants.
7. Remediation of Contaminated Soil: Plasma gasification is used to treat contaminated soil, breaking down harmful substances and reducing the environmental impact.
8. Metal Recovery: Plasma gasification can aid in the recovery of metals from various waste streams, allowing for recycling and reducing the need for raw materials [47].

4.3.2.2.4 Hydrothermal Gasification

Hydrothermal gasification is a method in which biomass is treated using water under elevated temperatures and pressure. The specific outcomes of this process are contingent on factors like temperature, pressure, and treatment duration. Hydrothermal gasification can be conducted in either supercritical or subcritical water conditions. This technique presents several benefits, including the efficient utilization of water as a reaction medium and the potential to generate a diverse array of valuable products from biomass. It holds significant promise as a technology for converting moist biomass feedstocks, such as sewage sludge and algae, into valuable energy resources [48].

4.3.2.3 Combustion

Combustion is the method of obtaining energy mainly in the form of heat energy by burning biomass in the presence of oxygen. Feedstock for this process is usually agricultural waste materials like husk and so on. Unlike other thermochemical processes, the feedstock used is unselective and can be of any size and shape. The amount of energy released during combustion depends upon the carbon content present in the feedstock. The combustion process can be broadly divided into fixed bed combustion and fluidized bed combustion based on the method by which the combustion process is carried out [49, 50].

4.3.3 Biochemical Process

Biochemical conversion involves the use of enzymes of bacteria or other microorganisms to break down biomass. Biochemical conversion can offer several advantages, such as the ability to produce high-value products and the potential for lower greenhouse gas emissions compared to some thermochemical conversion processes. However, it can also be more complex and expensive than thermochemical conversion, and the biomass' availability and characteristics can affect the process' feasibility and economics [51]. Some of the important biochemical processes are: Anaerobic digestion and Fermentation.

4.3.3.1 Anaerobic Digestion

Anaerobic digestion is the process by which organic matter or biomass is converted to fertilizers, biogas, or biologically useful products with the help of bacteria under anaerobic conditions, as shown in Figure 4.3. This process can also be utilized for the production of heat, methane, hydrogen,

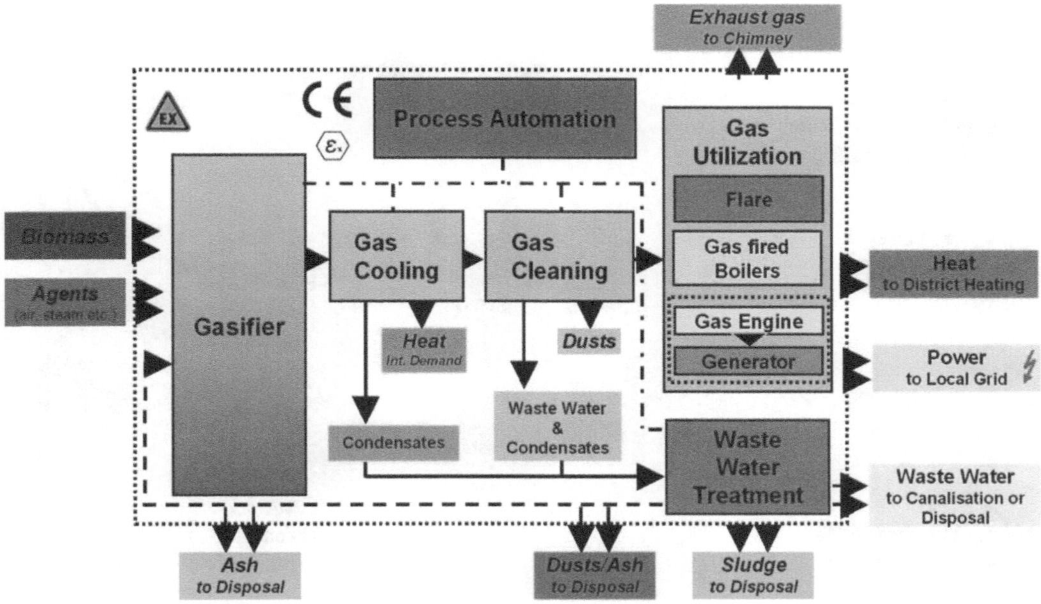

FIGURE 4.3 Biomass conversion using gasification.

Process schematic of a biomass gasification system including feedstock input, gasifier, gas cooling & cleaning, gas utilization (flare, gas-fired boiler, gas engine/generator), wastewater treatment and ash/sludge disposal.

Adapted from **"Biomass Gasification Process," by Salman Zafar, in BioEnergy Consult (2023).**

etc., under specific conditions. Compared to thermochemical processes, biochemical processes such as anaerobic digestion take several days for the final product formation. Anaerobic digestion can be carried out at a wide range of temperatures from 0°C to 60°C, but temperatures ranging from 30–35°C to 50–55°C are mostly preferred. The amount of energy produced and the quality of the subproducts of the anaerobic digestion are mainly determined by the quality of the feedstock and the quality of the bacteria used for the process of digestion. A good bacterium can produce a high yield and can convert up to 90% feedstock into biogas. The sludge produced is usually odorless and non-toxic. The major disadvantage of anaerobic digestion is that it causes health hazards to people and livestock. It also requires continuous monitoring of factors such as temperature, pH, etc. Anaerobic digestion plants are being used on an industrial scale; at the same time, compost pits and small biogas-producing plants can be used as anaerobic digesters on a small scale and in homes (Figure 4.4).

Some of the major health concerns during anaerobic digestion are:

1. Gas exposure: Methane (CH_4) and Hydrogen Sulfide (H_2S): Biogas produced in anaerobic digestion contains methane and hydrogen sulfide, both of which can be harmful in high concentrations. Methane is flammable, and hydrogen sulfide is toxic and can cause respiratory issues.
2. Pathogen exposure: Digestate: The leftover organic material after anaerobic digestion, called digestate, can contain pathogens such as bacteria, viruses, and parasites from organic waste. Direct contact can lead to the spread of diseases.
3. Confined space: Manure Pits and Digesters: Workers entering confined spaces like manure pits or digesters face risks of low oxygen levels, toxic gas buildup, and potential engulfment in the waste material.

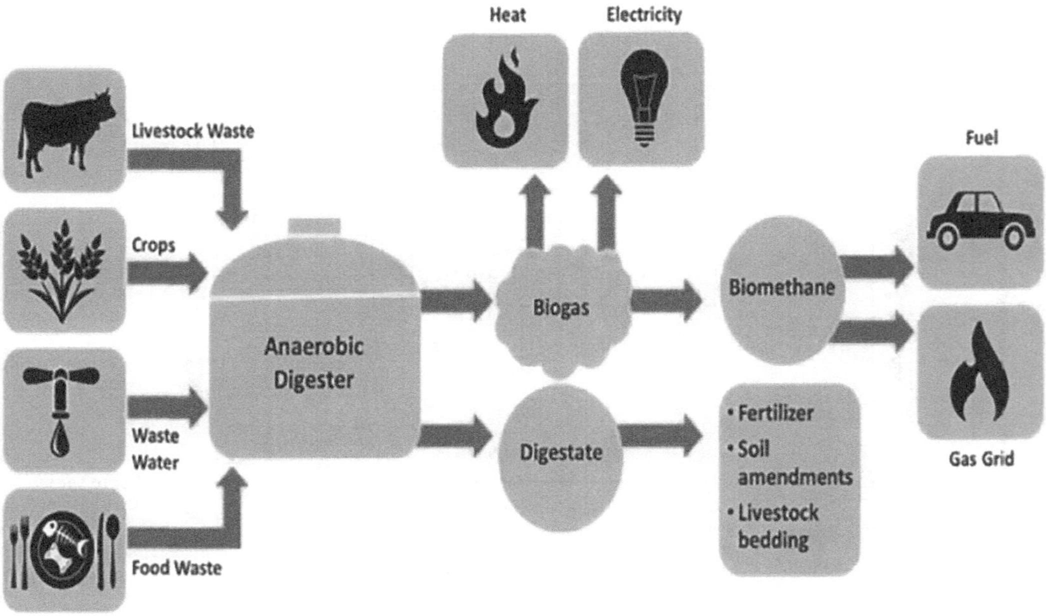

FIGURE 4.4 Anaerobic digestion system.

Process flow of an anaerobic digestion system: livestock, crops, waste-water, and food waste feed into the digester; biogas is used for heat, electricity or fuel; digestate is used as fertilizer/soil amendment.

Adapted from U.S. Environmental Protection Agency, "How does anaerobic digestion work?".

Safety precautions to reduce these risks are:

1. Gas detection system: Industrial-scale facilities install gas detection systems to monitor methane and hydrogen sulfide levels. Alarms are triggered if gas concentrations reach unsafe levels, alerting workers to evacuate the area.
2. Ventilation system: Proper ventilation systems are installed to remove gases from confined spaces, ensuring a constant flow of fresh air and preventing the buildup of toxic gases.
3. Personal protection equipment: Workers are provided with appropriate PPE, including gas masks with filters for toxic gases, safety harnesses, and communication devices for working in confined spaces.
4. Training and safety protocols: Employees receive thorough training in handling gases, operating equipment, and understanding the potential risks associated with anaerobic digestion. Strict safety protocols are enforced and regularly reviewed.
5. Automation and remote monitoring: Automation reduces the need for human entry into dangerous areas. Remote monitoring systems allow operators to assess conditions inside digesters and tanks without physical presence.
6. Regular maintenance: Proper maintenance of equipment and digesters reduces the risk of leaks and malfunctions, ensuring that gases are contained within the system.
7. Emergency response plan: Facilities have well-defined emergency response plans, including procedures for gas leaks, fires, and worker injuries. Regular drills are conducted to prepare workers for emergency situations.
8. Waste pre-treatment: Proper pre-treatment of organic waste, such as removing hazardous materials before digestion, reduces the potential for harmful gas production and exposure to toxins [52, 53].

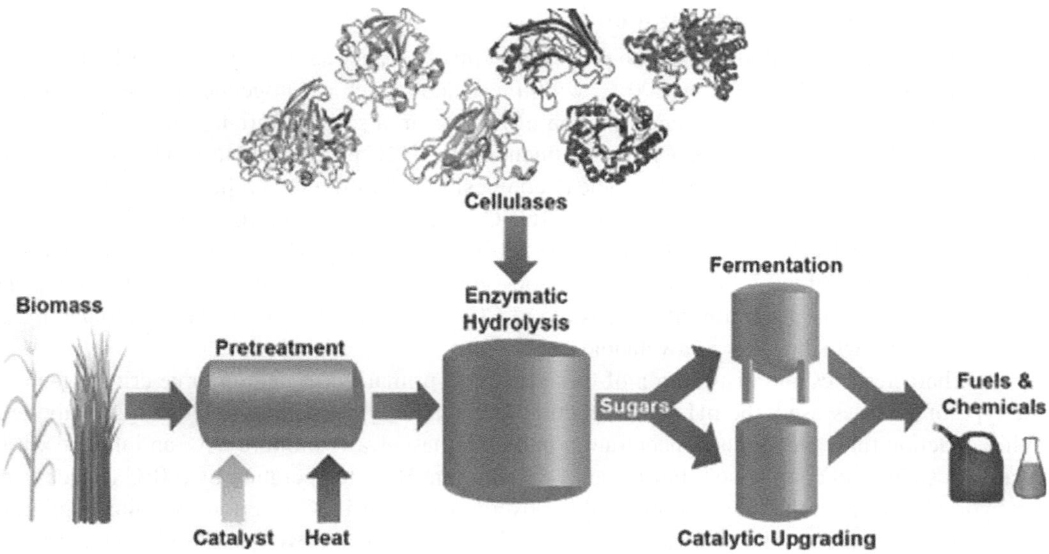

FIGURE 4.5 Fermentation system for biomass conversion.

Conceptual process flow diagram for conversion of lignocellulosic biomass: pretreatment → enzymatic hydrolysis → sugar production → catalytic upgrading/fermentation → fuels and chemicals.

Adapted from "Process Design and Economics for the Conversion of Lignocellulosic Biomass to Hydrocarbon Fuels" (p. XX), by R. Davis et al., 2015, National Renewable Energy Laboratory. Copyright 2015 by NREL.

4.3.3.2 Fermentation

Fermentation is the process of converting biomass into ethanol by means of biochemical pathways. The ethanol produced can be utilized for the production of energy, biodiesel, bio-oil, etc. The ethanol produced is usually considered as a petroleum substitute. Biomass selection plays an important role in determining the yield of the process. Biomass containing high amounts of carbohydrates or sugar is being used in the fermentation process. Unlike other processes, fermentation takes a comparatively long period of time for the production of ethanol [54] (Figure 4.5).

4.3.4 Hydrothermal Processing

A major percentage of biomass available is wet biomass whose moisture content can range up to 95%. Pyrolysis is one of the important and frequently used biomass conversion processes, but biomass containing a moisture content of more than 30% is required to be dried before the pyrolysis steps. This makes the energy required for processing much greater than that of the energy obtained from the processed biomass and makes the usage of wet biomass less efficient. In this case, hydrothermal processing plays an important role. Hydrothermal processing converts wet biomass (moisture content 70% or more) into carbonaceous biofuel in the presence of water and by the application of heat and pressure to the biomass.

In hydrothermal processing, water acts as a solvent and reactant. It utilizes either subcritical or supercritical water to transform biomass into final products without the presence of atmospheric oxygen. One of the biggest positives of this process is that it doesn't erudite any energy intended for drying before or during the process. Based on temperature, pressure, and residence time required, hydrothermal processing can be classified into three processes: Hydrothermal carbonization (HTC), Hydrothermal liquefaction (HTL), and Hydrothermal gasification (HTG) [55].

4.3.4.1 Hydrothermal Carbonization

Hydrothermal Carbonization (HTC) is a process that involves heating biomass feedstock in water at low temperatures, typically below 200°C, within a sealed vessel at autogenous pressure. The primary outcome of this process is the formation of hydrochar, a solid material with a wide range of applications. Understanding the key factors influencing the HTC process and its products is essential for optimizing its efficiency and utility. The decomposition of different components during HTC is hemicellulose starts breaking down at around 160°C, while cellulose and lignin follow suit at temperatures between 180°C and 200°C and above 220°C, respectively. The three primary products of HTC are hydrochar (the solid fraction), an aqueous solution (comprising bio-oil mixed with water), and a small volume of gas, primarily consisting of CO_2. The main objective of HTC is to maximize hydrochar yield, which serves as a valuable end product.

The characteristics and distribution of these products primarily depend on three critical factors: the type of biomass used, the pH of the reaction medium, and the maximum temperature applied. While reaction time and solids concentration in the biomass–water mixture have an influence, it is relatively smaller in comparison. For instance, a minimum HTC temperature of 160°C is necessary for hydrochar formation from glucose, and the overall extent of HTC reactions remains negligible below these conditions. The highest hydrochar yield is typically achieved at 200°C, with a gradual decrease in yield at higher temperatures due to the prevalence of gasification reactions that result in the formation of volatile compounds.

Hydrothermal Carbonization is a versatile and eco-friendly technology with a wide range of applications, from renewable energy production to waste management and carbon sequestration. Its ability to convert biomass into valuable hydrochar, while producing fewer harmful emissions, positions it as a promising solution for addressing sustainability challenges and reducing our reliance on fossil fuels [56].

4.3.4.2 Hydrothermal Liquefaction

Hydrothermal Liquefaction (HTL) is a thermochemical process that operates at high temperatures and pressures, leading to the conversion of biomass feedstock into a liquid product. The temperatures during the process range from 200 to 350°C, and pressures between 5 and 20 MPa. In HTL, biomass feedstock undergoes a transformation primarily into a liquid product that is aqueous soluble. This conversion mainly depends upon the combination of free radical and ionic reactions taking place. In HTL, water is used as a reaction medium. Water's behavior changes significantly under the elevated temperature and pressure conditions of HTL. The dielectric constant of water, which measures its ability to form hydrogen bonds, decreases rapidly with increasing temperature.

The dissociation of water into its constituent ions becomes more pronounced with increasing temperature. This indicates that as the temperature rises, water becomes less stable in its molecular form and more prone to ionization. As the temperature rises, water becomes less viscous, which has a significant impact on mass transfer processes within the HTL system. The reduction in water viscosity enhances mass transfer rates, allowing for faster transport of reactants and products within the reaction medium. This, in turn, accelerates any mass-transfer-limited chemical reactions, improving overall process efficiency [57].

4.3.4.3 Hydrothermal Gasification

Hydrothermal Gasification (HTG) is an advanced thermochemical process that operates at high temperatures and pressures, using supercritical water to efficiently convert biomass into a valuable mixture of non-condensable gases, including hydrogen-enriched syngas. It operates at temperatures near or above the critical point of water, typically in the range of 400–600°C, and under high pressures, typically between 23 and 45 MPa. In HTG, biomass feedstock is primarily converted into a mixture of non-condensable gases, including hydrogen, carbon monoxide, methane, and carbon dioxide. One of the distinguishing features of HTG is its ability to produce syngas enriched with hydrogen, making it a valuable technology for renewable energy and chemical production.

HTG is well-suited for wet biomass feedstock, making it a valuable technology for utilizing biomass with high moisture content, exceeding 80%. Conventional thermal gasification technologies often rely on partial oxidation using air, resulting in the production of tar and the dilution of syngas with nitrogen. In contrast, syngas produced from HTG is devoid of tars, or if they are produced, they remain in the liquid phase and do not dilute the gas phase. This makes the syngas from HTG cleaner and more suitable for a wide range of applications, including energy production and chemical synthesis. HTG is preferred over HTC and HTL. HTC and HTL face the problem of the production of undesired byproducts, but in the case of HTG, those byproducts are not produced or are released in the form of gaseous products rather than being dissolved in water [58].

4.4 ADVANTAGES AND DISADVANTAGES OF BIOMASS CONVERSION TECHNOLOGIES

Advantages:

- Renewable and eco-friendly: Biomass energy is derived from organic materials, making it a renewable energy source. It produces significantly less pollution compared to fossil fuels, contributing to cleaner air and reduced greenhouse gas emissions.
- A biomass energy plant contributes to improved sanitation in both rural and urban areas. Biomass energy is inherently carbon-neutral, thus aiding in the mitigation of global warming. When biomass energy is produced in a decentralized manner, it can offer high-quality, consistent, and reliable electricity to villages, supporting various essential needs such as lighting, clean water supply, irrigation, milling, and more.
- Biomass energy is reliable, cost-effective, and efficient.
- Biomass energy can be easily integrated with solar and wind energy systems, creating hybrid setups for enhanced energy generation and reliability.
- Biomass energy promotes the recycling of organic waste, reducing pollution and limiting the spread of diseases associated with waste accumulation.
- Improves sustainability.
- Reduce fossil fuel dependency.
- Reduces global warming to a greater extent.

Disadvantages:

- Requires continuous biomass supply.
- High initialization costs.
- Inefficiency due to improper construction.
- Environmental impact: Clearing large forest areas for biomass fuel can be environmentally problematic, although local sourcing of fuel can lower emissions compared to fossil fuels.
- Requires a large space for installation.
- Odor concerns [59].

4.4.1 LCA Analysis of Biomass Processing Technologies

The objective of this study is to appraise the sustainability, both environmentally and economically, of various biomass processing technologies. These technologies encompass biomass direct combustion power generation, biomass gasification power generation, biomass mixed combustion power generation, and biomass biogas power generation. To achieve this, the study employs the life cycle assessment (LCA) and dynamic analysis methods to measure the environmental impacts and economic advantages of these technologies in contrast to coal-fired power generation. The study focuses on key operational stages, including raw material production, processing, transportation, and power generation. Notably, the analysis excludes stages related to equipment manufacturing, recovery, depreciation, and plant construction. The unit capacities for biomass power generation technologies are as follows: 30 MW for biomass direct combustion, 4 MW for biomass gasification, 300 MW for biomass mixed combustion, and 2 MW for biomass biogas. For comparison purposes, the unit capacity for coal-fired power generation is set at 1320 MW.

The inventory analysis primarily examines the release of gas pollutants during the operational phase of power plants. Emission data for both biomass power generation and coal-fired power generation is sourced from literature and field investigations. Nevertheless, owing to limitations in biomass mixed-combustion technology at the industrial level, emission data for biomass mixed-combustion power generation is approximated by substituting a portion of biomass direct combustion for the traditional coal-fired power generation. The environmental impact assessment is executed following the framework of the LCA impact assessment stage. The data gathered during the inventory analysis is utilized to calculate the potential environmental impact for each impact category, encompassing global warming potential, acidification potential, photochemical ozone creation potential, human toxicity potential, and soot potential. These impact potentials are standardized and assigned weights to derive the overall environmental burden.

4.4.1.1 Economic Analysis of Biomass Processing Technologies

The economic assessment of the biomass power plant project involves an examination of economic factors during both the construction and operational phases. The initial investment during the construction phase encompasses expenses related to civil construction, equipment acquisition and installation, equipment commissioning and delivery, as well as other associated costs. This phase is characterized as the stage of capital expenditure. The operational phase extends from the commencement of formal production to the conclusion of the project's life cycle. Capital outflows in this phase encompass expenses for raw materials, fuel procurement and transportation, labor management, equipment depreciation, and other operating costs. Conversely, the stage of capital inflow includes revenues generated from electricity sales and other derived products.

The economic analysis of domestic biomass power generation technologies is conducted from three perspectives:

1. **Analysis of capital inflow and outflow**: This analysis considers the cash flows during the construction and operation periods, identifying the expenses and income associated with the biomass power plant project.
2. **Economic evaluation of biomass power generation**: This evaluation assesses the financial viability of the biomass power plant project by considering factors such as the net present value (NPV), internal rate of return (IRR), payback period, and profitability index. These indicators help determine the economic feasibility and attractiveness of the project.
3. **Analysis of economic evaluation results**: This analysis compares the economic evaluation results of different biomass power generation technologies, providing insights into the economic benefits and potential risks associated with each technology [60].

4.5 CONCLUSION

Biomass is a heterogeneous blend of carbon, oxygen, and hydrogen-based polymers. Additionally, it contains an assortment of macronutrients, micronutrients, trace elements, and heavy metals, leading to a notably diverse composition. Using it directly without any form of conversion can often lead to pollution and waste of energy. Hence, biomass conversion technologies are very crucial in the proper utilization of energy and in maintaining a sustainable environment. To make an efficient and effective conversion, knowing the biomass composition is crucial. There is a wide range of biomass conversion processes available, but the most preferred one is the thermochemical process due to its wide range of applications and the efficiency in producing energy. Thermochemical conversion is considered to be the most cost-efficient technology among biomass conversion technologies.

Biomass conversion technologies are very useful and help in the sustainable development of the nation. The whole world is facing a critical energy crisis due to the reduction in fossil fuel supply and other non-renewable energy sources. Conversion of biomass helps in managing organic waste and also serves as a great source of energy. Biomass conversion technologies have been performing for decades on a small scale. During the past few years, all these technologies have gained industrial importance and have become popular all over the world. Over the advantages, a few biomass conversion processes like combustion pose a threat to global warming as it produces a large amount of greenhouse gases. So, we can conclude that biomass conversion technologies are advantageous to society, but at the same time, a few of them can cause small threats too.

REFERENCES

[1] Kumar, A., Kumar, N., Baredar, P., & Shukla, A. (2015). A Review on Biomass Energy Resources, Potential, Conversion, and Policy in India. *Renewable and Sustainable Energy Reviews*, 45, 530–539.

[2] Chaursiya, K., & Dubey, A. (2018). Biomass Utilization Focused on Environment Protection, Energy Generation, and Conversion Technologies. *Journal of Advancements in Material Engineering*, 3(2), 1–6. © MAT Journals 2018. All Rights Reserved.

[3] Williams, C. L., Westover, T. L., Emerson, R. M., Tumuluru, J. S., & Li, C. (2016). Sources of Biomass Feedstock Variability and the Potential Impact on Biofuels Production. *BioEnergy Research*, 9(1), 1–14. DOI: 10.1007/s12155-015-9694-y

[4] Carrillo-Parra, A., Foroughbakhch-Pournavab, R., Bustamante-García, V., & Garza-Ocañas, F. (2014). Biofuels: Plants and Trees as Source of Bioenergy. *Applied Botany*.

[5] Unal, H., & Alibas, K. (2007). Agricultural Residues as Biomass Energy. *Energy Sources, Part B*, 2, 123–140.

[6] Stąsiek, J., & Szkodo, M. (2020). Thermochemical Conversion of Biomass and Municipal Waste into Useful Energy Using Advanced HiTAG/HiTSG Technology. *Energies*, 13(16), 4218.

[7] Kitani, O., Roy, K., Yoshida, M., & Endo, R. (2005). Control of Water Environment of Plants using Waste Biomass. *Agricultural Engineering International: The CIGR Ejournal*, Vol. VII, Manuscript LW 05 007.

[8] Rahman, R. N. U. A., Ismail, M., Rasid, R. A., & Ahamad Nordin, N. I. (2019). Torrefaction of Food Waste as a Potential Biomass Energy Source. *Indonesian Journal of Chemistry*, 19(4), 993–999. DOI: 10.22146/ijc.40871

[9] Carlin, N. T. (2007). *Thermo-Chemical Conversion of Dairy Waste Based Biomass Through Direct Firing* (Doctoral dissertation, Texas A&M University).

[10] Oladokun Sulaiman, O., Saman Abd Kader, A., & Azman, S. (2015). Study of Macro Algae as Marine Biomass Energy Source. *Journal of Aquaculture & Marine Biology*, 2(1), 00012.

[11] Zehnsdorf, A., Moeller, L., Stabenau, N., Bauer, A., Wedwitschka, H., Gallegos, D., Stinner, W., & Herbes, C. (2018). Biomass Potential Analysis of Aquatic Biomass and Challenges for Its Use as a Nonconventional Substrate in Anaerobic Digestion Plants. *Engineering in Life Sciences*, 18(6), 492–497. DOI: 10.1002/elsc.201800032

[12] Ramola, S., Belwal, T., & Srivastava, R. K. (2020). Thermochemical Conversion of Biomass Waste-Based Biochar for Environment Remediation. In O. V. Kharissova et al. (Eds.), *Handbook of Nanomaterials and Nanocomposites for Energy and Environmental Applications*. DOI: 10.1007/978-3-030-11155-7_122-2

[13] Lee, S. Y., Sankaran, R., Chew, K. W., Tan, C. H., Krishnamoorthy, R., Chu, D. T., & Show, P. L. (2019). Waste to Bioenergy: A Review on the Recent Conversion Technologies. *BMC Energy*, 1(1), 4.

[14] Claro, J. C. A. R., & Costa-Gonzalez, D. (2014). Biomass Pellet Production with Industrial and Agro-industrial Waste. *WIT Transactions on State of the Art in Science and Engineering*, 83, 37–42.

[15] Godinho João, K. A. (2013). *Pre-treatment of Different Types of Lignocellulosic Biomass Using Ionic Liquids*. Universidade Nova de Lisboa, Faculdade de Ciências e Tecnologia, Departamento de Química.

[16] Williams, C. L., Westover, T. L., Emerson, R. M., Tumuluru, J. S., & Li, C. (2016). Sources of Biomass Feedstock Variability and the Potential Impact on Biofuels Production. *BioEnergy Research*, 9(1), 1–14. DOI: 10.1007/s12155-015-9694-y

[17] Kanase-Patil, A. B., Saini, R. P., & Sharma, M. P. (2007). Biomass-Based Electricity Production Technologies. Paper presented at the *23rd National Convention of Chemical Engineers on Recent Trends in Chemical Engineering*.

[18] Sridhar, G., Sridhar, H. V., Dasappa, S., Paul, P. J., Subbukrishna, D. N., & Rajan, N. K. S. (2012). Green Electricity from Biomass-fueled Producer Gas Engine. Combustion Gasification and Propulsion Laboratory, Centre for Sustainable Technologies, Department of Aerospace Engineering, Indian Institute of Science, Bangalore, India.

[19] Ali, A. (n.d.). *Electricity generation from renewable biomass energy* (7 pp.). Academia.edu. (2017).

[20] Stevanović, S., & Stevanović, S. (2022). Biomass as a Renewable Energy Source. *Title of the Journal*. DOI: 10.5937/ekoPolj2201195S

[21] Danso-Boateng, E., & Achaw, O.-W. (2022). Bioenergy and Biofuel Production from Biomass Using Thermochemical Conversions Technologies—A Review. *AIMS Energy*, 10(4), 585–647. DOI: 10.3934/energy.2022030

[22] Chandraratne, M. R., & Daful, A. G. (2021). Recent Advances in Thermochemical Conversion of Biomass. *Recent Perspectives in Pyrolysis Research*. Intechopen, September.

[23] Overend, R. P. (2002). Direct Combustion of Biomass. In R. P. Overend (Ed.), *Renewable Energy Sources Charged with Energy from the Sun and Originated from Earth-Moon Interaction - Vol. I*. Encyclopedia of Life Support Systems (EOLSS).

[24] Bridgwater, T. (1992). Thermochemical and Biochemical Biomass Conversion Activities. *Biomaw and Biomergy*, 2(1–6), 307–318.

[25] Rajesh Banu, J., Godvin Sharmila, V., Ushani, U., Amudha, V., & Kumar, G. (2020). Impervious and Influence in the Liquid Fuel Production from Municipal Plastic Waste Through Thermo-chemical Biomass Conversion Technologies - A Review. *Science of the Total Environment*, 718, 137287.

[26] Moldoveanu, S. C. (2019). General Information About Pyrolysis. In S. C. Moldoveanu (Ed.), *Pyrolysis of Organic Molecules (Second Edition): Applications to Health and Environmental Issues* (pp. 1–33). Elsevier. DOI: 10.1016/B978-0-444-64000-0.00001-9

[27] Adhikari, S., Nam, H., & Chakraborty, J. P. (2018). Conversion of Solid Wastes to Fuels and Chemicals Through Pyrolysis. In *Waste Biorefinery* (pp.239–263). Elsevier B.V. DOI: 10.1016/B978-0-444-63992-9.00008-2

[28] Verma, M., Godbout, S., Brar, S. K., Solomatnikova, O., Lemay, S. P., & Larouche, J. P. (2012). Biofuels Production from Biomass by Thermochemical Conversion Technologies. *International Journal of Chemical Engineering*, 2012, Article ID 542426, 18. doi:10.1155/2012/542426

[29] Bridgwater, A. V. (1994). Catalysis in Thermal Biomass Conversion. *Applied Catalysis A: General*, 116, 5–47.

[30] Dhakal, B. (2021). *Intermediate Pyrolysis of Wheat Straw and Softwood Pellets* (Master's thesis). University of Alberta, Canada.

[31] Zanzi Vigouroux, R. (2001). *Pyrolysis of Biomass: Rapid Pyrolysis at High Temperature, Slow Pyrolysis for Active Carbon Preparation* (Report No. TRITA-KET R 144). Department of Chemical Engineering and Technology, Royal Institute of Technology, Stockholm, Sweden.

[32] Das, P., Chandramohan, V. P., Mathimani, T., & Pugazhendhi, A. (2021). Recent Advances in Thermochemical Methods for the Conversion of Algal Biomass to Energy. *Science of the Total Environment*, 766, 144608.

[33] Serio, M. A., Cosgrove, J. E., Wójtowicz, M. A., Wignarajah, K., & Fisher, J. (2011). Microwave-Assisted Pyrolysis of Solid Waste. Paper presented at *the 41st International Conference on Environmental Systems*, Portland, Oregon.

[34] Demirbas, A., & Arin, G. (2002). An Overview of Biomass Pyrolysis. *Energy Sources*, 24(5), 471–482. DOI: 10.1080/00908310252889979

[35] Han, Y., Gholizadeh, M., Tran, C. C., Kaliaguine, S., Li, C.-Z., Olarte, M., & Garcia-Perez, M. (2019). Hydrotreatment of Pyrolysis Bio-oil: A Review. *Fuel Processing Technology*, 195, 106140.

[36] Lee, X. J., Ong, H. C., Gan, Y. Y., Chen, W. H., & Mahlia, T. M. I. (2020). State of Art Review on Conventional and Advanced Pyrolysis of Macroalgae and Microalgae for Biochar, Bio-oil and Bio-syngas Production. *Energy Conversion and Management*, 210, 112707.

[37] Borsodi, N., Szentes, A., Miskolczi, N., Wu, C., & Liu, X. (2016). Carbon Nanotubes Synthesized from Gaseous Products of Waste Polymer Pyrolysis and Their Application. *Journal of Analytical and Applied Pyrolysis*, 120, 304–313.

[38] Guliyev, N. G., Ibrahimov, H. J., Alekperov, J. A., Amirov, F. A., & Ibrahimova, Z. M. (2018). Investigation of Activated Carbon Obtained from the Liquid Products of Pyrolysis in Sunflower Oil Bleaching Process. *International Journal of Industrial Chemistry*, 9(3), 277–284.

[39] Pang, S. (2019). Advances in Thermochemical Conversion of Woody Biomass to Energy, *Fuels and Chemicals. Biotechnology Advances*, 37(4), 589–597.

[40] Molino, A., Chianese, S., & Musmarra, D. (2016). Biomass Gasification Technology: The State of the Art Overview. *Journal of Energy Chemistry*, 25, 10–25.

[41] Delgado, E., Avilés, O., & Aperador, W. (2014). Gasification of Biomass in a Fixed Bed Reactor. In *Advanced Materials Research Vols. 875–877* (pp. 1831–1836). Trans Tech Publications.

[42] Gao, N., Li, A., Quan, C., & Gao, F. (2008). Hydrogen-rich Gas Production from Biomass Steam Gasification in an Updraft Fixed-bed Gasifier Combined with a Porous Ceramic Reformer. *International Journal of Hydrogen Energy*, *33*, 5430–5438. DOI: 10.1016/j.ijhydene.2008.07.021

[43] Bhavanam, A., & Sastry, R. C. (2011). Biomass Gasification Processes in Downdraft Fixed Bed Reactors: A Review. *International Journal of Chemical Engineering and Applications*, 2(6), 42.

[44] Huang, S., Wu, S., Wu, Y., & Gao, J. (2017). Structure Characteristics and Gasification Activity of Residual Carbon from Updraft Fixed-Bed Biomass Gasification Ash. *Energy Conversion and Management*, 136, 108–118.

[45] Subbaiah, B. S., Murugan, D. K., Deenadayalan, D. B., & Dhamodharan, M. I. (2014). Gasification of Biomass Using Fluidized Bed. *International Journal of Innovative Research in Science, Engineering, and Technology*, 3(2), 8995.

[46] Puig-Arnavat, M., Bruno, J. C., & Coronas, A. (2010). Review and Analysis of Biomass Gasification Models. *Renewable and Sustainable Energy Reviews*, 14, 2841–2851.

[47] Luche, J., Falcoz, Q., Bastien, T., Leninger, J. P., Arabi, K., Aubry, O., Khacef, A., Cormier, J. M., & Lédé, J. (2012). Plasma Treatments and Biomass Gasification. *IOP Conference Series: Materials Science and Engineering*, 29(1), 012011. doi: 10.1088/1757-899X/29/1/012011

[48] Kruse, A. (2009). Hydrothermal Biomass Gasification. *Journal of Supercritical Fluids*, *47*, 391–399.

[49] Zhang, L., Xu, C., & Champagne, P. (2010). Overview of Recent Advances in Thermo-Chemical Conversion of Biomass. *Energy Conversion and Management*, 51, 969–982.

[50] Adams, P., Bridgwater, T., Lea-Langton, A., Ross, A., & Watson, I. (2018). Biomass Conversion Technologies. In *Greenhouse Gas Balances of Bioenergy Systems* (p. 107). Elsevier Inc.

[51] Devi, S., Suhag, M., Dhaka, A., & Singh, J. (2011). Biochemical Conversion Process of Producing Bioethanol from Lignocellulosic Biomass. *International Journal of Microbial Resource Technology*, 1(1), 28–32.

[52] Roder, M., & Welfle, A. (2019). Bioenergy. In *Managing Global Warming*. Elsevier Inc.

[53] Sharma, S., Meena, R., Sharma, A., & Goyal, P. K. (2014). Biomass Conversion Technologies for Renewable Energy and Fuels: A Review Note. *IOSR Journal of Mechanical and Civil Engineering (IOSR-JMCE)*, 11(2), 28–35.

[54] Chan, W. N., & Holtzapple, M. T. (2003). Conversion of Municipal Solid Wastes to Carboxylic Acids by Thermophilic Fermentation. *Applied Biochemistry and Biotechnology*, 111(1–3), 93.

[55] Van Doren, L. G., Posmanik, R., Bicalho, F. A., Tester, J. W., & Sills, D. L. (2016). *Prospects for Energy Recovery During Hydrothermal and Biological Processing of Waste Biomass*. Elsevier.

[56] Heilmann, S. M., Davis, H. T., Jader, L. R., Lefebvre, P. A., Sadowsky, M. J., Schendel, F. J., … Valentas, K. J. (2010). Hydrothermal Carbonization of Microalgae. *Biomass and Bioenergy*, 34(6), 875–882.

[57] Tushar, M. S. H. K., Joarder, M. S. A., Raj, R. F., Sarker, M. A., & Mahmud, M. R. (2019). Hydrothermal Carbonization and Hydrothermal Liquefaction of Water Hyacinth. In *International Conference on Mechanical Engineering and Renewable Energy 2019 (ICMERE2019)*, 11–13 December 2019, Chittagong, Bangladesh (ICMERE2019-PI-242).

[58] Prifti, K., Basso, D., Pavanetti, R., & Manenti, F. (2021). Improving Hydrothermal Carbonization (HTC) Processes by Hydrochar Gasification. *Chemical Engineering Transactions*, 86, 925–930.

[59] Tekale, A., God, S., Bedre, B., Madake, G., Vaghela, P., & Labade, S. (2017). Energy Production from Biomass: Review. *International Journal of Innovative Science and Research Technology*, 2(10), 25. ISSN 2456-2165.

[60] Chen, S., Feng, H., Zheng, J., Ye, J., Song, Y., Yang, H., & Zhou, M. (2020). Life Cycle Assessment and Economic Analysis of Biomass Energy Technology in China: A Brief Review. *Processes*, 8(9), 1112.

5 Life Cycle Sustainability Analysis of Energy Production from Wood Waste through Pyrolysis and Integrated Energy Systems

Aneesh Raj and Praveen Kumar Ghodke

5.1 INTRODUCTION

The energy crisis of the 1970s placed serious doubts on the world's reliance on fossil fuels, prompting the scientific community to intensify its quest for sustainable alternatives [1]. Biomass-based energy is considered carbon neutral, as CO_2 emissions during combustion are balanced by photosynthesis [2]. Biomass is a cleaner and more sustainable alternative to fossil fuels that plays a crucial role in meeting rising global energy demand and aligning with sustainable energy development objectives. In recent years, biomass utilization for biofuel production has garnered considerable interest. As concerns over the environmental impact of fossil fuels increase due to climate change and the COVID-19 pandemic, organizations increasingly favor a shift toward green and diverse energy sources [3]. In addition to providing thermal and electrical energy, biomass serves as a precursor for valuable chemicals such as furfural, valerolactone, and levoglucosan [4, 5]. The thermochemical pathways, particularly pyrolysis, are gaining increased interest due to their potential to convert biomass into high-value products without the use of additional chemicals or the emission of pollutants. In addition, in pursuit of a transition to a more sustainable and circular economy, waste-to-energy (WTE) practices have gained prominence, with a focus on enhancing energy supply and waste management. Due to its potential to convert waste materials into energy and valuable resources while mitigating their environmental impact, pyrolysis has attracted significant research interest among various WTE technologies [6].

Although the pyrolysis process shows promise as a renewable energy option to meet rising demand, much more work is needed to bring it to market, particularly in the areas of technology, economics, and regulations. These efforts are geared toward realizing pyrolysis' full potential as an environmentally responsible and economically viable biomass conversion process that can provide reliable alternative energy for a sustainable future. It is important to address a variety of challenges, including shifts in biomass feedstock, system boundaries, analysis methods, and regional factors, when constructing a comprehensive and consistent LCSA consensus for biomass pyrolysis [7, 8].

LCSA has the potential to shed light on the environmental implications of the pyrolysis process across the board, from plantation to product upgrading and end-of-life disposal. LCSA analyzes how well-proposed solutions work and how they affect the environment. The results of the LCSA have applications in policymaking and the advancement of sustainable pyrolysis technology [8, 9].

DOI: 10.1201/9781003732907-5

5.2 SOURCES OF WOOD WASTE

In order to preserve the environment and promote the economy, wood waste can be used to generate renewable energy. Industrial and urbanization processes generate wood residue, which is a renewable and sustainable feedstock. The different sources of wood waste are highlighted below, together with their relevance, potential, and options for use as an environmentally friendly fuel.

5.2.1 Forestry Residues

Significant volumes of wood waste are produced during forestry operations in the form of branches, treetops, and other leftovers. These materials are sometimes discarded in ways that affect the environment or are left unused. These leftovers from processes like logging and thinning have an enormous potential for energy conversion. They can be gathered, chipped, and processed to provide electricity, heat, or biofuels, making them a more environmentally friendly fuel source than conventional fossil fuels [10].

5.2.2 Construction and Demolition Debris

The production of surplus materials, offcuts, and discarded lumber during construction projects and demolition makes the construction industry a substantial contributor to wood waste. Environmental concerns are exacerbated by the frequent disposal of this wood waste in landfills. The demand for virgin timber resources can be reduced by reusing it for energy production or other purposes, which eases the burden on landfills. Construction waste can be converted into useful energy using technologies like biomass boilers and gasification systems, advancing the ideas of the circular economy [11].

5.2.3 Urban Green Waste

By maintaining parks, gardens, and other public places, urban regions produce substantial amounts of wood debris. Waste management systems are strained as a result of the accumulation of fallen trees, clipped branches, and abandoned wooden objects like furniture and pallets. Communities can take advantage of this wood waste's potential for composting or the production of biomass energy by keeping it out of landfills and burning. Reusing urban green trash also encourages the use of nearby resources and lowers emissions associated with transportation [11].

5.2.4 Agricultural Residues

Wood waste can develop in agricultural settings as a result of procedures including vineyard management, crop harvesting, and orchard upkeep. Wood residues devoid of pesticides can be used for energy production without posing any contaminant risks. Additionally, by incorporating these residues into anaerobic digestion processes, biogas that can be used for both electricity and heating. Including agricultural wood waste in renewable energy plans can improve farming practices' sustainability while generating the least amount of trash [12].

5.2.5 Wood Processing and Manufacturing

Sawdust, shavings, and offcuts are common waste products produced by the wood processing sector, which includes sawmills, pulp and paper mills, and furniture producers. These leftovers can be converted into important products for use in energy production, such as particleboard or wood pellets. Additionally, innovative techniques like pyrolysis can transform these waste streams into biochar and bio-oil, opening up a variety of opportunities for energy production, soil improvement, and chemical synthesis [13].

5.2.6 Landscaping and Tree Maintenance

Wood debris is created as a result of tree removal, trimming, and maintenance activities in urban and suburban settings. This waste can be converted into biomass pellets, gasification, or combustion to provide energy. Local governments, tree care businesses, and energy providers may work together to make it easier to collect, process, and use this resource, which will cut down on waste and carbon emissions [6].

As a result, wood waste is shown as a flexible and plentiful resource with the potential to transform efforts to promote sustainability, waste management, and energy production in a variety of fields. Societies may make considerable progress in reducing dependency on limited fossil fuels, mitigating environmental effects, and creating a more resilient and environmentally friendly future by recognizing and utilizing the value of wood waste from various sources.

5.3 WOOD WASTE TO ENERGY – AVAILABLE TECHNOLOGIES

In view of the urgent need to transition away from fossil fuels and toward renewable energy sources, wood waste has emerged as a potential feedstock for the generation of sustainable energy. Urban green trash, construction debris, and other types of wood waste present a great chance to harness energy using a number of innovative technologies. This chapter thoroughly examines the available technologies for converting wood waste into energy, paying particular attention to their benefits, limitations, and prospective impacts on the state of renewable energy sources today. This chapter concentrates on the pyrolysis process, which is the most suited conversion method for the wood biomass waste under discussion [10, 14]. The available technologies are outlined in Figure 5.1.

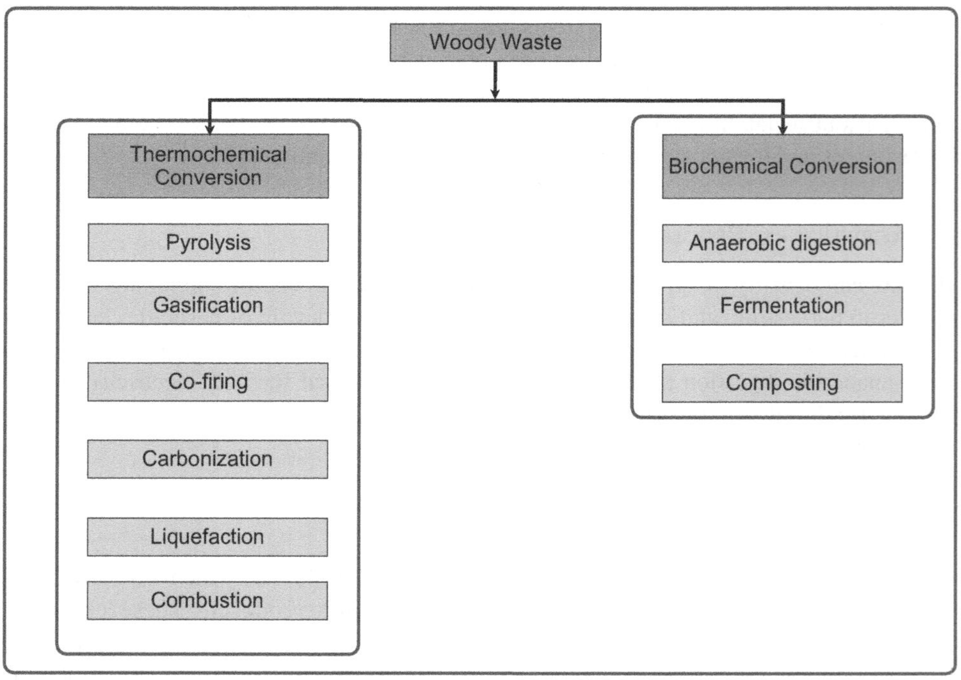

FIGURE 5.1 Available conversion technologies for woody waste biomass.

5.3.1 Pyrolysis

Biochar, bio-oil, and non-condensable gases are produced through the pyrolysis process, in which biomass substrates are heated to temperatures between 400 and 1200°C in an oxygen-free environment [15]. Figure 5.2 shows that pyrolysis can be classified in several ways according to the residence time, heating rate, and oxidizer presence or absence [9].

Pyrolysis can be categorized into slow, fast, and flash pyrolysis based on the heating rate and residence time. For optimal biochar production, use a heating rate of 0.1–10°C/s and a residence time of up to 550 s during slow pyrolysis, which occurs between 400 and 900°C. A larger bio-oil yield up to 75 percent by weight is possible with fast pyrolysis due to the shorter vapor residence time of 0.5–10 s and the faster heating rate of 10–200°C/s. The temperature of operation will be in the range of 400–900°C. Flash pyrolysis, on the other hand, takes place at even higher temperatures of 600–1200°C, with an exceptionally fast heating rate of 10^3–10^4°C/s, yielding a bigger percentage of non-condensable gases, up to 50–60 wt.%. Biochar, the byproduct of pyrolysis, can be substituted for coal in solid fuel boilers. Other possible uses for bio-char include carbon nanotube production, activated carbon production, soil enhancer production, fertilizer production, and so on. Power facilities can switch to using the bio-oil made from pyrolysis instead of fossil fuels in their boilers and furnaces. The possibility exists to convert bio-oil into a variety of valuable alternative fuels. Gases produced by pyrolysis are not condensable; therefore, they can be used for heating, cooking, and power generation in both commercial and residential applications [9, 10, 13, 16, 17].

Allothermal and autothermal pyrolysis are defined by the presence or absence of an oxidizing agent, respectively. Thermal degradation of biomass feedstock in an inert environment is what allothermal pyrolysis is all about. Being an endothermic process that necessitates external heat, allothermal pyrolysis makes it difficult to increase production volume. In order to complete the pyrolysis cycle and meet the heat demands, a small amount of oxygen is introduced during autothermal pyrolysis to partially oxidize a portion of the pyrolysis product. Combustion of organic compounds and oxidation of volatile substances are examples of exothermic reactions that can start in the presence of an oxidant. Thermal deterioration of feedstocks involves endothermic events, which are counteracted by these exothermic reactions [18–20].

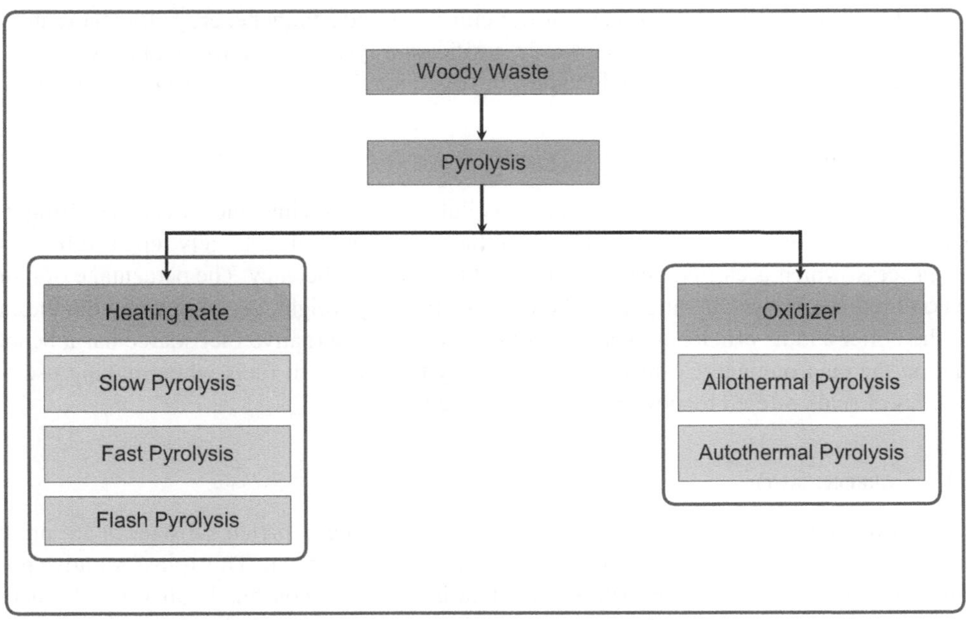

FIGURE 5.2 Classification of pyrolysis.

Although pyrolysis shows great promise as a biomass conversion process, it is important to consider the many possible downsides and obstacles before commercializing it. Considerations such as these include product and market acceptability, consistency and availability of feedstock, operating and capital costs, process complexity, and technological obstacles. Social and stakeholder considerations are also a part of them. Going from small-scale experiments in a lab to large-scale commercial applications of pyrolysis technologies can be costly. The construction and operation of large-scale pyrolysis plants necessitate substantial investments in infrastructure, machinery, and personnel. In the absence of subsidies or incentives from the government, it can be especially difficult to keep these costs in check and bring the technology up to cost parity with others used to produce energy and petroleum. The market popularity and demand for pyrolysis products such as biochar, bio-oil, and biogas can be influenced by their functional efficacy and competitive pricing in relation to comparable items. It is crucial to precisely control a lot of variables, such as temperature, heating rate, residence time, and pressure, in order to optimize product yields and characteristics during pyrolysis. Managing these characteristics on a commercial scale can be difficult, and process deviations can impact product performance and quality. To address potential issues with corrosion, pollution, and emissions control in large-scale pyrolysis processes, more technical knowledge and financial resources may be required [9, 21, 22].

As a whole, pyrolysis shows a lot of promise as an alternative energy source and might drastically change the way we meet our energy demands. Full realization of pyrolysis' potential as a sustainable and economically feasible biomass conversion process, however, requires ongoing study and development. A number of considerations, such as technological advancement, financial feasibility, conformity with regulations, and involvement of stakeholders, are necessary to accomplish this. A more sustainable energy future may be within reach if these initiatives help pyrolysis reach its full potential as an alternative energy source.

5.3.2 Gasification

Advanced thermochemical technology called gasification transforms wood waste into synthesis gas (syngas), which contains hydrogen, carbon monoxide, and other trace gases. Syngas can be used to generate power, heat buildings, or undergo additional processing to yield biofuels and chemicals. Compared to combustion, gasification has lower emissions and a higher energy efficiency, and it can handle a larger variety of feedstocks. However, gasification systems are intricate, necessitating exact control over operating parameters and the purging of contaminants from the syngas [19, 23].

5.3.3 Co-firing

One effective and eco-friendly way to turn lignocellulosic biomass into energy is by co-firing it with coal, a relatively inexpensive process. The basic fuel, which is coal, is largely replaced by biomass in this process, which is carried out in a boiler that has a high efficiency. The percentage of biomass that is co-fired in a boiler can range anywhere from 5 to 20 by weight, depending on both the capacity of the boiler and its efficiency. The use of biomass as an alternative fuel source has a beneficial impact on the environment. Co-firing is another way to save fossil fuels while making use of the economies of scale offered by power plants [22, 24, 25].

5.3.4 Carbonization

Carbonization is a thermal decomposition process in which carbon-rich substances are produced by heating organic materials in a controlled, low-oxygen environment. This transformative process is commonly used to create charcoal, activated carbon, and carbon black, all valuable materials with diverse applications. Initially, organic materials are selected, dried to reduce their moisture content, and then inserted into a chamber that has been specially designed. The material is heated progressively within the chamber, typically reaching temperatures between 400°C and 900°C. The lack of oxygen precludes combustion and induces pyrolysis, which breaks the material down into volatile gases, tars, and charcoal. Charcoal, which is primarily constituted of carbon, is the primary product. After chilling, the material is gathered and processed as required. Additionally, byproducts such as creosote and gases can be utilized for various industrial purposes. Carbonization is an efficient and environmentally favorable process for converting organic matter into valuable carbon-based materials with minimal waste [14, 26].

5.3.5 Liquefaction

Due to the importance of water in this thermochemical transition, this process is also called hydrothermal liquefaction. Organic materials undergo decomposition and reformation when water, in its dual role as reactant and catalyst, is heated to a high enough temperature to introduce hydrogen ions into hydrocarbons. Unlike pyrolysis and gasification, liquefaction does not require dried biomass as the initial feedstock (drying is an energy-intensive process) and reduces the number of unit operations required to convert biomass to liquid fuels. This method converts biomass directly into liquid fuels [22, 25, 27].

5.3.6 Combustion

One of the earliest and most fundamental methods for converting waste timber into energy is combustion. In this process, waste wood is incinerated to produce heat, which is then utilized to generate steam and drive turbines that generate electricity. Despite being a relatively ancient technology, combustion has undergone efficiency and pollution reduction enhancements. However, problems still exist, such as the emission of air pollutants and the requirement for circumspect ash residue management. Combustion of biomass in an oxygen-rich environment is the most basic method of thermochemical conversion. Heat, electricity, and combined heat and power are the main aspects of directly burning lignocellulosic biomass. A sequence of both homogeneous and heterogeneous reactions forms the basis of this process. Particle size, feedstock characteristics, combustion temperature, and combustion atmosphere composition are the primary determinants of biomass combustion. Combustion leads to excessive emissions of nitrogen oxides, carbon dioxide, particulate matter, and ash [18, 25, 28].

5.3.7 Anaerobic Digestion

Through microbial decomposition, anaerobic digestion is a biological process that transforms organic materials, particularly wood waste, into biogas (methane and carbon dioxide). It can handle both wood and organic trash, while organic waste is its main application. Wet and biodegradable wood debris is ideally suited for anaerobic digestion. It generates nutrient-rich digestate that can be utilized as a soil amendment in addition to biogas for energy [29, 30].

5.3.8 Fermentation

In the fermentation process, to make the cellulose available to the microbial activity, pretreatment is necessary. Complex sugars linked together in long chains make up cellulose fibers, which are covered in a lignin coating. There have been studies on a wide range of pretreatment methods, including biological, physical, and chemical pretreatments. Until recently, the main choice for commercial-scale biomass-to-ethanol systems was steam pretreatment. In the steam pretreatment process, high-pressure steam is used to heat the biomass to temperatures between 160 and 240°C for a set duration. The majority of the hemicellulose and a little amount of the lignin disintegrate when the pressure is relieved. Even when the cellulose is not dissolved, the enzymes can still easily access it.

Enzymatic hydrolysis converts the non-hydrolyzed cellulose and hemicellulose into carbohydrates. Cellulases are a complex mixture of several kinds of enzymes that break down the glucose-glucose linkages in cellulose. Because of its structural differences from cellulose and its composition largely of carbohydrates other than glucose, hydrolysis of hemicellulose requires a large number of enzymes (hemicelluloses). Aqueous solutions produced by enzymatic hydrolysis can be fermented to provide biochemicals, methanol, butanol, and ethanol [31].

The simple carbohydrates created by hydrolysis are converted into ethanol by yeast. Common baker's yeast is the most prevalent kind of microbe, and it has a reputation for being robust, safe, and ethanol-tolerant (*Saccharomyces cerevisiae*). However, baker's yeast can only ferment six-carbon-atom carbohydrates, that is, glucose. Yeast cannot metabolize carbohydrates containing five carbon atoms unless it undergoes genetic modification or is replaced with other microbes, like bacteria (such as xylose and arabinose). The amount of water needed for fermentation is more than for other processes [22, 31–33].

5.3.9 Composting

Composting is a method for producing fertilizer by decomposing organic waste in an aerobic setting. Nevertheless, the lignin layer found in lignocellulosic biomass prevents bacteria and enzymes required for the rapid breakdown of complex biomass from accessing cellulose and hemicellulose. Decomposition of lignocellulose biomass is thus slowed down. Fungi, in addition to microbes, are added to the compost pile to speed up the decomposition process. The composting rate is influenced by several factors, including the carbon-to-nitrogen ratio (20:1 to 40:1), carbon-to-phosphorus ratio (75:1 to 150:1), biomass porosity, availability of air, moisture content (40–60 percent), temperature (20–70°C), particle size, and inoculum (Pleurotus spp. and Trichoderma). At temperatures of 55°C and 62°C, respectively, weed seeds can grow, and the elimination of diseases is the primary benefit of decomposition. Regular stirring and aeration of the compost pile allow for temperature management. With a C/N ratio of 15:1, the compost that is produced at the end of the process can be used as a direct fertilizer. There hasn't been much enthusiasm for composting lignocellulose biomass since it decomposes more slowly than organic materials with lower lignin concentrations. Existence of well-established technologies for the generation of biofuels and bioproducts, such as gasification and pyrolysis.

As a result of numerous technologies, wood waste has enormous potential as a source of renewable energy. The choice of technology is influenced by things like the availability of feedstock, the need for energy, and environmental concerns. Each technology has its own pros and disadvantages. As technology develops, wood waste could play a significant role in the shift to cleaner, more sustainable energy sources. This chapter delves into the LCSA of energy generation from wood waste using pyrolysis and integrated systems [10, 21, 29, 31].

5.4 LCSA OF ENERGY PRODUCTION FROM WOOD WASTE THROUGH PYROLYSIS

A life cycle sustainability assessment (LCSA) looks at how a product, system, or process impacts the environment at every stage of its existence. By carefully employing inputs, outputs, emissions, and information systems, LCSA has proven to be a useful approach for assessing product and service eco-profiles. With the data provided by LCSA, policymakers will be able to make informed decisions regarding environmentally friendly technology. ISO 14040–14049 are the standards and guidelines for LCSAs that have been published by ISO. The four primary steps of life cycle assessment (LCSA) are goal and scope determination, carrying out a life cycle inventory analysis (LCIA), performing an environmental impact assessment (EIA), and evaluating the results [9, 34] as shown in Figure 5.3.

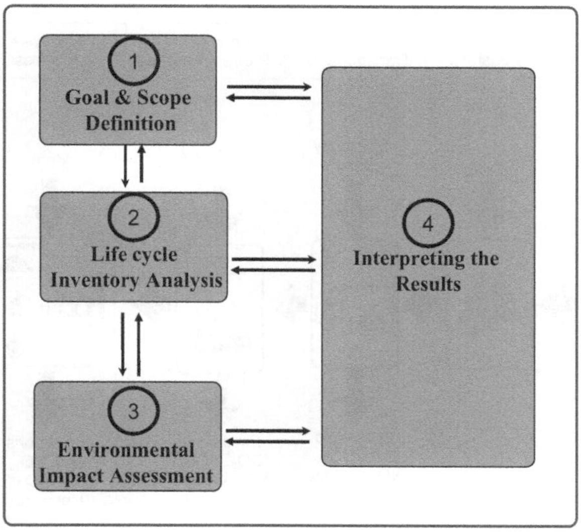

FIGURE 5.3 Stages of LCSA.

Defining the purpose and scope of a project is essential and includes defining the functional unit and system boundaries. Inputs and outputs, such as feedstock, energy, emissions, and refuse, are analyzed during the LCIA phase. For LCIA calculations, various databases are used. A more precise system boundary definition necessitates more information. Potential environmental consequences, including eutrophication, human toxicity, ecotoxicity, abiotic depletion, and global warming, are measured during the EIA phase. Concluding the process, pinpointing areas for improvement, and sharing findings with stakeholders are all part of the last stage. Here we'll go over each of these steps individually. As a result, after the results are generated, the functional unit is required to do unit computations. It also establishes the premise for conducting analysis. The number of resources used and the amount of discharges (waste and emissions) produced during production can be ascertained by inventory analysis. Indicators and the project's basis can both be laid out with the help of life cycle impact assessment [17, 35–37].

5.4.1 Defining the Goal and Scope

The LCSA's aim and scope are established during this preliminary phase, along with the system's boundaries and the probability of any unintended consequences. By identifying the functional unit and associated processes, it establishes the framework for the evaluation. This step should specifically explain the intended application, the goal of the research, the target audience, etc., in accordance with ISO 14040 and 14044. The focus of current standards and recommendations, however, is mostly on defining phases and does not provide sufficient direction. While carrying out the goal and scope phase, the guides for defining goals and scope make important methodological decisions in light of the stated LCSA standards, guidelines, and research. The main responsibilities of this step are to define the functional unit, locate the system border, and distribute the environmental burden [6, 37].

5.4.2 System Boundary

The system boundary definition plays a crucial function in the LCSA analysis. The pretreatment, processing, product utilization, and product end of life after use are all included in the LCSA

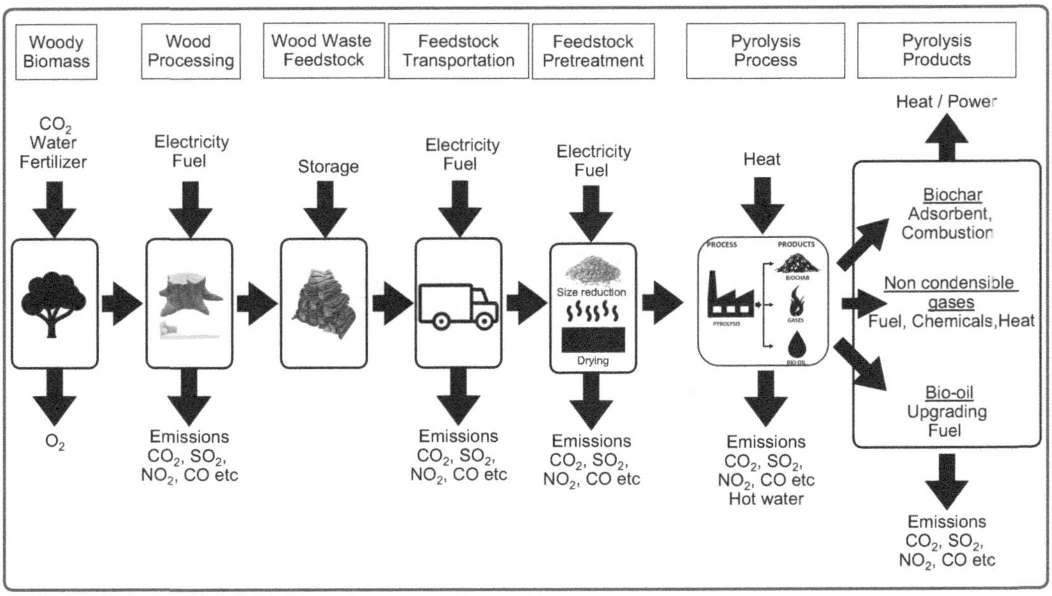

FIGURE 5.4 System boundary for energy production from wood waste.

system boundary framework for the pyrolysis of wood waste. A comprehensive analysis of the land changes caused by feedstock farming, fertilizer use, feedstock transportation to the plant site, biomass conversion, and product disposal is provided by the LCSA study. A possible approach to establishing the boundary is to consider the "cradle-to-grave" system, which entails cultivating feedstock to dispose of the product. In the gate-to-gate system boundary analysis, the LCSA results reveal the possibility of global warming from the time the feedstock enters the processing unit's gate until the finished goods depart the unit. A typical cradle-to-grave LCSA system border is shown in Figure 5.4.

Transporting biomass from the site of wood processing to the pyrolysis facility is a crucial component of the system boundary and is required for calculating the environmental impact using the LCSA method. To reduce environmental impact and transportation distance, it is always preferable for the plantation site and wood processing unit to be close to the pyrolysis facility [17, 37, 38].

5.4.3 Functional Unit

The functional unit, which measures the final product's functionality, is an important part of the Life Cycle Sustainability Assessment (LCSA). Functional units can be selected based on the system boundaries. Choosing a functional unit is essential since it makes it easier to assess and compare how different options will affect the environment. The choice of an appropriate functional unit is crucial to the objective and scope specification. Comparisons between many alternative scenarios or products are a necessary component of the LCSA results since the functional unit serves as the standard basis for calculation and reference. Which functional units are chosen depends on the system's final output and the scope that has been defined. Standard functional unit categories are often defined by mass (1 kg/1 ton feedstock/product) and energy (1 MJ/1 kWh/1 MW). Applying the same functional unit to compare and contrast different scenarios and projects could provide a useful foundation. Various researchers selected different functional units, which are listed in Table 5.1. Area and distance are not commonly used in biomass conversion systems, but they will be if the product's function can be demonstrated convincingly [6, 37].

TABLE 5.1
Functional Unit for Different LCSA Studies [8]

Feedstock	End Product	Functional Unit
Pine wood waste	Bio oil, Biochar, and incondensable gas	1 MJ
Eucalyptus tree	Bio oil, Biochar, and incondensable gas	1 MJ
Forest residue	Bio gasoline and biodiesel	1 km
Chips from branches and tops 500	Renewable diesel	1 MJ

5.4.4 Allocation of Environmental Burden

In stage 1, Life Cycle Sustainability Assessment (LCSA) must determine how to distribute the environmental load among a production system's goods and co-products. The literature identifies different LCSA techniques for solving this challenging environmental allocation. In the first technique, "no allocation," the product bears the entire environmental responsibility. This approach puts the product through the full environmental impact of the production system without considering co-products. Despite its simplicity, this strategy may misrepresent environmental consequences. LCSA investigations often use the second method, "economic allocation," which is popular. It suggests distributing environmental impacts by product value. LCSA practitioners sometimes allocate based on end-user costs rather than immediate post-production values, which can lead to erroneous results. The third method, "biophysical allocation," divides environmental costs among co-products and outputs by mass and energy. Some academics believe energy distribution is more exact because it matches economic factors, since energy content often determines market value. Economic allocation, which fluctuates with market pricing, is less stable than mass allocation. The fourth strategy, the "attributional approach," divides co-product environmental impact by mass, energy, or economic worth. It distributes environmental burdens across co-products based on criteria. The fifth method for assessing co-product environmental impact is "system expansion." This method calculates environmental burden by subtracting saved emissions from total emissions and considering co-product displacement. This method addresses co-products accurately, making it useful for consequential LCSA. Most LCSA studies combine these allocation methods to assess a product's environmental impact. ISO standards encourage avoiding allocation and increasing system boundaries; however, finding co-product alternatives is difficult [6, 39].

5.4.5 Life Cycle Inventory Analysis

A crucial component in evaluating the environmental impact of diverse processes and systems is life cycle inventory, or LCI. It involves meticulously compiling inventories of all the inputs, outputs, and products connected to a specific lifetime. The analysis goes into detail when calculating the physical quantities of inputs, such as energy and material consumption, as well as assessing direct emissions like N_2O, CO, CO_2, CH_4, NO_x, and NH_3, as well as the management of wastes at each stage, in order to quantify the LCI of bioenergy production systems using pyrolysis. Since these calculations are tied to a specific functional unit, they enable accurate comparisons and assessments. However, the quality of the data gathered has a significant impact on the dependability and accuracy of LCI analysis. In the data gathering process, which is a vital stage in the larger context of Life Cycle Sustainability Assessment (LCSA), duplicate data must be carefully avoided. In light of this, it is clear that gathering accurate, precise, and realistic data is not only desired but also necessary for creating reliable LCSA models.

The lack of unit process data is one of the main obstacles to creating life cycle inventories for assessment. LCSA practitioners use primary and secondary sources to close this data gap. These

rankings. To evaluate the reliability of decisions based on LCSA and to direct future attempts to decrease uncertainty, data on the sensitivity of model results is helpful. Furthermore, there are a number of crucial points that must be considered when gathering data for sensitivity analysis. However, if sensitivity analysis shows that specific actions and results are crucial to the study, re-evaluation and re-analysis will be necessary. Thus, the LCSA model relies heavily on sensitivity analysis for data collection and boundary setting [9].

5.5 LCSA INTEGRATED SYSTEM

The tool life cycle assessment (LCA) is used to analyze the environmental effects of biomass-based biofuel production chains as well as material and energy-related emissions. Biomass pyrolysis is a thermochemical conversion method that converts biomass at high temperatures and low oxygen concentrations into bio-oil, charcoal, and syngas. By substituting biofuels for conventional fuels, biomass pyrolysis can lower greenhouse gas (GHG) emissions and fossil fuel usage. Life cycle sustainability assessment (LCSA) is a holistic approach that considers not only environmental, but also social and economic aspects of biofuel production. An integrated system of biomass pyrolysis is a concept that involves the integration of different processes and technologies to optimize the efficiency and sustainability of biofuel production from biomass. The steps for the LCSA of an integrated system will be the same as discussed above, with the change in system boundary for different integrated systems [29]. Listed below are some examples of integrated systems for biomass pyrolysis.

5.5.1 Pyrolysis with Gasification

This system utilizes a combination of pyrolysis and gasification to convert biomass into syngas and charcoal. Biochar can be used as a soil supplement or as a carbon sequestration agent, while syngas can be utilized for power generation or further processed into liquid fuels. Syngas, a combination of hydrogen, carbon monoxide, and carbon dioxide, is produced by the gasification process by combining organic or fossil-based materials with oxygen and steam at high temperatures. The new technology of pyrolysis with gasification (PG) combines the benefits of both processes into one, resulting in increased energy efficiency, decreased emissions, and improved product quality. Life-cycle assessment (LCA) is a method for calculating how a product or procedure affects the environment from the time it is first conceived until it is finally discarded. Energy use, greenhouse gas emissions, water use, land use, toxicity, and human health are only some of the aspects of environmental performance that may be compared between waste-to-energy technologies using LCA. Using LCA, you can pinpoint the precise feedstock, pretreatment, product upgrade, and end-use scenarios that have the biggest impacts on the environment. The environmental sustainability of PG and other waste-to-energy technologies can be enhanced with the help of insights and recommendations provided by LCA. Figure 5.5 shows a hypothetical system boundary for a combined pyrolysis and gasification setup.

5.5.2 Pyrolysis with Biorefinery

Pyrolysis and biorefinery are combined in this system to create a variety of bio-based products from biomass, including bio-oil, charcoal, bioethanol, biogas, and biochemicals. While biochar can be applied in a variety of ways, bio-oil can be converted into chemicals or fuels for transportation. LCA has recently been extensively used in the realm of biofuel production to compare various systems. Studies that compared the production of biofuels and fossil fuels in their main focus confirmed that the majority of biofuel production from forestry and agricultural leftovers has remarkable carbon reduction potential when compared to fossil fuels. When some writers analyzed the GWP (global warming potential) of pyrolytic biofuel made from various types of forest resources (such as logging residues, willows, and poplars), they discovered that it was significantly lower than that of fossil

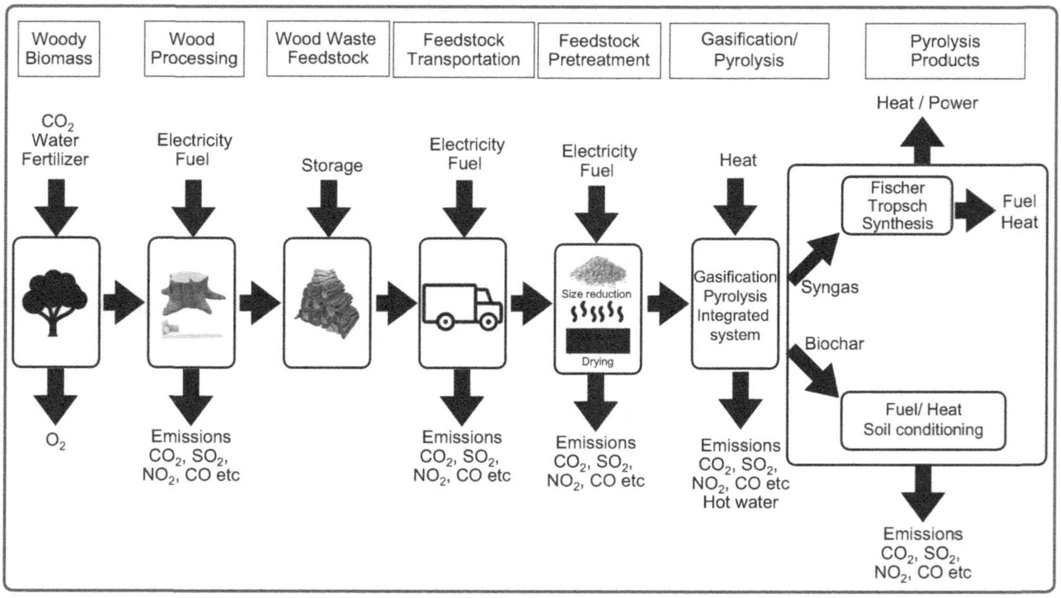

FIGURE 5.5 System boundary for integrated gasification pyrolysis process.

fuels. Some writers have looked at the benefits of poplar fast pyrolysis for the production of bio-oil and have found that it reduces greenhouse gas emissions by 54.5 percent when compared to traditional fossil fuels. The various biorefinery activities, such as extraction, biochemical processing, etc., are merged with the thermochemical process pyrolysis to obtain the products in an integrated pyrolysis with biorefinery concept [38]. A sample system boundary with an integrated pyrolysis biorefinery process is shown in Figure 5.6.

The life cycle assessment of integrated biomass pyrolysis with biorefinery can help to compare the environmental performance of different biofuel production pathways and identify the hotspots and improvement opportunities. When evaluating the environmental, social, and economic impacts of biofuel generation from biomass, a life cycle evaluation that incorporates biorefinery technology into integrated biomass pyrolysis is useful. Ongoing life cycle assessment of integrated biomass pyrolysis with biorefinery faces several challenges, including improving bio-oil quality and stability, creating selective and efficient catalysts, optimizing process parameters and configurations, and evaluating economic and environmental implications.

5.5.3 Pyrolysis with Carbon Capture and Storage (CCS)

The carbon dioxide (CO_2) released during the pyrolysis of biomass can be captured and stored using this technique. This has the potential to generate a negative emission technology and increase the GHG mitigation potential of biofuel production. The carbon dioxide (CO_2) released during biomass pyrolysis or gasification can be captured and stored via CCS. By using biofuels instead of traditional fossil fuels, pyrolysis with CCS is a negative emissions method that accomplishes both goals. It can also create a solid carbon product that can be used for various applications. LCA can help to compare the environmental performance of different biofuel production pathways and identify the hotspots and improvement opportunities. LCA can also help to assess the sustainability of biofuel production from biomass by considering not only environmental, but also social and economic aspects. LCA of pyrolysis with CCS is still ongoing and faces some challenges, such as improving the quality and stability of bio-oil, developing efficient and selective catalysts, optimizing the process conditions and configurations, and evaluating the economic and environmental impacts. A sample system boundary with integrated pyrolysis and carbon capture is shown in Figure 5.7.

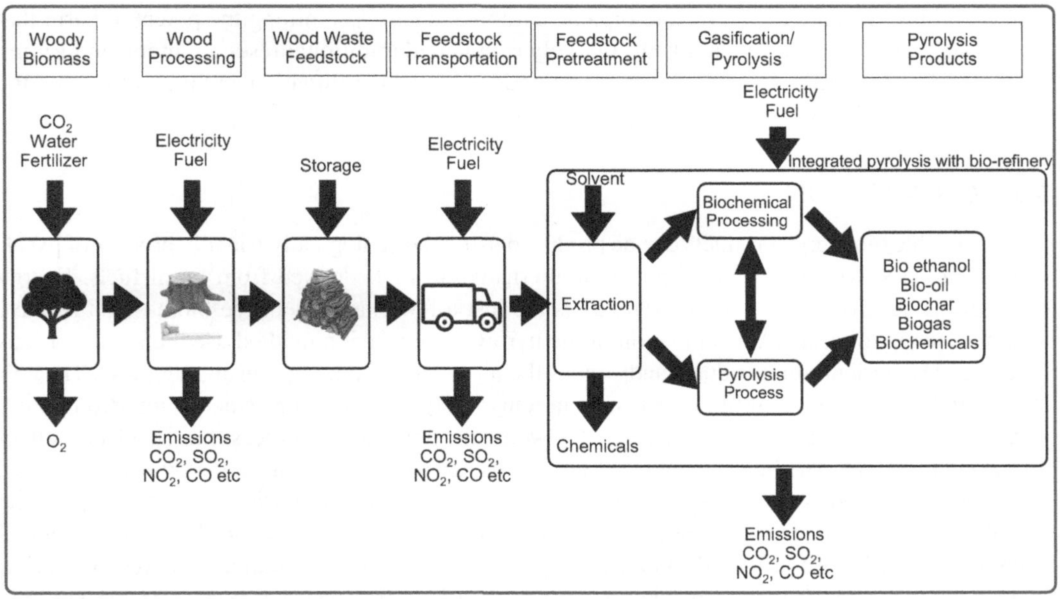

FIGURE 5.6 System boundary for integrated pyrolysis and biorefinery process.

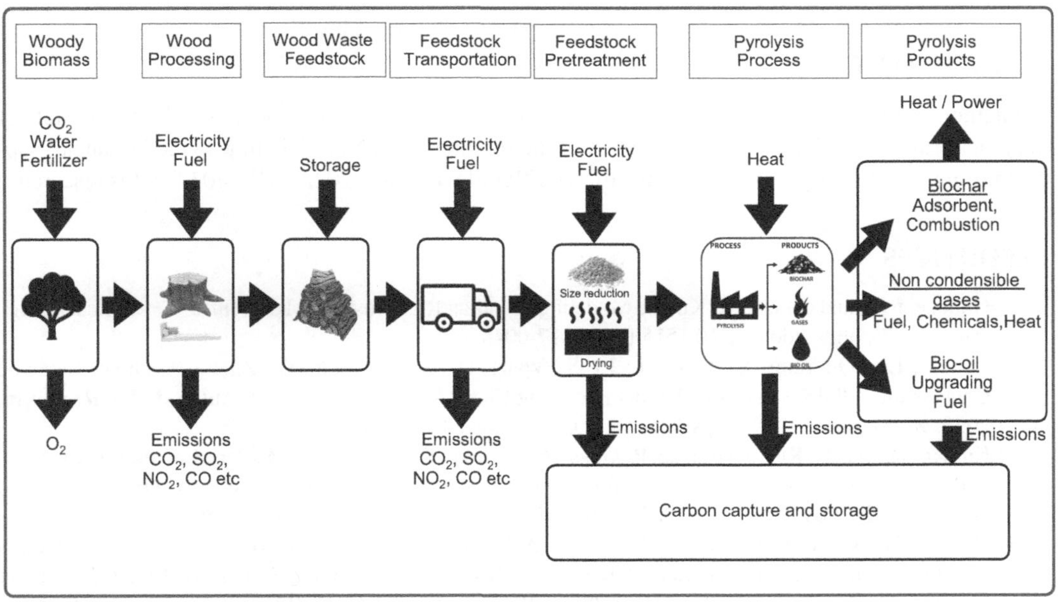

FIGURE 5.7 System boundary for integrated pyrolysis, carbon capture and storage.

5.6 RESEARCH OPPORTUNITIES

The pyrolysis process, which yields bio-oil, has been the subject of numerous life cycle assessments (LCAs), but the upgrading process, which turns bio-oil into items with added value, has received far less attention. The bio-oil undergoes deoxygenation as part of the upgrading process, with catalyst deactivation posing the greatest challenge. How catalytic synthesis would change things is still up in the air. Very few academics have provided any substantial discussion of the life cycle assessment (LCA) of biochar and its impacts on soil applications, including amendment, water holding capacity, structure, and organic carbon and soil. Examining the potential harm that may have been avoided if biochar hadn't largely supplanted traditional resource inputs in farming is crucial. It is important to study different methods of biomass cultivation and their effects on the overall pyrolysis LCA.

5.7 CONCLUSION

Many valuable bioenergy products can be produced using the ecologically friendly process of pyrolysis of biomass, which provides a potential alternative to fuels derived from fossil fuels. Energy crops, forest leftovers, and agricultural wastes are just a few examples of the biomasses that have been the subject of an LCA examination of the pyrolysis conversion method used to create various products. The agricultural residue stands out as the most eco-friendly option among these. The collection and transportation of biomass results in remarkably low emissions when compared to other pyrolysis methods. The use of chemicals and power in the pretreatment process has an adverse effect on the environment. The biggest source of greenhouse gas emissions during the pyrolysis process, according to the cradle-to-grave system boundary, is electricity use. To further lessen the impact, the use of more environmentally friendly energy sources, such as natural gas or electricity generated from biomass, may be considered. Biomass was identified as the most sustainable source of electricity in a scenario analysis that assessed the effects of electricity from various sources. To have as little influence on the environment as possible, bio-oil must be upgraded to high-quality fuel. Ecosystem, human health, and resource depletion are all included in the full studies of LCA, but possible global warming is the most researched environmental effect category.

ACKNOWLEDGMENT

We authors are thankful to the Chemical Engineering Department, NIT Calicut, and the Directorate of Technical Education, Kerala, for the support of the research. The NIT Calicut provided financial support through the Faculty Research Grant (FRG), FRG file no. FRG/2022/CHED/03 for this research.

REFERENCES

[1] Ghodke, P.; Mandapati, R. N. Kinetic Modeling of Indian Rice Husk Pyrolysis. *Int. J. Chem. React. Eng.*, 2018, *16* (2). https://doi.org/10.1515/ijcre-2017-0048

[2] Zhu, Y.; Liu, Q.; Yang, X.; Wang, J.; Li, S.; Wang, D.; Ye, Y.; Wang, D.; Zheng, Z. Direct Catalytic Conversion Cellulose Pyrolysis Vapors into Long Chain Alkanes (LCAs) over Au/TS-1. *J. Energy Inst.*, 2021, *98* (March), 11–19. https://doi.org/10.1016/j.joei.2021.05.003

[3] Kumar Sharma, A.; Kumar Ghodke, P.; Goyal, N.; Nethaji, S.; Chen, W. H. Machine Learning Technology in Biohydrogen Production from Agriculture Waste: Recent Advances and Future Perspectives. *Bioresour. Technol.*, 2022, *364* (August), 128076. https://doi.org/10.1016/j.biortech.2022.128076

[4] Chen, W. H.; Naveen C; Ghodke, P. K.; Sharma, A. K.; Bobde, P. Co-Pyrolysis of Lignocellulosic Biomass with Other Carbonaceous Materials: A Review on Advance Technologies, Synergistic Effect, and Future Prospectus. *Fuel*, 2023, *345* (December 2022), 128177. https://doi.org/10.1016/j.fuel.2023.128177

[5] Ashokkumar, V.; Venkatkarthick, R.; Jayashree, S.; Chuetor, S.; Dharmaraj, S.; Kumar, G.; Chen, W.-H.; Ngamcharussrivichai, C. Recent Advances in Lignocellulosic Biomass for Biofuels and Value-Added Bioproducts - A Critical Review. *Bioresour. Technol.*, 2022, *344*, 126195. https://doi.org/10.1016/j.biortech.2021.126195

[6] Cusenza, M. A.; Longo, S.; Cellura, M.; Guarino, F.; Messineo, A.; Mistretta, M.; Volpe, M. Environmental Assessment of a Waste-to-Energy Practice: The Pyrolysis of Agro-Industrial Biomass Residues. *Sustain. Prod. Consum.*, 2021, *28*, 866–876. https://doi.org/10.1016/j.spc.2021.07.015

[7] Vuppaladadiyam, A. K.; Varsha Vuppaladadiyam, S. S.; Sikarwar, V. S.; Ahmad, E.; Pant, K. K.; Murugavelh, S.; Pandey, A.; Bhattacharya, S.; Sarmah, A.; Leu, S.-Y. Y. A Critical Review on Biomass Pyrolysis: Reaction Mechanisms, Process Modeling and Potential Challenges. *J. Energy Inst.*, 2023, *108* (March), 101236. https://doi.org/10.1016/j.joei.2023.101236

[8] Yu, Z.; Ma, H.; Liu, X.; Wang, M.; Wang, J. Review in Life Cycle Assessment of Biomass Conversion through Pyrolysis-Issues and Recommendations. *Green Chem. Eng.*, 2022, *3* (4), 304–312. https://doi.org/10.1016/j.gce.2022.08.002

[9] Akinpelu, D. A.; Adekoya, O. A.; Oladoye, P. O.; Ogbaga, C. C.; Okolie, J. A. Machine Learning Applications in Biomass Pyrolysis: From Biorefinery to End-of-Life Product Management. *Digit. Chem. Eng.*, 2023, *8* (March), 100103. https://doi.org/10.1016/j.dche.2023.100103

[10] Sharma, A. K.; Ghodke, P. K.; Goyal, N.; Bobde, P.; Kwon, E. E.; Lin, K. Y. A.; Chen, W. H. A Critical Review on Biochar Production from Pine Wastes, Upgradation Techniques, Environmental Sustainability, and Challenges. *Bioresour. Technol.*, 2023, *387* (June), 129632. https://doi.org/10.1016/j.biortech.2023.129632

[11] Mohan, D.; Pittman, C. U.; Steele, P. H. Pyrolysis of Wood/Biomass for Bio-Oil: A Critical Review. *Energy & Fuels*, 2006, *20* (3), 848–889. https://doi.org/10.1021/ef0502397

[12] Collins, S.; Ghodke, P. Kinetic Parameter Evaluation of Groundnut Shell Pyrolysis through Use of Thermogravimetric Analysis. *J. Environ. Chem. Eng.*, 2018, *6* (4), 4736–4742. https://doi.org/10.1016/j.jece.2018.07.012

[13] Hoang, A. T.; Ong, H. C.; Fattah, I. M. R. R.; Chong, C. T.; Cheng, C. K.; Sakthivel, R.; Ok, Y. S. Progress on the Lignocellulosic Biomass Pyrolysis for Biofuel Production toward Environmental Sustainability. *Fuel Process. Technol.*, 2021, *223* (June), 106997. https://doi.org/10.1016/j.fuproc.2021.106997

[14] Lee, J.; Kim, S.; You, S.; Park, Y. K. Bioenergy Generation from Thermochemical Conversion of Lignocellulosic Biomass-Based Integrated Renewable Energy Systems. *Renew. Sustain. Energy Rev.*, 2023, *178* (June 2022), 113240. https://doi.org/10.1016/j.rser.2023.113240

[15] Bridgwater, A. V.; Peacocke, G. V. C. Fast Pyrolysis Processes for Biomass. *Renew. Sustain. Energy Rev.*, 2000, *4* (1), 1–73. https://doi.org/10.1016/S1364-0321(99)00007-6

[16] Liu, J.; Chen, X.; Chen, W.; Xia, M.; Chen, Y.; Chen, H.; Zeng, K.; Yang, H. Biomass Pyrolysis Mechanism for Carbon-Based High-Value Products. *Proc. Combust. Inst.*, 2022, *39* (3), 3157–3181. https://doi.org/10.1016/j.proci.2022.09.063

[17] Gahane, D.; Biswal, D.; Mandavgane, S. A. Life Cycle Assessment of Biomass Pyrolysis. *Bioenergy Res.*, 2022, *15* (3), 1387–1406. https://doi.org/10.1007/s12155-022-10390-9

[18] Luo, G.; Wang, W.; Zhao, Y.; Tao, X.; Xie, W.; Wang, K. Autothermal Pyrolysis of Lignocellulosic Biomass: Experimental, Kinetic, and Thermodynamic Studies. *J. Anal. Appl. Pyrolysis*, 2023, *171*. https://doi.org/10.1016/j.jaap.2023.105972

[19] Ghodke, P. K.; Sharma, A. K.; Jayaseelan, A.; Gopinath, K. P. Hydrogen-Rich Syngas Production from the Lignocellulosic Biomass by Catalytic Gasification: A State of Art Review on Advance Technologies, Economic Challenges, and Future Prospectus. *Fuel*, 2023, *342* (February), 127800. https://doi.org/10.1016/j.fuel.2023.127800

[20] Bieniek, A.; Reinmöller, M.; Küster, F.; Gräbner, M.; Jerzak, W.; Magdziarz, A. Investigation and Modelling of the Pyrolysis Kinetics of Industrial Biomass Wastes. *J. Environ. Manage.*, 2022, *319*, 115707. https://doi.org/10.1016/j.jenvman.2022.115707

[21] Mujtaba, M.; Fraceto, L.; Fazeli, M.; Mukherjee, S.; Savassa, S. M.; Araujo de Medeiros, G.; do Espírito Santo Pereira, A.; Mancini, S. D.; Lipponen, J.; Vilaplana, F. Lignocellulosic Biomass from Agricultural Waste to the Circular Economy: A Review with Focus on Biofuels, Biocomposites and Bioplastics. *J. Clean. Prod.*, 2023, *402* (March), 136815. https://doi.org/10.1016/j.jclepro.2023.136815

[22] Kargbo, H.; Harris, J. S.; Phan, A. N. "Drop-in" Fuel Production from Biomass: Critical Review on Techno-Economic Feasibility and Sustainability. *Renew. Sustain. Energy Rev.*, 2021, *135* (July 2020). https://doi.org/10.1016/j.rser.2020.110168

[23] Emmanouilidou, E.; Mitkidou, S.; Agapiou, A.; Kokkinos, N. C. Solid Waste Biomass as a Potential Feedstock for Producing Sustainable Aviation Fuel: A Systematic Review. *Renew. Energy*, 2023, *206* (February), 897–907. https://doi.org/10.1016/j.renene.2023.02.113

[24] Bridgwater, A. V. Review of Fast Pyrolysis of Biomass and Product Upgrading. *Biomass and Bioenergy*, 2012, *38*, 68–94. https://doi.org/10.1016/j.biombioe.2011.01.048

[25] Patel, M.; Zhang, X.; Kumar, A. Techno-Economic and Life Cycle Assessment on Lignocellulosic Biomass Thermochemical Conversion Technologies: A Review. *Renew. Sustain. Energy Rev.*, 2016, *53*, 1486–1499. https://doi.org/10.1016/j.rser.2015.09.070

[26] Poomsawat, W.; Poomsawat, S. Pyrolysis Kinetic Behavior of Composite Polypropylene-Biomass Solid Fuels Derived via Co-Hydrothermal Carbonization Process. *Therm. Sci. Eng. Prog.*, 2023, *43* (May), 101953. https://doi.org/10.1016/j.tsep.2023.101953

[27] Wu, Y.; Wang, H.; Li, H.; Han, X.; Zhang, M.; Sun, Y.; Fan, X.; Tu, R.; Zeng, Y.; Xu, C. C.; et al. Applications of Catalysts in Thermochemical Conversion of Biomass (Pyrolysis, Hydrothermal Liquefaction and Gasification): A Critical Review. *Renew. Energy*, 2022, *196*, 462–481. https://doi.org/10.1016/j.renene.2022.07.031

[28] Cheng, K.; Winter, W. T.; Stipanovic, A. J. A Modulated-TGA Approach to the Kinetics of Lignocellulosic Biomass Pyrolysis/Combustion. *Polym. Degrad. Stab.*, 2012, *97* (9), 1606–1615. https://doi.org/10.1016/j.polymdegradstab.2012.06.027

[29] Tijjani Usman, I. M.; Ho, Y. C.; Baloo, L.; Lam, M. K.; Sujarwo, W. A Comprehensive Review on the Advances of Bioproducts from Biomass towards Meeting Net Zero Carbon Emissions (NZCE). *Bioresour. Technol.*, 2022, *366* (October), 128167. https://doi.org/10.1016/j.biortech.2022.128167

[30] Opatokun, S. A.; Lopez-Sabiron, A. M.; Ferreira, G.; Strezov, V. Life Cycle Analysis of Energy Production from Food Waste through Anaerobic Digestion, Pyrolysis and Integrated Energy System. *Sustain.*, 2017, *9* (10). https://doi.org/10.3390/su9101804

[31] Awasthi, M. K.; Sar, T.; Gowd, S. C.; Rajendran, K.; Kumar, V.; Sarsaiya, S.; Li, Y.; Sindhu, R.; Binod, P.; Zhang, Z.; et al. A Comprehensive Review on Thermochemical, and Biochemical Conversion Methods of Lignocellulosic Biomass into Valuable End Product. *Fuel*, 2023, *342* (October 2022), 127790. https://doi.org/10.1016/j.fuel.2023.127790

[32] Rodionova, M. V.; Bozieva, A. M.; Zharmukhamedov, S. K.; Leong, Y. K.; Chi-Wei Lan, J.; Veziroglu, A.; Veziroglu, T. N.; Tomo, T.; Chang, J. S.; Allakhverdiev, S. I. A Comprehensive Review on Lignocellulosic Biomass Biorefinery for Sustainable Biofuel Production. *Int. J. Hydrogen Energy*, 2022, *47* (3), 1481–1498. https://doi.org/10.1016/j.ijhydene.2021.10.122

[33] Puschnigg, S.; Fazeni-Fraisl, K.; Lindorfer, J.; Kienberger, T. Biorefinery Development for the Conversion of Softwood Residues into Sustainable Aviation Fuel: Implications from Life Cycle Assessment and Energetic-Exergetic Analyses. *J. Clean. Prod.*, 2023, *386* (December 2022), 135815. https://doi.org/10.1016/j.jclepro.2022.135815

[34] Henkel, H. J. K. Editorial The Revision of ISO Standards 14040–3. *Int. J. Life Cycle Assess.*, 2005, *10* (3), 1

[35] Yadav, A.; Sharma, V.; Tsai, M. L.; Chen, C. W.; Sun, P. P.; Nargotra, P.; Wang, J. X.; Dong, C. D. Development of Lignocellulosic Biorefineries for the Sustainable Production of Biofuels: Towards Circular Bioeconomy. *Bioresour. Technol.*, 2023, *381* (April), 129145. https://doi.org/10.1016/j.biortech.2023.129145

[36] Sharifzadeh, M.; Sadeqzadeh, M.; Guo, M.; Borhani, T. N.; Murthy Konda, N. V. S. N.; Garcia, M. C.; Wang, L.; Hallett, J.; Shah, N. The Multi-Scale Challenges of Biomass Fast Pyrolysis and Bio-Oil Upgrading: Review of the State of Art and Future Research Directions. *Prog. Energy Combust. Sci.*, 2019, *71*, 1–80. https://doi.org/10.1016/j.pecs.2018.10.006

[37] Zhu, X.; Labianca, C.; He, M.; Luo, Z.; Wu, C.; You, S.; Tsang, D. C. W. Life-Cycle Assessment of Pyrolysis Processes for Sustainable Production of Biochar from Agro-Residues. *Bioresour. Technol.*, 2022, *360* (June), 127601. https://doi.org/10.1016/j.biortech.2022.127601

[38] Sharifzadeh, M.; Wang, L.; Shah, N. Integrated Biorefineries: CO2 Utilization for Maximum Biomass Conversion. *Renew. Sustain. Energy Rev.*, 2015, *47*, 151–161. https://doi.org/10.1016/j.rser.2015.03.001

[39] Mohamed, B. A.; O'Boyle, M.; Li, L. Y. Co-Pyrolysis of Sewage Sludge with Lignocellulosic and Algal Biomass for Sustainable Liquid and Gaseous Fuel Production: A Life Cycle Assessment and Techno-Economic Analysis. *Appl. Energy*, 2023, *346* (April), 121318. https://doi.org/10.1016/j.apenergy.2023.121318

[40] Stančin, H.; Strezov, V.; Mikulčić, H. Life Cycle Assessment of Alternative Fuel Production by Co-Pyrolysis of Waste Biomass and Plastics. *J. Clean. Prod.*, 2023, *414* (May). https://doi.org/10.1016/j.jclepro.2023.137676

6 Techniques for Processing Macroalgal Biomass in Energy and Food Production

Sowmiya Kadalmani, Geethalakshmi Ramakrishnan, and Renganathan Sahadevan

6.1 INTRODUCTION

The world's biggest difficulties in the coming decades will be sustainable food production and energy generation [1]. There are numerous uses for macroalgae in food, construction materials, chemicals, and pharmaceuticals. Seaweeds have been used as a vital component of nutrition and healthcare for more than 14,000 years [2], and the market is now worth more than USD 6 billion worldwide (FAO, 2018). Macroalgae are multicellular marine creature that can float or attach to solid surfaces like rocks [3]. Macroalgae forests play crucial roles in coastal ecosystems: they provide a habitat for a variety of species, influence coral-microbial ecology, shield creatures from storms, lessen acidification and deoxygenation, sustain biogeochemical cycles, and boost fisheries yield [4]. In this context, the generation of bioenergy from algal biomass appears promising; nevertheless, more research is needed because the efficiency of bioenergy production is controlled by several aspects. Two of the most important variables are the alga species—there are an estimated 150,000 known species—and the optimization of parameters to create bioenergy on a large scale in an economically competitive fashion [5].

At present, many nations have chosen first-generation feed stocks as their preferred option for use as an automotive co-fuel. Notwithstanding growing anxiety over the "Food and Fuel Issue," second-generation resources are used as the basic components for the manufacturing of biofuel [6]. To produce them, lignocellulosic materials like bagasse, cotton stalks, corn stover, wood chips, rice straw, organic waste, and plant remnants in decay are used [7]. The main drawback of second-generation feed stocks, however, is their high lignin content and the need for extensive agricultural space for their cultivation [8]. Algal biomass has been identified as a third-generation energy resource as a result of the above limitations of these feedstocks [9]. Because macroalgae grow in the sea, they are unique among energy crops. Seaweed works as a bio-filter during growth, bio extracting excess nitrogen, phosphate, carbon dioxide, and contaminants such as heavy metals [10]. When such compounds are gathered, they are removed from the aquatic system, improving the quality of coastal and marine waterways. The amount of nutrients in macroalgae varies by species, harvesting season, region of origin, and environmental factors [11]. Sustainability is currently putting pressure on the development of bioenergy sources and conversion pathways to establish a circular economy. However, there are still a lot of challenges in the way of commercializing macroalgal bioenergy, including environmental effects as well as technological, economic, and social issues. Recent research results have helped us focus on marine macroalgae biomass as a potentially sustainable feedstock for the manufacture of feed components for monogastric aquaculture and terrestrial animals [12]. Algal biomass normally comprises the three primary macromolecules—carbohydrates,

DOI: 10.1201/9781003732907-6

lipids, and proteins—along with a small quantity of minerals. Once the crucial macromolecules have been extracted from the algae, the leftover biomass can be used for the fermentation-based production of secondary products like biogas, butanol, ethanol, etc. [13]. Seaweed is a complex category of eukaryotic, multicellular, photosynthetic aquatic organisms classed as Phaeophyta (brown algae), Rhodophyta (red algae), and Chlorophyta (green algae). Only 221 of the 20,000 known marine macroalgal species are commercially important [14]. Marine macroalgae are a diversified collection of multicellular, plant-like protists that can be divided into brown (Phaeophyta), green (Chlorophyta), and red (Rhodophyta) algae. These are some of the polysaccharides that can be found in macroalgae: cellulose, starch, laminarin (the primary storage polysaccharide in brown seaweed), Fucoidans (a sulfated fucose-rich polysaccharide found in brown seaweed), carrageenan (a sulfated polysaccharide found in red seaweed), and alginate (a structural polysaccharide found in brown seaweed) [15].

Over the last five decades, there has been outstanding development in the global seaweed business, with annual global production of seaweeds estimated at 26 million tons fresh (FAO, 2014), with a value of USD 7.3 billion [16]. Seaweed polysaccharides such as agar, carrageenan, and alginate are commercially valuable compounds produced from seaweed resources, with a market worth of more than USD 1 billion [17].

6.2 LIFE CYCLE OF MACROALGAE

The most crucial component of the marine environment is macroalgae, which guard marine resources by preventing eutrophication and pollution. Because macroalgae lack roots, stems, and leaves, they are inferior to plants. Instead, they consist of thallus, which resembles a leaf, and occasionally stems and feet. Novel biomass feedstocks that can be grown responsibly and converted to sources of energy are being sought; algae is one example that is currently gaining popularity [18]. The marine macroalgae feedstock has the potential to be advantageous.

Most algal life cycles can be classified into one of three types: gametic (diplontic), zygotic (haplontic), and biphasic (sporic/haplo-diplontic) [5]. The biphasic life cycle is the most complex of the three [19]. It is characterized by an obligate cycling between separate, free-living, and independent haploid (gametophyte) and diploid (sporophyte) phases [20].

6.2.1 Haploid Gametophyte Phase

The life cycle begins with the discharge of haploid spores from adult sporophytes. These spores are frequently formed in sporangia or sori on the sporophyte. The spores are dispersed into the water and can be moved by currents [21]. When conditions are favorable, the spores settle and attach to a substrate.

6.2.2 Germination and Growth

The settling spore germinates and grows into a haploid gametophyte. The gametophyte is the main phase in several macroalgae species [19]. It develops and matures, often by cell division and expansion, until it reaches the reproductive stage.

6.2.3 Reproduction

During the gametophyte phase, specialized structures known as gametangia form. These gametangia generate gametes, which are haploid reproductive cells. Male gametangia generate sperm cells (spermatia), while female gametangia generate egg cells [22]. When sperm cells are released into the water, they find their way to egg cells, resulting in the formation of a diploid zygote.

6.2.4 Diploid Sporophyte Phase

The zygote grows into a diploid sporophyte. This phase often adheres to the substrate and can grow to a higher size than the gametophyte phase [23]. Through meiosis, the sporophyte develops spore-producing structures (sporangia or sori) carrying diploid spores. These spores are discharged into the water, completing the generational alternation and restarting the cycle.

6.3 CULTIVATION AND HARVESTING OF MACROALGAE

Several environmental factors influence macroalgae cultivation success. Water quality, photosynthetic active radiation (PAR), temperature, salinity, inorganic nutrient concentrations, especially CO_2, and the lack of environmental contaminants are all required for successful biomass accumulation via photosynthesis [18]. These environmental parameters are dynamic and depend on natural and human factors [24]. The Norwegian coastline spans more than ten degrees of latitude and provides a variety of abiotic and biotic conditions for successful macroalgal farming [25]. Although farming is still in its early stages, seaweed farmers along the Norwegian coast cultivated and gathered 176 tons ww of brown seaweeds [19]. The principal cost of off-bottom seaweed cultures, such as offshore Macrocystis and Laminaria rope farms, is connected with farm construction. To overcome this, the concept of cultivating floating seaweed in an enclosed structure similar to those used for fish cultivation in the ocean was devised [14]. Some seaweeds, such as Sargassum, may grow while floating on the ocean's surface and can be farmed in the same way.

6.3.1 Cultivation

Brown macroalgae and red macroalgae predominate over green macroalgae among cultivated macroalgae. Brown, red, and green macroalgae contributed 47.3%, 52.6%, and 0.05% of total global production in 2019, respectively [26]. Algae can be cultivated in practically any kind of water, even wastewater.

6.3.1.1 Site Selection

The location of cultivation is critical. Water quality, fertilizer availability, temperature, wave exposure, and substrate type must all be taken into perspective [27]. Coastal sites with suitable conditions are frequently chosen.

6.3.1.2 Collection of Spores

Macroalgae cultivation can begin with the collection of spores or propagules from wild or produced sources [28]. Sporophyte spores can be gathered or acquired from specialized vendors.

6.3.1.3 Seeding

The spores are often sown onto suitable substrates, such as ropes, nets, or frames, which are then immersed in water. These substrates act as anchors for the spores, allowing them to develop [29]. Attaching seaweeds to ropes, lines, or nets is a popular method of cultivation due to the minimal installation and maintenance expenses [24]. Cultivation can be accomplished by physically attaching seedlings to the ropes.

6.3.2 Cultivation Methods

The method of growing algae for raw materials is rather simple, and it can proceed with little or no supervision [30].

6.3.2.1 Long Line Cultivation Method

Ropes with connected spores are suspended in the water and tethered to the seabed by buoys. The algae grow around the ropes [9].

6.3.2.2 Raft Cultivation

Ropes or nets containing spores are tied to floating rafts, which allow them to be moved to appropriate places and depths [31].

6.3.2.3 Bottom Culture

Substrates are put directly on the bottom in shallow waters.

6.3.3 Maintenance

Monitoring water quality, nutrient levels, and potential diseases are all part of proper maintenance. Cleaning and eradication of fouling organisms may be required on a regular basis.

6.3.4 Harvesting

Macroalgae are normally harvested when they reach the required size and quality [32]. There are numerous techniques available, ranging from hand-picking or cutting subtidal thalli to bulldoze, tractor, or boat harvesting [33]. Skimmer boats have been constructed to gather macroalgae easily, far from the coast, such as Laminaria digitata.

6.3.4.1 Manual Harvesting

Hand harvesting is typical in smaller-scale farms. Harvesters manually cut or pluck adult algae from cultivation substrates.

6.3.4.2 Mechanical Harvesting

Mechanical equipment such as harvesters, conveyor belts, and pumps may be employed in larger-scale operations [34]. These methods are more efficient, but they must be handled with care to avoid damaging the algae.

6.3.4.3 Continuous Harvesting

In some cultivation techniques, algae are continually harvested by removing a piece of the biomass at regular intervals while leaving the rest to develop [35].

Manual harvesting for moderate-scale farming operations and mechanical or automated harvesting, comprising boat, crane, conveyor, suction tubes, dredge cutters, and revolving blades, for large-scale farming operations, are regarded as potential options [14]. The high moisture content (85 wt%) of fresh seaweed is a significant barrier to delivery to the conversion facility.

Green algae were the least grown, accounting for only 0.05% of total farmed seaweed, and were mostly utilized as food [36]. *Saccharina japonica* (*S. japonica*) and Eucheuma spp. were the most farmed among brown and red algae in 2019, with yearly production of 12.27 and 9.81 million wet tons, respectively, demonstrating that the cultivation infrastructure for these seaweed types is well-developed [14].

6.4 COMPOSITION OF MACROALGAE

Algae have a wide spectrum of bioactive characteristics due to the presence of certain chemicals that promote antibacterial, anticancer, antioxidant, antifungal, and antiviral capabilities [37]. Much research has been conducted on the primary phytochemicals contained in seaweed, with the goal of studying their bioactive potential. The chemical makeup of diverse macroalgae (Table 6.1) differs

TABLE 6.1
Components of Macroalgae [32, 38]

	Red Macroalgae	Green Macroalgae	Brown Macroalgae
Pigments	Chlorophyll a, phycobiliproteins, and carotenoids	Chlorophyll a & b, beta-carotene, xanthophyll	Chlorophyll a & c, fucoxanthin, and xanthophylls.
Cell wall	Composed of cellulose, agar, and carrageenan.	Composed of cellulose.	Composed of cellulose and alginates.
Storage polysaccharides	Floridean Starch	Starch	Laminarin, fucoidan
Secondary metabolites	Many macroalgae produce secondary metabolites of varying ecological and commercial relevance. These chemicals frequently have bioactive features such as antioxidant, antibacterial, antiviral, and antifungal activity.		
Nutritional Value	Macroalgae are high in dietary fiber, vitamins (such as vitamin C, vitamin K, and several B vitamins), and minerals (such as iodine, calcium, iron, and magnesium).They also provide vital fatty acids, such as omega-3 fatty acids (EPA and DHA).		

greatly from terrestrial plants [32]. They have higher levels of nitrogen and sulfur than land-based, lignocellulosic biomass, as well as lower levels of carbon, hydrogen, and oxygen. Compared to the 16% lignin found in corn stover, macroalgae contain almost little lignin [38]. Additionally, seaweed grows faster than terrestrial crops due to its increased photosynthetic efficiency [11]. Seaweeds are a promising source of protein, with protein levels ranging from 5 to 47% depending on species. Green and red seaweeds have higher protein content (10–47%) than brown seaweeds (5–24% DW) [39].

6.5 TECHNIQUE FOR THE MACROALGAE BIOMASS PROCESSING

The development of highly efficient biorefinery techniques is critical for large-scale bio-based product and bio-energy production. Because macroalgae have distinct chemical structures and compositions than terrestrial plants, macroalgae-based biorefinery has enormous promise for the creation of new bio-products and bio-energy generation [40].

6.5.1 Bioethanol Production from Macroalgae

Bioethanol/ethyl alcohol is a form of ethanol produced through the processing of any natural biological source. It might be used in place of fuel. The need for bioethanol as a transportation fuel grows over time [30]. The high carbon and low lignin content of macroalgae makes them suited for conversion into bioethanol (Figure 6.1). Several pretreatment processes are followed by microbial fermentation in the synthesis of bioethanol from macroalgae [41]. In general, bioethanol synthesis from biomass includes pretreatment, enzymatic hydrolysis, fermentation, and distillation. Before ethanol fermentation, macroalgae must be saccharified. Various physical, chemical, and biological pretreatments have been investigated in order to improve saccharification efficiency [42].

Bioethanol production from wet, dried, and residues (after hydrocolloid extraction) was attempted (Table 6.2) [43]. Residues from hydrocolloid extraction are high in cellulose and have been used to make bioethanol and followed by fermentation occurs in two stages: Separate Hydrolysis and Fermentation (SHF) and Simultaneous Saccharification and Fermentation (SSF). SHF entails sequential hydrolysis and fermentation, whereas SSF entails simultaneous hydrolysis and fermentation [44]. Finding enzymes and microorganisms that can use and convert all carbohydrates to ethanol is a serious issue in the conversion of seaweed to ethanol, with alginate being the key limiting component.

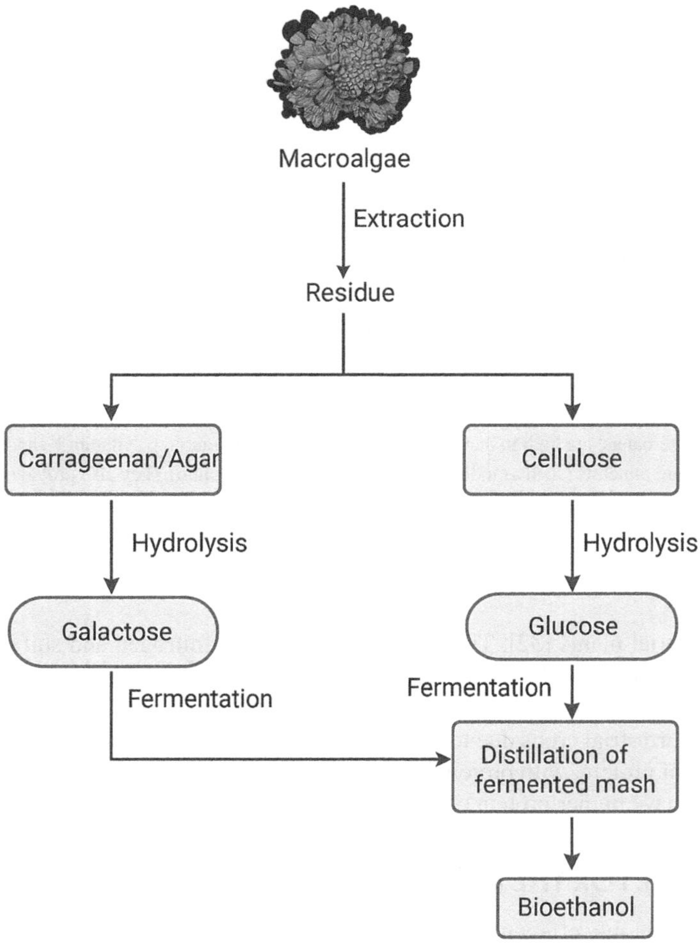

FIGURE 6.1 Extraction of bioethanol from macroalgae.

TABLE 6.2
Sugar Profile of Macroalgae [43]

Macroalgae Species	Mannitol (%)	Fructose (%)	Glucose (%)	Other Carbohydrates (%)
Green algae				
Ulva sp.	2–15	1–10	5–25	Cellulose, Hemicellulose
Enteromorpha sp.	2–10	1–7	5–20	Cellulose, Hemicellulose
Chlorella sp.	1–10	1–8	Low	Starch, Cellulose, Glucomannan
Spirogyra sp.	1–15	1–10	Low	Cellulose, Hemicellulose, Xylans
Red algae				
Palmaria palmata	1–5	0.5–2	5–20	Agar, Carrageenan
Porphyra sp.	1–8	0.5–4	5–15	Agar, Porphyran
Gracilaria sp.	1–10	0.5–5	Low	Agar, Carrageenan, Glucans
Gelidium sp.	1–10	0.5–5	Low	Agar, Carrageenan, Glucans
Brown algae				
Laminaria sp.	1–10	0.5–5	5–30	Alginates, Laminaran
Fucus vesiculosus	1–10	0.5–5	5–20	Fucoidan, Alginates
Saccharina sp.	1–10	0.5–5	5–20	Alginates, Laminaran
Sargassum sp.	1–10	0.5–5	5–20	Fucoidan, Alginates

6.5.2 Biogas

Biogas production from macroalgae is a promising avenue for sustainable energy production, but it also requires careful process optimization and consideration of local conditions and regulations [45]. Biogas can be made by converting organic waste derived from algal biomass. This process involves the breakdown of algal biomass into methane and carbon dioxide, which can be used in the home or as a fuel for automobile engines [30]. Anaerobic digestion of macroalgae produces biogas. In this process, which happens in the absence of oxygen, the biomass goes through four major sequential phases: hydrolysis, acidogenesis, acetogenesis, and methanogenesis [46]. This enables the conversion of organic matter into biogas, which primarily constitutes a mixture of methane (CH_4) and carbon dioxide (CO_2), as well as a semisolid fraction (residual biomass) known as digestate.

6.5.3 Bio Crude Oil and Biofuel

Biodiesel is simply a fatty acid methyl or ethyl ester generated by the trans-esterification (in the presence of alcohols) of triglycerides. So, biodiesel is alkyl fatty acids (with chain lengths ranging from C14 to C25), esterified in the presence of alcohols, primarily CH_3OH or C_2H_5OH [30]. Meanwhile, Indonesia is trying to achieve Net Zero Emissions by 2060. Increasing the share of renewable energy obtained from biomass in the country's overall energy mix is one effort [47]. Because of its organic content, macroalgae can be converted into bio-oil and solid fuel for energy sources; however, this application method must be included. Bio-oil, once developed, can replace fossil fuels in a variety of applications, including as a source of heat, electricity, and chemicals. Better-quality bio-oil can be utilized in boilers, furnaces, turbines, diesel engines, power plants, and other industrial processes [48]. Thermochemical liquefaction is a method of converting a wet algal biomass material into a liquid fuel.

Thermochemical techniques are much faster than biochemical processes and are widely used to improve biomass biofuel production. Among the thermochemical conversion strategies for biomass, the pyrolysis process is most suited for producing biofuels. Pyrolysis is one method for converting biomass thermochemically to produce energy and chemicals. Biomass can be used to generate gas, oil, and char in the absence of oxygen [49].

6.6 ADVANTAGES AND DISADVANTAGES OF TECHNIQUES USED FOR BIOMASS CONVERSION

In comparison to microalgae, using macroalgae for energy production provides benefits, including easy harvesting and affordable algae separation systems from the water culture environment. The microalgae harvesting sector typically accounts for 30–40% of the operational costs in systems where microalgae are utilized to produce energy [50]. On the other hand, macroalgae may survive in a variety of environmental situations [51]. The use of seawater for the formation of algal biomass has a higher potential for reducing the water problem, and it can also assist in reducing carbon dioxide in the atmosphere while providing oxygen to the ocean's living animals [52]. Aside from all of the benefits of incorporating algae in foods, some difficulties, such as sensory concerns (such as a greenish color and/or a strong fishy taste), are visible in these goods [5]. The biggest challenge to using algal biomass for biofuel production is breaching the stiff cell wall that limits the soluble organics [53]. Because the cell wall components of algal biomass differ depending on the species, obtaining a common pretreatment approach is nearly impossible. The need for drying high moisture feedstock from 85 to 90 wt% to 5–10 wt% and high ash content are the primary factors restricting the large-scale adoption of thermochemical processes [14]. Although hydrothermal liquification is appropriate for wet biomass, the primary obstacles are low technological maturity and the requirement for catalyst upgrade, biocrude extraction, organic solvent recovery, and biocrude upgradation.

6.7 APPLICATION OF MACROALGAL BIOMASS IN ENERGY AND FOOD PRODUCTION

Macroalgae have a wide range of applications and are utilized as a starting raw material in a variety of industries, including animal aquaculture, cosmetics, and the hydrocolloid industry [51]. Many household products, including cosmetics, meals like ice cream, and beverages like wine, contain seaweed. Seaweed is not only a whole food or food item in and of itself, but it is also a complementary supplement to other foods with the ability to provide key functional characteristics [39]. The macroalgae biomass market is growing in terms of market capitalization on an annualized basis, with data from the Food and Agriculture Organization of the United Nations (FAO) indicating that global macroalgal biomass production in 2016 amounted to approximately 30 million tons at a value of USD $11.6 billion. Asia is the world's top producer of macroalgae, with China leading the way with 14 million tons valued at USD$8.6 billion, followed by Africa with approximately 140,000 tons and the Americas with 15,634 tons. Several macroalgal species, including *Laminaria japonica*, Eucheuma spp., *Kappaphycus alverezii*, *Pyropiayezoensis*, *Undaria pinnatifida*, and *Graciliaria verrucosea*, have already been mass cultivated in Asia [24]. The classification of edible seaweeds numbers around 600 species. As a low-calorie food high in vitamins, minerals, and dietary fibers, edible seaweeds are now recognized to offer excellent nutritional value. Their nutritional content may vary based on the region, time of year, growth stage, and portion of the seaweed picked, among other factors. Seaweeds extract an amazing wealth of mineral components, macro elements, and trace elements from the sea. They are known to be an excellent source of vitamins and minerals, particularly potassium and iodine (i.e., *Palmaria palmata*, *Fucus vesiculosus*, Laminaria sp.), and potentially good sources of proteins (i.e., red algae such as *Pyropia tenera* (formerly *Porphyra tenera*), *Grateloupia filicina*), long-chain polysaccharides (i.e., *P. tenera*), and soluble and insoluble dietary fibers (i.e., *G. filicina*, *Chondrus crispus*, *Ulva lactuca*) [54]. The effects of food supplementation with macroalgae or macroalgal extracts on the immunological state and intestinal health of many monogastric farm animal species, including pigs, poultry and fish have been widely studied. Bioactive components from macroalgae, including Laminaria-derived laminarin and fucoidan, have been proposed as alternatives to in-feed antibiotics due to their health- and growth-promoting benefits [11].

Laminarin is one such chemical being researched as a human health supplement. It is a polymer that stores energy in the vacuoles of brown seaweed cells. Different species contain varying levels of laminarin, which is further regulated by harvest season and extraction procedures. According to research, it has promising applications in cancer therapy, tissue engineering, anti-inflammatory, and antioxidant compounds [39].

A sulfated polysaccharide called fucoidan is mostly present in the cell-wall matrix of brown seaweed species. Numerous research now support the advantages of fucoidans and fucose-rich sulfated polysaccharides, which also include antiviral, antitumor, anti-inflammatory, and immunomodulatory effects.

Due to rising rates of allergies related to certain food sources, heavy metal contamination of food (such as mercury in fish), and the persistence of other harmful chemicals in food and their link to human disease, industries such as seaweed aquaculture are under pressure to produce "clean and green" products [43]. Wheat and soybeans are commonly consumed commodities that have been linked to allergies; hence, a shift away from these monoculture crops is in high demand.

6.8 CONCLUSION

Seaweed is a promising food source for humans. Blue biomass can help customers increase their plant protein consumption while also improving their mineral intake, particularly iodine. Seaweed is a valuable source of functional biomolecules such as alginate, agar, and carrageenan for the food sector [39]. Bioactives such as polyphenols and carotenoids can be isolated and concentrated for use in the nutraceutical business to provide health advantages to consumers.

Third-generation biorefineries provide a diverse range of high-value products, including pharmaceuticals, nutraceuticals, energy, and other industrial commodities, but they are not yet commercially viable due to technical obstacles and production cost issues [5]. There is growing interest in the possible positive benefits of macroalgae bioactive substances, such as laminarin, fucoidan, and phlorotannins. Biorefinery techniques to establish cost-effective and environmentally friendly extraction technologies to produce interesting bioactive chemicals with quantifiable beneficial effects would benefit future commercialization. Protein levels and possibly limiting essential amino acids in macroalgae vary widely, and species-specific polysaccharides and phenolic chemicals may influence protein digestibility. Because of this, it is impossible to make generalizations regarding the value of entire macroalgae as a protein source. However, many species have insufficient amounts of digested protein to be desirable as substitute protein sources in animal feed [11].

Investments in macroalgae utilization are risky because of the uncertainties surrounding feedstock cultivation as well as processing technologies, contracts, and demand. When a new processing technique, like a new biofuel refining technology, is developed, it is especially important to take uncertainty into account [38]. Green energy and environmental protection can be accomplished by properly processing algal biomass. Because it emits no pollutants, bio-hydrogen will be used as bioelectricity in the near future. It will be significant in industrial operations since it will minimize the cost of the process.

Incorporating macroalgal biomass processing into existing energy and food production systems can help to create a more sustainable and resilient future. We can minimize our dependency on finite fossil fuels, cut greenhouse gas emissions, and relieve strain on traditional agriculture while also supplying nutritious food supplies by harnessing the power of macroalgae. As technology progresses and public awareness of the benefits rises, macroalgal biomass processing is positioned to play a critical role in creating the future landscape of energy and food production.

REFERENCES

[1] Soria-Lopez, A., Garcia-Perez, P., Carpena, M., Garcia-Oliveira, P., Otero, P., Fraga-Corral, M., ... Simal-Gandara, J. (2023). Challenges for future food systems: From the green revolution to food supply chains with a special focus on sustainability. *Food Frontiers*, 4(1), 9–20.

[2] Ivanov, V. M., Shevchenko, O., Marynin, A., Stabnikov, V., Stabnikova, E., Gubenia, O., ... Salyuk, A. (2021). Trends and expected benefits of the breaking edge food technologies in 2021–2030. *Ukrainian Food Journal*, 10(1), 7–36. DOI: 10.24263/2304-974X-2021-10-1-3

[3] Sudhakar, K., Mamat, R., Samykano, M., Azmi, W. H., Ishak, W. F. W., & Yusaf, T. (2018). An overview of marine macroalgae as bioresource. *Renewable and Sustainable Energy Reviews*, 91, 165–179.

[4] Li, S., Hu, M., Tong, Y., Xia, Z., Tong, Y., Sun, Y., ... He, P. (2023). A review of volatile compounds in edible macroalgae. *Food Research International*, 165, 112559.

[5] da Rosa, M. D. H., Alves, C. J., dos Santos, F. N., de Souza, A. O., Zavareze, E. D. R., Pinto, E., ... de Pereira, C. M. P. (2023). Macroalgae and microalgae biomass as feedstock for products applied to bioenergy and food industry: a brief review. *Energies*, 16(4), 1820.

[6] Dave, N., Selvaraj, R., Varadavenkatesan, T., & Vinayagam, R. (2019). A critical review on production of bioethanol from macroalgal biomass. *Algal Research*, 42, 101606.

[7] Worku, L. A., Bachheti, A., Bachheti, R. K., Rodrigues Reis, C. E., & Chandel, A. K. (2023). Agricultural residues as raw materials for pulp and paper production: Overview and applications on membrane fabrication. *Membranes*, 13(2), 228.

[8] Koul, B., Yakoob, M., & Shah, M. P. (2022). Agricultural waste management strategies for environmental sustainability. *Environmental Research*, 206, 112285.

[9] Bak, U. G., Mols-Mortensen, A., & Gregersen, O. (2018). Production method and cost of commercial-scale offshore cultivation of kelp in the Faroe Islands using multiple partial harvesting. *Algal Research*, 33, 36–47.

[10] Seghetta, M., Hou, X., Bastianoni, S., Bjerre, A. B., & Thomsen, M. (2016). Life cycle assessment of macroalgal biorefinery for the production of ethanol, proteins and fertilizers–a step towards a regenerative bioeconomy. *Journal of Cleaner Production*, 137, 1158–1169.

[11] Overland, M., Mydland, L. T., & Skrede, A. (2019). Marine macroalgae as sources of protein and bioactive compounds in feed for monogastric animals. *Journal of the Science of Food and Agriculture*, 99(1), 13–24.

[12] Pangestuti, M. B., Suhartini, S., & Hidayat, N. (2021, November). Life cycle assessment of bioenergy production from macroalgae: A review. In *IOP Conference Series: Earth and Environmental Science* (Vol. 924, No. 1, p. 012070). IOP Publishing.

[13] Dutta, N., Kundu, P., Lee, J. T. E., & Bhattacharya, S. (2023). Implementation and optimization of algal biomass in value-added products recovery: A step towards algae-based green economy. *Hydrobiology*, 2(2), 326–346.

[14] Liu, J. J., Dickson, R., Niaz, H., Van Hal, J. W., Dijkstra, J. W., & Fasahati, P. (2022). Production of fuels and chemicals from macroalgal biomass: Current status, potentials, challenges, and prospects. *Renewable and Sustainable Energy Reviews*, 169, 112954.

[15] Stiger-Pouvreau, V., Bourgougnon, N., & Deslandes, E. (2016). Carbohydrates from seaweeds. In J. Fleurence & I. A. Levine (Eds.), *Seaweed in Health and Disease Prevention* (pp. 223–274). Academic Press.

[16] Porse, H., & Rudolph, B. (2017). The seaweed hydrocolloid industry: 2016 updates, requirements, and outlook. *Journal of Applied Phycology*, 29(5), 2187–2200.

[17] Baghel, R. S., Trivedi, N., & Reddy, C. R. K. (2016). A simple process for recovery of a stream of products from marine macroalgal biomass. *Bioresource Technology*, 203, 160–165.

[18] Aitken, D., Bulboa, C., Godoy-Faundez, A., Turrion-Gomez, J. L., & Antizar-Ladislao, B. (2014). Life cycle assessment of macroalgae cultivation and processing for biofuel production. *Journal of Cleaner Production*, 75, 45–56.

[19] Chauton, M. S., Forbord, S., Mäkinen, S., Sarno, A., Slizyte, R., Mozuraityte, R., … Skjermo, J. (2021). Sustainable resource production for manufacturing bioactives from micro-and macroalgae: Examples from harvesting and cultivation in the Nordic region. *Physiologia Plantarum*, 173(2), 495–506.

[20] Thornber, C. S. (2006). Functional properties of the isomorphic biphasic algal life cycle. *Integrative and Comparative Biology*, 46(5), 605–614.

[21] Santelices, B. (1990). Patterns of reproduction, dispersal and recruitment inseaweeds. *Oceanography and Marine Biology: An Annual Review*, 28, 177–276.

[22] Krishnamurthy, V. (2001). Reproductive biology of eukaryotic algae. In B. M. Johri & P. S. Srivastava (Eds.), *Reproductive Biology of Plants* (pp. 57–95). Springer Berlin Heidelberg.

[23] Milledge, J. J., Smith, B., Dyer, P. W., & Harvey, P. (2014). Macroalgae-derived biofuel: a review of methods of energy extraction from seaweed biomass. *Energies*, 7(11), 7194–7222.

[24] Fernand, F., Israel, A., Skjermo, J., Wichard, T., Timmermans, K. R., & Golberg, A. (2017). Offshore macroalgae biomass for bioenergy production: Environmental aspects, technological achievements and challenges. *Renewable and Sustainable Energy Reviews*, 75, 35–45.

[25] Price, C. S., & Morris Jr, J. A. (2013). Marine cage culture and the environment: Twenty-first century science informing a sustainable industry (NOAA Technical Memorandum NOS NCCOS 164). National Oceanic and Atmospheric Administration, National Centers for Coastal Ocean Science.

[26] Anisha, G. S., Padmakumari, S., Patel, A. K., Pandey, A., & Singhania, R. R. (2022). Fucoidan from marine macroalgae: Biological actions and applications in regenerative medicine, drug delivery systems and food industry. *Bioengineering*, 9(9), 472.

[27] Kotta, J., Futter, M., Kaasik, A., Liversage, K., Rätsep, M., Barboza, F. R., … Virtanen, E. (2020). Cleaning up seas using blue growth initiatives: Mussel farming for eutrophication control in the Baltic Sea. *Science of the Total Environment*, 709, 136144.

[28] Forbord, S., Skjermo, J., Arff, J., Handå, A., Reitan, K. I., Bjerregaard, R., & Lüning, K. (2012). Development of Saccharina latissima (Phaeophyceae) kelp hatcheries with year-round production of zoospores and juvenile sporophytes on culture ropes for kelp aquaculture. *Journal of Applied Phycology*, 24, 393–399.

[29] Dawes, C. P. (1995). Suspended cultivation of Gracilaria in the sea. *Journal of Applied Phycology*, 7, 303–313.

[30] Akram, F., Saleem, B., Irfan, M., Shakir, H. A., Khan, M., Ali, S., … Franco, M. (2023). Recent trends for production of biofuels using algal biomass. In I. A. Abbott (Ed.) *Basic Research Advancement for Algal Biofuels Production* (pp. 27–58). Springer Nature Singapore.

[31] Radulovich, R., Neori, A., Valderrama, D., Reddy, C. R. K., Cronin, H., & Forster, J. (2015). Farming of seaweeds. In N. Srivastava & P. K. Mishra (Eds.), *Seaweed Sustainability* (pp. 27–59). Academic Press.

[32] Biris-Dorhoi, E. S., Michiu, D., Pop, C. R., Rotar, A. M., Tofana, M., Pop, O. L., … Farcas, A. C. (2020). Macroalgae—A sustainable source of chemical compounds with biological activities. *Nutrients*, 12(10), 3085.

[33] Kirkman, H., & Kendrick, G. A. (1997). Ecological significance and commercial harvesting of drifting and beach-cast macro-algae and seagrasses in Australia: a review. *Journal of Applied Phycology*, 9, 311–326.

[34] Grima, E. M., Belarbi, E. H., Fernández, F. A., Medina, A. R., & Chisti, Y. (2003). Recovery of microalgal biomass and metabolites: process options and economics. *Biotechnology Advances*, 20(7–8), 491–515.

[35] Christenson, L. B., & Sims, R. C. (2012). Rotating algal biofilm reactor and spool harvester for wastewater treatment with biofuels by-products. *Biotechnology and Bioengineering*, 109(7), 1674–1684.

[36] Taelman, S. E., Champenois, J., Edwards, M. D., De Meester, S., & Dewulf, J. (2015). Comparative environmental life cycle assessment of two seaweed cultivation systems in North West Europe with a focus on quantifying sea surface occupation. *Algal Research*, 11, 173–183.

[37] Ananya, A. K., & Ahmad, I. Z. (2014). Cyanobacteria "the blue green algae" and its novel applications: A brief review. *International Journal of Innovation and Applied Studies*, 7(1), 251.

[38] Palatnik, R. R., & Zilberman, D. (2017). Economics of natural resource utilization-the case of macroalgae. In *Modeling, Dynamics, Optimization and Bioeconomics II: DGS III, Porto, Portugal, February 2014, and Bioeconomy VII, Berkeley, USA, March 2014-Selected Contributions 3* (pp. 1–21). Springer International Publishing.

[39] Healy, L. E., Zhu, X., Pojić, M., Sullivan, C., Tiwari, U., Curtin, J., & Tiwari, B. K. (2023). Biomolecules from macroalgae—nutritional profile and bioactives for novel food product development. *Biomolecules*, 13(2), 386.

[40] Kraan, S. (2013). Mass-cultivation of carbohydrate rich macroalgae, a possible solution for sustainable biofuel production. *Mitigation and Adaptation Strategies for Global Change*, 18, 27–46.

[41] Li, K., Liu, S., & Liu, X. (2014). An overview of algae bioethanol production. *International Journal of Energy Research*, 38(8), 965–977.

[42] Veza, I., Hoang, A. T., Abbas, M. M., Tamaldin, N., Idris, M., Djamari, D. W., … Opia, A. C. (2022). Microalgae and macroalgae for third-generation bioethanol production. In C. R. Soccol, G. A. G. Pereira, C.-G. Dussap, & L. P. de Souza Vandenberghe (Eds.), *Liquid biofuels: Bioethanol*, (pp. 301–331). Springer International Publishing.

[43] Chen, J., Bai, J., Li, H., Chang, C., & Fang, S. (2015). Prospects for bioethanol production from macroalgae. *Trends in Renewable Energy*, 1(3), 185–197.

[44] Kádár, Z., Szengyel, Z., & Réczey, K. (2004). Simultaneous saccharification and fermentation (SSF) of industrial wastes for the production of ethanol. *Industrial Crops and Products*, 20(1), 103–110.

[45] Pardilhó, S., Boaventura, R., Almeida, M., & Dias, J. M. (2022). Marine macroalgae waste: A potential feedstock for biogas production. *Journal of Environmental Management*, 304, 114309.

[46] Barbot, Y. N., Al-Ghaili, H., & Benz, R. (2016). A review on the valorization of macroalgal wastes for biomethane production. *Marine Drugs*, *14*(6), 120.

[47] Farobie, O., & Hartulistiyoso, E. (2022). Palm oil biodiesel as a renewable energy resource in Indonesia: Current status and challenges. *Bioenergy Research*, 15(1), 93–111.

[48] Amrullah, A., & Farobie, O. (2023). Conversion and rate behavior of brown macroalgae in pyrolysis: Detailed effects of operating parameters. *Heliyon*, 9(7), e18350.

[49] Jahirul, M. I., Rasul, M. G., Chowdhury, A. A., & Ashwath, N. (2012). Biofuels production through biomass pyrolysis—a technological review. *Energies*, 5(12), 4952–5001.

[50] Liu, J., Zhou, F., Abed, A. M., Le, B. N., Dai, L., Ali, H. E., … Zhang, G. (2022). Macroalgae as a potential source of biomass for generation of biofuel: Artificial intelligence, challenges, and future insights towards a sustainable environment. *Fuel*, 336, 126826.

[51] Tan, I. S., Lam, M. K., Foo, H. C. Y., Lim, S., & Lee, K. T. (2020). Advances of macroalgae biomass for the third generation of bioethanol production. *Chinese Journal of Chemical Engineering*, 28(2), 502–517.

[52] Rajkumar, R., Yaakob, Z., & Takriff, M. S. (2014). Potential of micro and macro algae for biofuel production: a brief review. *Bioresources*, 9(1), 1606–1633.
[53] Kavitha, S., Gondi, R., Kannah, R. Y., Kumar, G., & Banu, J. R. (2022). A review on current advances in the energy and cost effective pretreatments of algal biomass: Enhancement in liquefaction and biofuel recovery. *Bioresource Technology*, 369, 128383.
[54] Leandro, A., Pereira, L., & Gonçalves, A. M. (2019). Diverse applications of marine macroalgae. *Marine Drugs*, 18(1), 17.

7 Nature's Powerhouse
Photobiological Hydrogen Production for a Cleaner Future

Poonam Kumari and S. Venkata Mohan

7.1 INTRODUCTION

Photosynthesis represents a photochemically driven oxidation-reduction (redox) process, recognized as the cornerstone of nearly all terrestrial life forms, with a profound impact spanning geological eras. This intricate mechanism orchestrates the continual cycling of carbon within the biosphere, primarily energized by the perpetual infusion of sustainable energy sources. (Sherif et al., 2014) Throughout the archives of industrial civilization, preserved remnants of photosynthetic products from ancient epochs have historically constituted the predominant energy reservoir. Nonetheless, the foreseeable future presents a formidable challenge in the form of dwindling fossil fuel reserves, coupled with the imminent global peril of climate change, thereby mandating a pivotal paradigm shift. Notably, as early as 1874, Jules Verne, in his novel "The Mysterious Island," presciently foretold the rise of hydrogen as an alternative to coal, opining that "I believe that water will one day be harnessed as a fuel, for the hydrogen and oxygen it comprises, whether utilized separately or in concert, shall provide an inexhaustible font of heat and illumination" (Verne 2004). Now, almost a century and a half later, in the throes of a climate crisis, hydrogen has emerged as the foremost candidate for sustainable energy solutions.

The synthesis pathways of hydrogen energy (H_2) can be classified into four distinct categories: Brown hydrogen, generated as a byproduct of industrial operations; Gray hydrogen, obtained from fossil fuel sources; Blue hydrogen, created from fossil fuels with integrated carbon capture technology; and Green hydrogen, extracted from renewable resources. Among these, Brown and Gray methodologies typically exhibit higher carbon intensity, while Blue hydrogen presents a comparatively lower carbon footprint, and Green hydrogen is often perceived as having a carbon-neutral profile (Dahiya et al., 2023). Green hydrogen production primarily hinges on two principal techniques: water electrolysis and biological production. Historically, fossil-based electricity sources have been utilized for water electrolysis. Nonetheless, this approach raises environmental concerns and contributes to carbon emissions associated with the combustion of fossil fuels. Such practices not only contradict the objective of generating clean hydrogen but also lead to increased operational expenses. Despite the environmental benefits of employing sustainable electricity sources, the cost of producing hydrogen via water electrolysis remains a limiting factor. Additionally, the efficiency of the process poses a significant challenge. Water electrolysis necessitates a specific amount of electrical energy to break down water molecules, and the ongoing task lies in optimizing this process for maximum efficacy. Excessive energy consumption can escalate production costs, rendering hydrogen less competitive in the energy market (Pilon and Berberoğlu, 2014). Biological H_2 production can be broadly classified into two categories: light-dependent and light-independent processes (Suresh et al. 2023). Light-dependent processes encompass activities such as photo-fermentation and bio-photolysis, which

DOI: 10.1201/9781003732907-7

occur in organisms capable of photosynthesis, including anoxygenic phototrophic bacteria, cyanobacteria, and microalgae (Pandey et al., 2019). Of particular interest is the potential for photo-biological hydrogen production by photosynthetic organisms, a concept that has garnered substantial attention due to its potential to generate clean, carbon-free, and renewable energy using sunlight and water, two of nature's most abundant resources (Pathy et al., 2022). Within the realm of microorganisms, hydrogen metabolism plays a pivotal role, where hydrogen often serves as the final product or a reactant in separate and distinct processes.

The ability of various microbes to store hydrogen differs depending on the mechanism at play (Cheng et al., 2011). The two reaction centers in plants and algae, PSI and PSII, have close links from an evolutionary standpoint. The horizontal arrows in Figure 7.1 each exhibits a unique relationship with various kinds of more archaic bacterial reaction centers. The different electron transport carriers are represented by the acronyms in the diagram. (Blankenship 1992) and (Allen 2014). NADPH or NADH levels determine the entry of electrons into the Benson–Calvin cycle, which fixes carbon dioxide (Mukherjee and Srivastava, 2023). Proton transport related to electron transfer moves from top to bottom across the membrane, albeit it is not shown in the schematic (Figure 7.1). An ATP-producing proton gradient is created as a result of this proton transport, which also leads to proton transfer. NAD^+ reduction in purple bacteria is associated with a reversible electron transfer mechanism (Mukherjee and Srivastava, 2023). Through effective and coordinated research endeavors, we possess the chance to fully realize the capabilities of this technology and move it from isolated lab environments to a broad range of applications in society. Within this chapter, we embark on a scientific odyssey into the realm of photosynthetic biohydrogen pathways, delving into the intricate shades and challenges that await exploration.

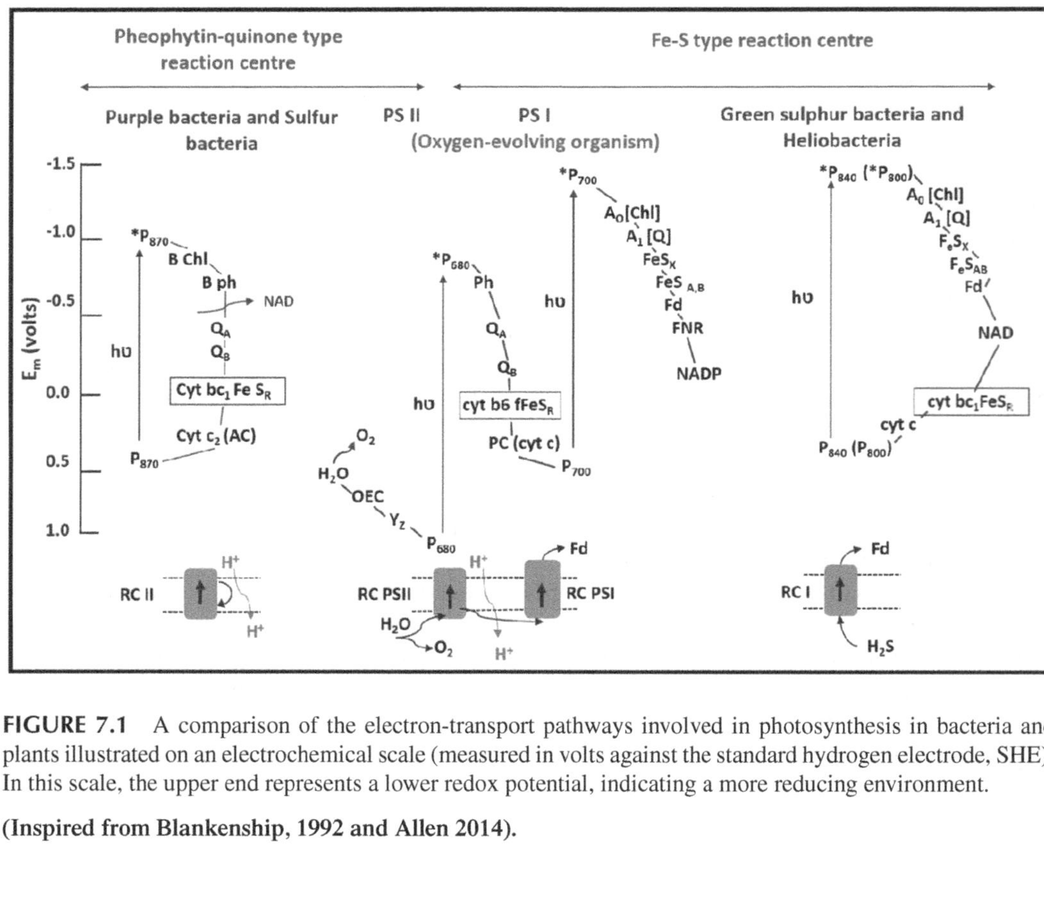

FIGURE 7.1 A comparison of the electron-transport pathways involved in photosynthesis in bacteria and plants illustrated on an electrochemical scale (measured in volts against the standard hydrogen electrode, SHE). In this scale, the upper end represents a lower redox potential, indicating a more reducing environment.

(Inspired from Blankenship, 1992 and Allen 2014).

7.2 FUNDAMENTALS OF PHOTOBIOLOGICAL HYDROGEN PRODUCTION

Photobiological H_2 generation offers the potential to produce a renewable fuel using two of nature's most abundant resources: light and water. This process is environmentally sustainable as it utilizes naturally occurring carbon dioxide to produce oxygen and biomass. This innovation is built upon the adaptation of autotrophic photosynthetic mechanisms, which have been reconfigured to enable the generation of hydrogen gas in conjunction with the synthesis of carbon-based biomass. Photosynthesis, a fundamental biological process, heavily relies on light energy. However, in natural environments, the distribution of light is non-uniform, characterized by variations in intensity and duration. To address these fluctuations in light availability, the machinery and electron transport systems within photosynthetic complexes have evolved a diverse range of adaptations. The microbial metabolism involved in H_2 production is influenced by various factors, as depicted in Figure 7.2. These factors include carbon sources, energy sources, electron sources, and nitrogen sources. All of these adaptations and metabolic processes are rooted in the fundamental act of charge separation that occurs within pigment-protein complexes (Blankenship, 2021). Microalgae and Cyanobacteria utilize photosynthetically produced carbohydrates as their energy source for H_2 production, which is accomplished through indirect bio-photolysis under anoxic conditions. In contrast, microalgae use light energy to directly produce H_2 from water in the presence of oxygen. The generation of photobiological H_2 can be achieved through three fundamental processes: direct bio-photolysis, indirect bio-photolysis, and photo-fermentation (Giri et al., 2022). A summary of the processes involved in photobiological H_2 production, along with their associated strengths and limitations, can be found in Table 7.1 (Boichenko et al., 2004).

7.2.1 Direct Bio-photolysis

The process of turning water into hydrogen with solar energy is called bio-photolysis. This results in the production of H_2 and molecular oxygen (Figure 7.3). This method can be applied to individual cellular components or microalgal cultures (Schlegel and Barnea, 2013). Direct bio-photolysis and the photosynthetic mechanism employed by algae and plants with pigments that absorb light are extremely similar. The process involves Photosystem I and II (PSI/700 and PSII/680), which generate potent oxidants that can convert water into protons, electrons, and oxygen, or their equivalents (Oh et al., 2013). Then, with the aid of an electron transport chain, the ferredoxin oxidoreductase

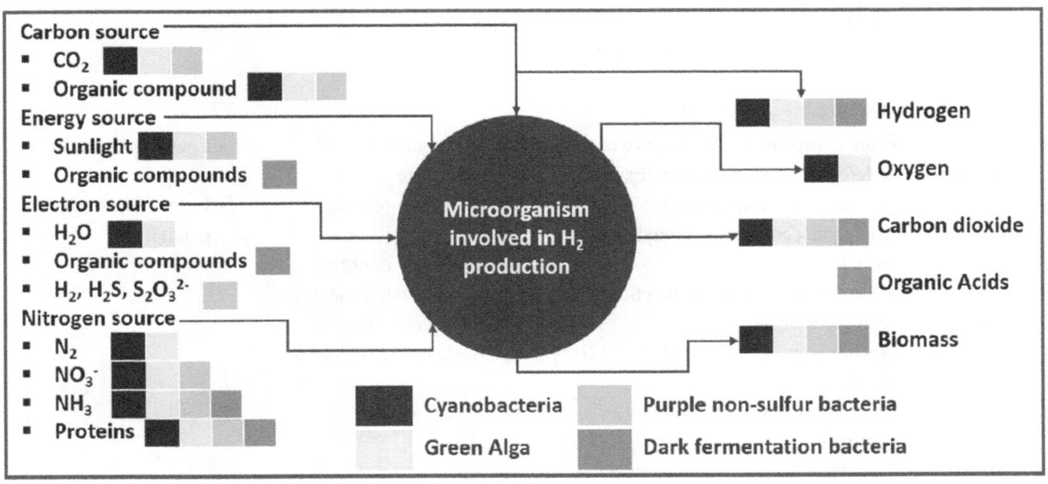

FIGURE 7.2 Overview representation of carbon, energy, electron, and nitrogen sources of microbial H_2 metabolism.

TABLE 7.1
Summary of Processes Involved in Photobiological H_2 Production and Associated Key Advantages and Limitations

Process	General Features	Advantages	Limitations
Direct bio-photolysis	**Basic principle**: In the presence of a direct light source such as the sun, microorganisms containing photosynthetic pigments initiate a series of electron transfer processes to produce hydrogen gas from water. **Microorganism**: Green algae, Cyanobacteria **Mechanism**: $2H_2O$ + Ferredoxin(oxidized) + light → $4H^+$ + Ferredoxin(reduced) ($4e^-$) + O_2 $4H^+$ + Ferredoxin(reduced) ($4e^-$) → $2H_2$ + Ferredoxin(oxidized)	• Simple cultivation process • CO_2 sequestration • Production utilizing nature's copious resources: solar energy and water	• O_2 sensitivity • Thresholds of light intensity and decrease in energy efficiency • Simultaneous H_2 and O_2 production, increasing downstream cost
Indirect bio-photolysis	**Basic principle**: In distinct cellular compartments or during specific time intervals, typically in situations characterized by a lack of sulfur and the inactivity of photosystem II (PSII), utilize complex carbohydrates or pyruvate, which have been synthesized from water and solar energy, to generate hydrogen gas. **Microorganism**: Cyanobacteria and blue-green algae under anoxic conditions **Mechanism**: $6H_2O + 6CO_2$ + light → $C_6H_{12}O_6 + 6O_2$ $C_6H_{12}O_6 + 2H_2O \rightarrow 4H_2 + 2CH_3COOH + 2CO_2$ $2CH_3COOH + 4H_2O$ + light → $8H_2 + 4CO_2$ $N_2 + 8H^+$ + Ferredoxin (reduced) ($8e^-$) + 16ATP → $2NH_3 + H_2$ + Ferredoxin (oxidized) + 16ADP + 16Pi	• Production utilizing nature's copious resources: solar energy and water • H_2 production O_2 production non-simultaneous (spatial or temporal or spatiotemporal separation) • Simultaneous N_2 fixation for microbial growth	• H_2 uptake by hydrogenase • Higher energy demand, particularly for nitrogen-fixing cyanobacteria • Limitations in H_2 evaluation due to nitrogen substrate
Photo-fermentation	**Basic principle**: In the presence of light, microorganisms transform organic biomass into simpler organic or inorganic compounds through various metabolic processes. **Microorganism**: Photosynthetic bacteria and green algae **Mechanism**: $C_6H_{12}O_6 + 6H_2O \rightarrow 12H_2 + 6CO_2$ Pyruvate + CoA → Acetyl-CoA + Formate Pyruvate + CoA + 2Ferredoxin(ox) → Acetyl-CoA + CO2 + 2Ferredoxin(red) ($2e^-$)	• Spectrum of valuable metabolites • Utilization of wide spectrum • Utilization of organic substrate (small acids to acidogenic effluent) • Waste bio-remediation	• Lower light conversion efficiency • Homogenous light requirement

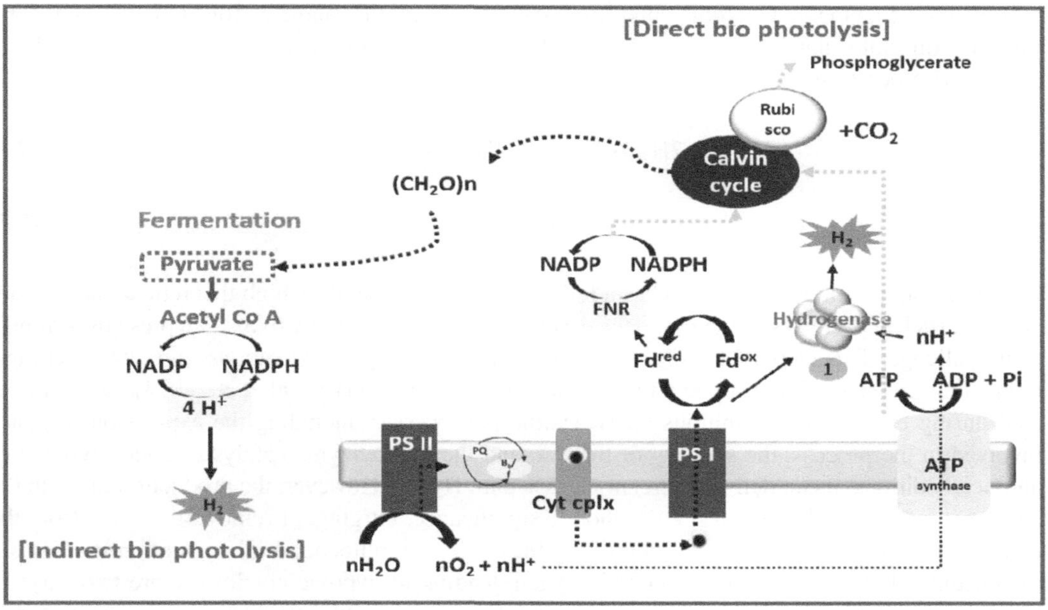

FIGURE 7.3 Bio-photolysis, direct and indirect mechanisms of H_2 production. Gray bar indicating the lipid bilayer of the thylakoid membrane of organism. Dotted black arrows represent the electron transfer. Black arrows represent the multiple possible reactions.

enzyme (FNR) assists the electrons in reducing ferredoxin. Then, using the reduced ferredoxin as an electron source, the hydrogenase enzyme catalyzes the reduction of protons, ultimately leading to the evolution of H_2. The direct bio-photolysis mechanism is used by *Chlamydomonas reinhardtii*, one of the most researched bacteria (Mukherjee and Srivastava, 2023). Direct bio-photolysis' outcome can be characterized as:

$$2H_2O + \text{Light energy} \rightarrow 2H_2 + O_2 \tag{7.1}$$

Light-harvesting complex proteins (LHC proteins) absorb sunlight energy during the first stage of photosynthesis in microalgae. The splitting of water into H^+, electrons (e^-), and oxygen (O_2) is subsequently triggered by this absorbed energy (Eq. 7.2). Photons captured by LHC-II (a light-harvesting complex composed of chlorophyll a, chlorophyll b, and carotenoids) within a protein framework in the thylakoid membrane, with a primary interaction with Photosystem II subsequently propel the linear flow of electrons along the electron transport chain, proceeding from PSII-LHC II super complex to plastoquinone (PQ), cytochrome b6f (Cyt), plastocyanin (PC), and PSI-LHC I super complex to ferredoxin (FD) (Oey et al., 2016). The ferredoxin-$NADP^+$ oxidoreductase (FNR) enzyme catalyzes the last process. These electrons are ultimately utilized to decrease $NADP^+$ to NADPH, which works with ATP in later metabolic reactions to change CO_2 into sugars, starches, and other biomolecules, producing biomass as a whole (Mukherjee and Srivastava, 2023). An electrochemical H^+ gradient is created across the thylakoid membrane by the release of protons into the lumen during the photosynthetic water-splitting process. This gradient is utilized to transport and utilize electrons for CO_2 fixation. This gradient encourages the return of extra H^+ to the stroma and causes the ATP synthase to produce ATP. Under typical aerobic photosynthetic circumstances, CO_2 is converted to carbohydrates, which are then utilized for respiration and cellular development in the mitochondria. The reduced ferredoxin, on the other hand, acts as an electron donor to the hydrogenase enzyme in anaerobic conditions, causing stromal H^+ and electrons to combine reversibly to form molecular H_2 (Eq. 7.3) (Oey et al., 2016). Following is a representation of the direct photolysis process' sequential reactions:

$$2H_2O + \text{light} \rightarrow 4H^+ + 4e^- + O_2 \tag{7.2}$$

$$4H^+ + 4e^- \rightarrow 2H_2 \tag{7.3}$$

Direct bio-photolytic H_2 generation has a lot of promise because of its high theoretical photon conversion efficiency. However, the hydrogenase enzyme's high sensitivity to oxygen presents a significant challenge. This enzyme is only active in anaerobic or hypoxic situations; in aerobic settings, oxygen progressively builds up and renders the enzyme inactive (Biswal et al., 2023). Oxygen created during photosynthesis inhibits all H_2 synthesis processes, including the expression of genes involved in the process, the stability of mRNA, and the final enzyme catalysis. Under hypoxic or anoxic conditions, it can deliver a stream of 95% pure H_2 gas. However, the production of H_2 in the environment is irregular and of little economic significance. Purging, or removing oxygen from the reaction mixture with inert gas, or controlling the cell's own mitochondrial respiration mechanism to consume photosynthetically generated oxygen, leading to hypoxic conditions, are two ways to boost the efficiency of H_2 synthesis. For its bio-photolytic H_2 generation mechanisms, *Chlamydomonas reinhardtii* is one of the most researched algae species (Suresh et al., 2023). Direct bio-photolysis' main advantage is its self-sufficiency because no extra substrates or nutrients are needed. The main source of electrons required for the generation of H_2 gas is water; the primary inputs required for growing the microorganisms carrying hydrogenase enzymes are solar energy and atmospheric CO_2. In order to achieve clean and effective H_2 production, ideal circumstances for direct bio-photolysis must be managed. These aspects will be discussed in more detail in the next section.

7.2.2 Indirect Bio-photolysis

Indirect bio-photolysis, which can happen in both cyanobacteria and microalgae, consists of two independent stages that are naturally separated in space or time. Oxygen is created as a byproduct of the first cycle, which transforms solar energy into chemical energy that is stored as carbohydrates (Pandey et al., 2019) (Eq. 7.4). These carbohydrates are used as substrates for the generation of H_2 in the second step (Eq. 7.5 and 7.6). Because the molecular energy contained within the carbohydrates is harnessed, the process is referred to as "indirect" because it uses biomass or carbs as the source of H^+ and e^- as opposed to water.

The general reactions for indirect bio-photolysis are as follows:

$$\text{First stage: } 6H_2O + 6CO_2 + \text{light} \rightarrow C_6H_{12}O_6 + 6O_2 \tag{7.4}$$

$$\text{Second stage: } C_6H_{12}O_6 + 2H_2O \rightarrow 4H_2 + 2CH_3COOH + 2CO_2 \tag{7.5}$$

$$2CH_3COOH + 4H_2O + \text{light} \rightarrow 8H_2 + 4CO_2 \tag{7.6}$$

The pyruvate ferredoxin oxidoreductase enzyme catalyzes the decarboxylation of pyruvate, which is produced following carbohydrate oxidation, into acetyl Co-A during the second stage, which involves the generation of H_2. Ferredoxin is reduced by NADPH, which results in the production of H_2. While microalgae only use hydrogenase to produce H_2, cyanobacteria use both hydrogenase and nitrogenase enzymes in the evolution of H_2 (Kose and Oncel, 2016). Although nitrogenase has the advantage of being unidirectional, it is less energy-efficient than hydrogenase-based indirect photolysis, requiring only 4 ATP to produce 1 H_2. Both enzymes must be kept apart from the reactions that produce H_2 and O_2 since they are both oxygen-sensitive. In filamentous cyanobacteria, spatial separation is achieved when nitrogen fixation and H_2 evolution reactions take place in specialized cells called heterocysts with functional PSI but no PSII activity. Photosynthetic reactions and oxygen evolution reactions take place in vegetative cells. The level of H_2 production is constrained by the low heterocyst to vegetative cell ratio (about 1:10), which calls for future genetic engineering improvement (Oncel et al., 2015). While most cyanobacteria are filamentous and nitrogen-fixing (genera like *Nostoc, Calothrix*, etc.), certain cyanobacteria (like *Synechocystis*) are not. Oxygen evolution and H_2 evolution occur at different times in microalgae. While H_2 is produced at night, the photosynthetic conversion of CO_2 to carbohydrates and lipids occurs during the day. This suppresses the hydrogenase enzyme inhibition caused by oxygen buildup during the night (Pilon and Berberoğlu, 2014). The reaction for H_2 evolution, along with N_2 fixation, is as follows:

$$N_2 + 8e^- + 8H^+ + 16ATP \rightarrow 2NH_3 + H_2 + 16ADP + 16Pi \quad (7.7)$$

The reaction for H_2 evolution without N_2 fixation is:

$$8H^+ + 8e^- + 16ATP \rightarrow 4H_2 + 16ADP + 16Pi \quad (7.8)$$

7.2.3 Photo-fermentation

Numerous types of photosynthetic bacteria are capable of photo-fermentation, which can take place in anaerobic environments (Figure 7.4). Purple non-sulfur (PNS) bacteria stand out among these microorganisms as being particularly promising in this procedure. They use sunlight as their source of energy and use small organic molecules as their main sources of CO_2 and H_2, often organic acids like succinate, malate, and lactate. Through the action of the nitrogenase enzyme in a nitrogen-deficient environment, PNS bacteria can create H_2 by employing these small organic molecules as carbon sources and solar energy as an energy source. PNS bacteria are highly efficient in producing H_2 because, among other things, they can work in anaerobic environments, avoid the oxygen sensitivity problems that nitrogenase and enzymes have, and they can absorb and use both visible and near-infrared solar radiation (400–950 nm) (Li et al., 2022).

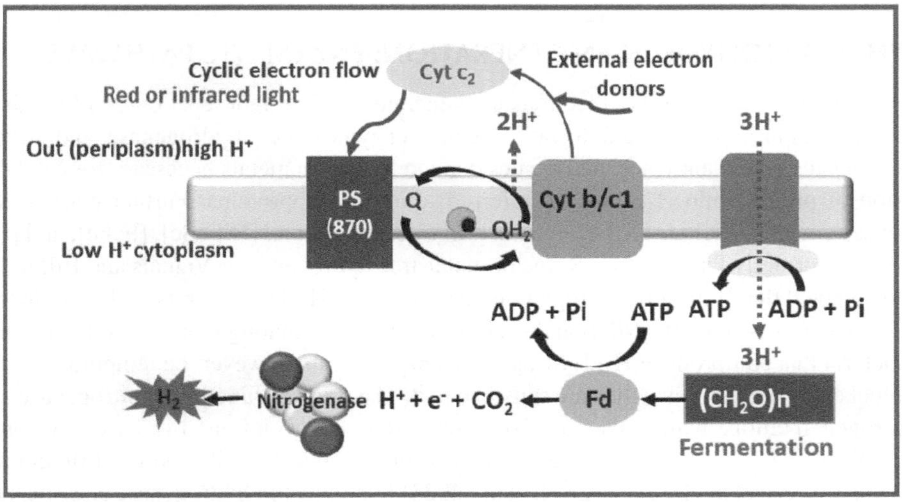

FIGURE 7.4 Bio-photolysis, photo-fermentation mechanism of H_2 production. Gray bar indicating the lipid bilayer of the thylakoid membrane of organism. Black arrows represent the multiple possible reactions.

In the photo-fermentation process, microorganisms use the sun's energy they absorb to create ATP, which speeds up the oxidative photo catabolic breakdown of organic substrates and releases high-energy electrons. The PQ pool at PSI and PSII receives these electrons after that. The enzyme NADPH-plastoquinone oxidoreductase is also involved in this step. The electrons are subsequently used to decrease ferredoxin by the action of hydrogenase or nitrogenase enzymes (Mosebach et al., 2017). Ultimately, the reduced ferredoxin transfers its electrons to protons to create H_2. N_2 undergoes this transformation into ammonium ions at high or moderate N_2 partial pressures. The organic substrates act as a source of electrons, which are then provided to ferredoxin by reverse flow for reduction. The H_2 generation by the photo-fermentation process (Baeyens et al., 2020) is shown in Eq. 7.9:

$$16ATP + N_2 + 16H_2O + 10H^+ + 8e^- \rightarrow 16ADP + 2NH_4^+ + 16Pi + H_2 \tag{7.9}$$

In this route, green algae obtain electrons through endogenous substrate catabolism and heterotrophic fermentation in place of water. Through photo-catabolism, electrons are extracted from the organic substrate products, with the help of light energy and oxidative carbon metabolism. These endogenous substrate-derived electrons then pass through the PQ pool, which is found in PSI and PSII. In the chloroplasts of various species of vascular plants, NADPH plastoquinone oxidoreductase is linked to the transport of electrons to the PQ pool (Dalena et al., 2017). In doing so, PSI absorbs light energy and produces electrons. Through their transfer to the equivalent levels of hydrogenase and ferredoxin, these electrons' redox potential can be increased. A large yield of biohydrogen is produced, thanks to hydrogenase being induced by the dark and anoxic environment (Ahmed et al., 2021). Theoretically, during the photo-fermentation process, organic acid substrates might be stoichiometrically converted into H_2. However, scientists have found that the production volume and light conversion efficiency are frequently both quite low. These restrictions might be circumvented by combining a series of photo-dark fermentation stages (Chandra and Mohan, 2014).

When compared to bio-photolysis, the photo-fermentation method has the benefit of being able to utilize organic waste products. This feature offers great potential for producing renewable hydrogen while fostering environmental sustainability. The PSB class of photosynthetic bacteria, which flourish in anaerobic or microaerophilic environments, is another interesting group that actively participates in photo-fermentation. PSB use reduced sulfur compounds, such as hydrogen sulfide, as donors rather than organic acids as substrates (Mohan et al., 2019). They can therefore use the photo-fermentation process to produce H_2 as well as granules of elemental sulfur.

7.3 PHOTO-BIOHYDROGEN GENERATION: ENZYMATIC PATHWAYS

In order to produce molecular H_2, electrons are generated during light reactions, carbohydrate degradation and respiration, directed toward specific enzymes, i.e., hydrogenase and nitrogenase enzymes. Because these enzymes' active sites contain transition metals necessary for catalyzing the conversion of protons into H_2, they are referred to as metalloproteins. Further classifications of hydrogenase enzymes include [NiFe]-uptake, [NiFe]-bidirectional, [NiFeSe], [FeFe], and [Fe]-only hydrogenase (formerly known as Fe-S free or metal-free hydrogenase) (Vignais and Billoud, 2007). It should be noted that [Fe]-only hydrogenase does not make H_2, but rather uses H_2 to reduce CO_2 to form CH_4. The most common [NiFe] and [FeFe] hydrogenases among them are found in photosynthetic bacteria that can produce H_2. Hydrogenase enzymes can, however, be temporarily or permanently blocked by oxygen since they are sensitive to it. The primary function of nitrogenase enzymes is in nitrogen fixation, which also involves concurrent H_2 evolution. Mo-based, V-based, and Fe-based nitrogenase enzymes are the three main varieties, and they all produce different amounts of hydrogen (Suresh et al., 2023). (Suresh et al., 2023) In contrast to hydrogenase enzymes, nitrogenase enzymes are oxygen-sensitive and need a substantial quantity of ATP in order to produce H_2. Additionally, the stoichiometric conversion of H^+ to H_2 is constrained by the conversion of N_2 into

NH_3 or NH_4^+. In order to increase H_2 generation in nitrogen-limited conditions, researchers in the field of biohydrogen production are working to improve nitrogenase enzyme efficiency.

7.3.1 Hydrogenase

Depending on the organism, hydrogenase enzymes are essential elements present in a variety of cellular compartments. They can be found in the periplasm of photosynthetic bacteria, the thylakoids of microalgae, and the stroma of cyanobacteria. The reversible reaction illustrates how these adaptable enzymes are involved in either hydrogen intake or evolution:

$$2H^+ + e^- \rightarrow H_2 \tag{7.10}$$

The hox gene encodes [NiFe]-uptake hydrogenase, [NiFe]-bidirectional hydrogenase, [NiFeSe] hydrogenases (where selenocysteine replaces Ni-coordinated cysteine residues), [FeFe] hydrogenases, and [Fe]-only hydrogenases, among other types of hydrogenase enzymes (Mulder et al., 2011). These include [NiFe] hydrogenases, which are typically present in cyanobacteria and bacteria, and [FeFe] hydrogenases, which are typically present in green microalgae and obligate anaerobic fermentative bacteria (Srivastava et al., 2020). Methanogenic bacteria and archaea contain [Fe]-only hydrogenases, which have mononuclear Fe-based active sites but lack the Fe-S cluster observed in [FeFe] hydrogenases. Green microalgae's chloroplasts include [FeFe] hydrogenase enzymes, which are in charge of catalyzing the conversion of H^+ to H_2 during hypoxic photosynthetic circumstances. These enzymes weigh 50 kDa on an average molecular weight and can be found as monomers or dimers. They contain a 6Fe-6S cluster or H-cluster, which is made up of a 4Fe-4S cubane subcluster covalently connected to a 2Fe-2S subcluster. Comparing this H-cluster in green microalgae to other hydrogenases, including the cyanobacterial [NiFe] hydrogenase, it shows a stunning 100-fold higher proton reduction activity. However, because of the strong bond that forms between the Fe atoms in the active site and O_2, oxygen irreversibly inhibits the [FeFe] hydrogenases in green microalgae under oxygenic photosynthetic conditions (Pathy et al., 2022). Only in hypoxic or anoxic conditions can these enzymes catalyze the reversible conversion of H^+ to H_2, with ferredoxin acting as an electron carrier. Only cyanobacteria and photosynthetic nitrogen-fixing bacteria have [NiFe]-uptake hydrogenases; these organisms also have nitrogenases that can produce H_2 (Srivastava et al., 2020). These [NiFe]-uptake hydrogenases, often referred to as unidirectional hydrogenases, use H_2 as a source of high-energy electrons, and the following is a representation of their catalytic reaction:

$$H_2 \rightarrow 2H^+ + e^- \tag{7.11}$$

[NiFe]-uptake enzymes are composed of two subunits, hupS and hupL, which stand for the small and large subunits, respectively. Within the roughly 30-kDa small subunit hupS, there are three 4Fe-4S clusters. In the electron transport chain, these clusters transfer electrons from hupL to electron-acceptor molecules. The main subunit, hupL (approximately 60 kDa), contains the Ni-Fe bimetallic active site. Here, Ni interacts with four cysteine residues, two of which act as bridging ligands to the $Fe(CN)_2CO$ unit. [NiFe]-bidirectional hydrogenases, as their name suggests, are mostly found in cyanobacteria and are capable of producing and consuming hydrogen in response to cellular redox conditions. Following another exposure to light, there is a brief burst of H_2 evolution that eventually leads to H_2 uptake. The process of producing H_2 mostly occurs in the dark. These enzymes are susceptible to substances such as O_2, CO, the bioenergetics of the cell, metabolic flux, and light exposure, all of which together lower the rates and yield of H_2 generation. Given the potential for large-scale, commercially viable H_2 production, it is imperative to explore the genetic, substrate, and environmental manipulation of [NiFe]-bidirectional and [FeFe] hydrogenase enzymes.

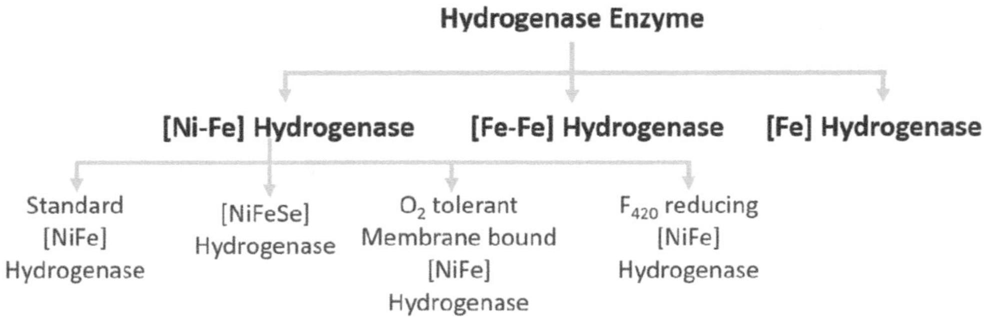

FIGURE 7.5 Classification of hydrogenase enzyme.

Although [FeFe] hydrogenases have a faster turnover rate for producing H_2, they are more oxygen-sensitive (Lau et al., 2015). However, [NiFe] hydrogenases have a higher oxygen tolerance, and their H_2 production could be improved by adjusting the N_2 partial pressure or exposure to light (Figure 7.5).

7.3.2 Nitrogenase

Different prokaryotic species, such as PNS bacteria, cyanobacteria, and green sulfur bacteria, depend on the presence of nitrogenase enzymes. Their major function is to accelerate N_2 fixation, with NH_3 being the end product (Madigan et al., 1997). However, as demonstrated in the irreversible reaction below, this mechanism also produces H_2 as a byproduct:

$$N_2 + 8e^- 8H^+ + 16ATP \rightarrow H_2 + 2NH_3 + 16(ADP + Pi) \tag{7.12}$$

In the absence of N_2, the reaction simplifies to:

$$2e^- 2H^+ + 4ATP \rightarrow H_2 + 4(ADP + Pi) \tag{7.13}$$

These equations demonstrate that a large amount of energy must be expended in the form of ATP during this process, especially the hydrolysis of 4 moles of ATP for each mole of H_2 produced. Dinitrogenase, also known as Fe-Mo protein, and dinitrogenase reductase, also known as Fe protein, make up the two-protein complex that makes up the nitrogenase enzyme (Brown et al., 2016). Dinitrogenase reductase is a homodimeric Fe-S protein with a molecular weight of around 65 kDa that facilitates the passage of electrons from an external electron source to the dinitrogenase complex. The dinitrogenase complex, also referred to as the MoFe protein, has a molecular weight of 230 kDa. It catalyzes the transformation of N_2 into NH_3 gradually. The MoFe cofactor is involved in both the production of NH_3 and the simultaneous evolution of H_2. The catalytic region of nitrogenase enzymes has been shown to contain either Fe or V; however, Mo-nitrogenase is the form that is most frequently identified in N_2-fixing prokaryotes (Einsle and Rees, 2020). As a result, there are three different types of nitrogenases dependent on the metal cofactor, each of which produces H_2 at various stoichiometric ratios, as seen in the following reactions:

$$\text{Mo based nitrogenase}: N_2 + 8H^+ + 8e^- \rightarrow 2NH_3 + H_2 \tag{7.14}$$

$$\text{Fe based nitrogenase}: N_2 + 21H^+ + 21e^- \rightarrow 2NH_3 + 7.5H_2 \tag{7.15}$$

$$\text{V based nitrogenase}: N_2 + 12H^+ + 12e^- \rightarrow 2NH_3 + 3H_2 \tag{7.16}$$

TABLE 7.2
Associated Enzymes in Photobiological H_2 Production

Enzymes	Microorganisms	Advantages	Disadvantages
Hydrogenase	Green algae, Cyanobacteria, and Purple-nonsulfur bacteria	• ATPs are not required • Higher efficiency rate approx. 41% • Very active and have larger turnover rate	• Higher sensitivity to O_2 • Limitation of generating H_2 under higher partial pressure
Nitrogenase	Photosynthetic bacteria and Cyanobacteria	• Robust production of H_2 • Effective under higher partial pressure	• Lower efficiency approx. 16% • Small turnover rate 2ATP/electron

The H_2 yield is naturally decreased when N_2 is present in the medium. It has been anticipated and experimentally proven that nitrogenase enzymes can only create H_2 in the absence of N_2 (Table 7.2). This field of study is still being actively investigated. However, a fundamental issue with nitrogenase-based H_2 synthesis is its high energy requirement, which reduces the process' total energy effectiveness. However, nitrogenase has the benefit of irreversible H_2 creation, which leads to the development of high-pressure H_2, making it appropriate for production on an industrial scale. Furthermore, this method does away with the potential for explosive H_2/O_2 mixture generation, which is frequently seen in photosynthetic bacteria that use hydrogenase enzymes (Eroglu and Melis, 2011).

7.4 FACTORS TO MODULATE THE HYDROGEN PRODUCTION

The process by which bacteria create photobiological hydrogen is influenced by a variety of circumstances. Several factors influence light-harvesting antenna pigments, like chlorophyll or bacteriochlorophyll, including their size, alternative electron consumption pathways, food selection, culture conditions, light intensity, and O_2 and CO_2 concentrations. They also cover how sensitive the oxygen-consuming enzymes nitrogenase and hydrogenase are. Many of these factors can be subject to manipulation through genetic and metabolic engineering techniques, offering opportunities to enhance the overall efficiency of photobiological hydrogen production. In the following section, we will provide an overview of these influencing factors and insights into how they can be adjusted to improve the yield and rate of hydrogen production.

7.4.1 Modulations in Metabolism

While genetic modification has potential for developing a more efficient and labor-intensive approach for photo-biohydrogen synthesis, commercialization remains a barrier. In this regard, more research and discovery of diverse microorganisms are required. Metabolic modulation offers an additional approach to regulate competitive pathways and H_2 metabolism.

1. Sulfur deprivation has emerged as a potentially effective technique. Sulfur deficiency affects the formation of the PSII reaction center protein D1, which contains sulfur-containing amino acids such as methionine. This decrease in D1 protein concentration causes photodamage to PSII as well as a decrease in its water-splitting capacity, resulting in decreased O_2 generation. When the rate of O_2 production is less than that of mitochondrial respiration, anoxia or hypoxia ensues. Enzyme hydrogenase expression and activity are increased as a result. According to a study (Nagy et al., 2018), ascorbate accumulation brought on by a sulfur deficit can render the manganese cluster in Photosystem II (PSII) inoperable.

Consequently, the reaction center deteriorates, low oxygen (hypoxia) develops, and H_2 gas is produced (Dahiya et al., 2021).

2. In large-scale H_2 production systems, the buildup of O_2 in the culture medium might inhibit the activity of hydrogen-producing enzymes, requiring the removal of O_2 (Babu et al., 2009). Several methods can be used to do this, including the use of inert gases such as argon, the inclusion of oxygen scavenging chemicals such as sodium sulfite, and the use of reducing agents such as ascorbate and glucose oxidase.
3. Magnesium deficiency is a relatively new technique for metabolic regulation. This method reduces photosynthesis while increasing respiration, allowing for long-term H_2 production (Li et al., 2022).

7.4.2 Substrate and Nutrient Optimization

It is critical to deliver the necessary nutrients in appropriate proportions to grown microorganisms in order to sustain both growth and H_2 production. Metal ions, such as Fe_2^+, are essential for the function of [FeFe] and [NiFe] hydrogenases. Furthermore, Fe and Mo are required for the metal cofactors of Fe or Mo (Fe) nitrogenases, whereas V is required for the V-only nitrogenase. Fe is required as a cofactor by a large number of proteins and enzymes, including cytochromes and ferredoxin, that are involved in the electron transport chain in both photosynthesis and respiration. Other metal ions, like as Mg_2^+ and Mn_2^+, must also be present in sufficient concentrations to maintain chlorophyll activities (Eroglu and Melis, 2011). Another critical aspect in nutrient supply is the carbon-to-nitrogen (C/N) ratio (Mohan et al., 2016). The presence of nitrogen-containing substrates or ammonia has a major influence on nitrogenase activity. As previously stated, nitrogen deficit promotes increased hydrogen generation when nitrogenase enzymes are involved. It is worth noting that nitrogenase enzymes are ammonium ion sensitive. When the quantity of ammonia salts in the culture medium is reduced, the efficiency of nitrogenase enzymes in the forward reaction of H_2 generation improves (Savasturk et al. 2018). When optimizing nutritional conditions for microorganisms involved in H_2 generation, not only the C/N ratio, but also the supply of nitrogen, must be considered.

7.4.3 Genetic Approaches

1. **Improving Light Absorption**: For growth and development, photosynthetic microorganisms, such as green microalgae, photosynthetic bacteria, and cyanobacteria, need an abundance of sunshine. Large light-harvesting antennas have developed in these creatures to enhance sunlight absorption even in low-light conditions (Kamshybayeva et al., 2022; Kosourov et al., 2020). These enormous antennas, however, can cause excessive light absorption in industrial H_2-generating systems requiring dense microbe cultures and bright sunshine, resulting in energy loss as heat or fluorescence. To remedy this inefficiency, the photosynthetic equipment can be genetically modified to lower antenna size. Truncating antenna size in high-density cultures of green algae and some bacteria dramatically increases H_2 generation compared to wild-type strains under similar conditions, according to research (Melis, 2012). Reduced pigment concentrations can also improve light penetration inside the culture, lowering solar energy waste.
2. **Reducing Oxygen Sensitivity**: During photosynthesis, wild-type cyanobacteria and green microalgae create oxygen, which suppresses hydrogenase and nitrogenase activities (Kosourov et al., 2021). This oxygen sensitivity makes photobiological hydrogen synthesis difficult, particularly in aerobic or open-system cultures. This issue can be solved in a number of ways, such as by adjusting the ratio of photosynthesis to respiration to produce hypoxic or anoxic conditions and by deactivating the PSII D1 reaction center protein to decrease oxygen generation and extend the production of H_2 (Wang et al., 2020).

3. **Uptake Hydrogenase Inactivation**: Cyanobacteria and photosynthetic bacteria express uptake hydrogenases, which absorb and recycle the hydrogen created during nitrogen fixation, lowering overall molecular H_2 output. Several approaches, including gene deletions and site-directed mutagenesis, have been investigated to inactivate or remove absorption hydrogenase, resulting in higher H_2 evolution (Raleiras et al., 2016).
4. **Mitigating Competitive Pathways**: Pathways competing with H_2 generation, such as nitrate absorption, respiratory electron transport chains, and carbon fixation, can limit hydrogen yield. For instance, the synthesis of H_2 and polyhydroxybutyrate (PHB) is in competition in certain bacteria. It has been shown that increasing the capacity for H_2 generation involves deleting the genes involved in PHB synthesis (Oyserman et al., 2016).
5. **Sulfur Deprivation for Hypoxic Conditions**: Genetic alteration can also diminish sulfate permease activity in chloroplasts, causing sulfur deprivation and restricting PSII D1 protein biosynthesis (Melis, 2007). This results in a hypoxic (or anoxic in sealed cultures) environment that is favorable for H_2 generation.

The lack of genome sequence information for many microbes, as well as limited transformation technologies, is one of the key obstacles in genetic engineering for enhanced H_2 generation. Exceptions, such as *Chlamydomonas reinhardtii*, which possesses a well-defined genetic system, have paved the path for considerable genetic engineering research.

7.4.4 Integrated Approaches

Integrated systems offer a viable way for leveraging the benefits and mitigating the constraints of various photobiological H_2 production processes (Table 7.1). These integrated systems are designed to reduce total energy usage, substrate needs, H_2 generation output, H_2 evolution, and waste treatment. There are two basic techniques for developing such linked systems:

1. **Integrating Photo and Dark Fermentation**: Anaerobic organisms present in organic waste products (Mohan, Babu, and Sarma, 2007b), such as anaerobic compost, sewage sludge, and garden soils, are the main agents of dark fermentation. These chemical compounds may be employed as dark fermentative H_2 generation substrates (Mohan et al., 2005). Total H_2 generation efficiency (Dahiya et al. (2020a, 2020b) and Mohan (2020) may be raised by combining non-photosynthetic dark fermentative bacteria with photo-fermentative microorganisms (Nikhil et al., 2017; Ramprakash et al., 2022). H_2, CO_2, and small organic acids are produced uring the dark fermentation process of carbohydrates (such as glucose) (Mohan et al., 2019). The biomass produced by dark fermentation is used as a growth medium for photosynthetic bacteria that perform photo-fermentation (Chandra and Mohan, 2011). In the presence of light, these photosynthetic bacteria utilize the organic acids in the biomass as substrates to produce hydrogen (Chandra et al., 2015). Combining these two processes potentially provides 12 moles of H_2 from 1 mole of hexose, with tiny organic acids acting as intermediates between the two (Dahiya et al., 2020b).
2. **Combining Different Photosynthetic Microorganisms**: Combining two photosynthetic microbial systems to boost H_2 generation efficiency is another interesting technique. One efficient combination is to include photosynthetic bacteria and green microalgae/cyanobacteria in the design of a photobioreactor. The effective absorption of visible light (400–700 nm) by microalgae is responsible for around 45% of the solar radiation energy that reaches Earth. Contrarily, anoxic photosynthetic bacteria absorb near-infrared light, which ranges in wavelength from 700 to 950 nm and makes up around 25% of the solar spectrum. These integrated systems use a wider range of the solar energy spectrum to increase the total light conversion efficiency for photobiological H_2 production (Mohanakrishna and Mohan, 2013).

Moreover, the integration of a three-component system, comprising dark fermentative bacteria along with two photosynthetic organisms, holds the potential for economical organic substrate synthesis aimed at facilitating H_2 evolution. It is essential to acknowledge, however, that these methodologies are currently in the experimental phase and have not yet been effectively implemented on a commercial scale.

7.5 TECHNICAL ASSETS AND LIABILITIES ASSOCIATED WITH PHOTOBIOLOGICAL H_2 SYNTHESIS

7.5.1 Associated Assets

Photobiological H_2 synthesis presents several notable advantages; this method harnesses abundant sunlight, making it a cost-effective alternative to electrochemical H_2 production. It leverages natural sunlight for energy-efficient H_2 generation. Photobiological technologies utilize renewable resources, relying on sunlight and water, eliminating the need for non-renewable energy sources. The process produces pure and clean H_2 while avoiding the emission of greenhouse gases, aligning with environmental sustainability goals (Dahiya et al., 2020a). An abundance of microorganisms involved, such as green algae, cyanobacteria, and photosynthetic bacteria, are readily available and adaptable to controlled cultivation. Many photosynthetic bacteria can efficiently utilize a broad spectrum of light and organic waste as substrates for H_2 production, enhancing versatility. Under anaerobic conditions, the process yields valuable byproducts like lactic acid, butyric acid, and acetic acid, adding further value. Photobiological H_2 synthesis demonstrates high photon conversion efficiency, maximizing solar energy utilization. The process contributes to carbon sequestration by converting CO_2 into biomass during the aerobic phase and generating H_2 during the anaerobic phase. A novel CO_2 sequestration technique transforms CO_2 into biomass during the aerobic phase, promoting sustainable carbon management.

7.5.2 Liabilities Associated

Despite its advantages, photobiological H_2 synthesis faces certain challenges. Current strategies suffer from low solar to hydrogen conversion efficiency ($\approx 1\%$), particularly under outdoor conditions (Ogbonna et al., 1999). Major issues are predominantly associated with light transfer limitations, mass transfer and hydrodynamic limitations, contamination of monocultures, photobioreactor cost and temperature management cost. The reliance on sunlight introduces intermittency, dependent on weather conditions and daylight hours. Excessive irradiance often leads to photooxidative damage; on the other hand, local irradiance within the photobioreactor due to microbes, media and bubbles may remain below the required levels, resulting in hindrance in productivity (Melis, 2002). So, there always remains the challenge of uniform redistribution throughout the bioreactor. Byproduct buildup, low interfacial area concentration, and sedimentation of microbes and nutrients are the issues related to the mass transfer and hydrodynamics. Oxygen-sensitive hydrogenase and nitrogenase enzymes necessitate anaerobic or hypoxic conditions for efficient H_2 production. Scaling up for commercial use remains challenging, as the reactor is scaled up. The surface area available for gas transfer per unit volume of the reactor decreases in non-sparged reactors, which may result in the buildup of byproducts hindering large-scale adoption. Achieving optimal H_2 production often requires complex and costly genetic modifications of microorganisms. The process' efficiency varies based on factors like light intensity, temperature, and nutrient availability (Chiranjeevi and Mohan 2016). The monocultures often suffer the challenges of contamination within photobioreactors, competing for light and nutrients, and adversely affecting the performance. The utilization of organic waste as a substrate may compete with other waste-to-energy or recycling processes. Photobiological H_2 synthesis offers significant environmental benefits but poses challenges related to scalability, oxygen sensitivity, and process optimization. Ongoing research and development are essential to fully realize its potential as a sustainable and clean H_2 production method.

7.6 CONCEPTUAL CIRCULAR FRAMEWORK FOR PHOTOBIOLOGICAL H_2 TECHNOLOGY

There are significant challenges in harnessing technology for various microorganism systems due to their diverse metabolism, resulting in low substrate efficiency and increased economic costs. Traditional processes suffer from inherent flaws such as poor resource and production efficiency, an inability to fully utilize slower material flows, and the necessity for high resource extraction to remain viable. In the pursuit of industrial feasibility, considerable efforts have been made to enhance productivity, improve light conversion efficiency, and maximize substrate utilization (Chandra and Mohan, 2014), as discussed earlier. Furthermore, there is ample room for incorporating a circular framework into the process to advance the technology. A circular framework for photobiological hydrogen production represents a transformative approach to harnessing the potential of microorganisms like algae, cyanobacteria, and photobacteria for H_2 generation (Figure 7.6). (Dahiya et al. (2020b). This innovative concept focuses on optimizing resource utilization, reducing waste, and enhancing overall efficiency throughout the entire hydrogen production process. These issues can be addressed by adopting a circular framework in the process. Algae, cyanobacteria, and photobacteria all have the potential for efficient H_2 production and excellent sustainability performance (Katakojwala and Mohan, 2021).

One of the central elements of this circular framework is the integration of wastewater treatment (Mohan, Mohanakrishna, et al., 2007a) and biogenic municipal waste (Sarkar et al., 2023) into the hydrogen production system along with CO_2 sequestration (Mohan et al., 2019). Microorganisms have a remarkable capacity to thrive in wastewater and organic waste (Iragavarapu et al., 2023) environments, utilizing pollutants and nutrients as substrates. By coupling hydrogen production with wastewater treatment, the process not only contributes to environmental remediation but also minimizes the need for additional nutrients and water resources. This dual-purpose approach addresses the issue of resource scarcity and water scarcity while simultaneously mitigating pollution, resulting in the production of biomass enriched with value-added products like lipids, carbohydrates, biomolecules, proteins, and more. These value-added products have diverse applications in the market, including biodiesel (Mohan et al., 2015), food supplements (Kuravi and Mohan, 2022), animal fodder (Mohan et al., 2020), bioplastics (Kumari et al., 2022; Mohan et al., 2010), carotenoids (Suresh et al., 2017), and feedstocks (Kiran and Mohan, 2022), among others. Utilizing waste streams as feedstock offers significant advantages, including reduced greenhouse gas emissions, lower production costs, and waste valorization (Akca et al., 2021). By capturing and redirecting CO_2 emissions into the photobiological hydrogen production system, this circular approach not

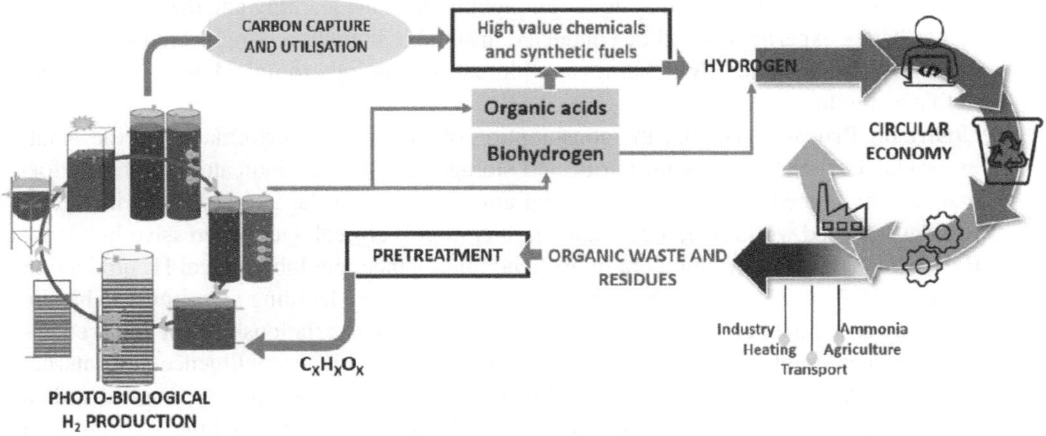

FIGURE 7.6 Conceptual circular framework of pho-biological H_2 technology.

only reduces the carbon footprint but also enhances the efficiency of the process. Establishing a sustainable process that effectively reduces, recycles, and reuses these wastes is crucial for achieving energy efficiency in a biorefinery focused on producing food, pharmaceuticals, and other value-added products (Sarkar et al., 2021). The ultimate goal is to create a zero-waste system. This approach not only advances hydrogen production technology but also contributes to a more sustainable and circular economy (Chandra et al., 2019; Mohan et al., 2016).

7.7 PROSPECTS: TOWARD SUSTAINABLE PHOTOBIOLOGICAL HYDROGEN PRODUCTION

1. **Mainstream Prospect**: Photo-biological processes depend upon the capacity of photosynthetic organisms to work as a biological catalyst for the production of H_2 from sunlight, water, and enzymes. The physiological diversity in microbial metabolism of microorganisms, each one stands with its own advantages and limitations. From the perspective of engineering, the process stands out with a potential advantage of a low-cost catalyst (biological catalyst) and minimum energy-consuming operating conditions (mesophilic) than all present processes of H_2 making in industries. Though certain biological and engineering challenges are the bottlenecks for this promising technology to become a practical reality. Cellular metabolism, with basic bio-chemistry of the process, needs to be understood well, along with fundamental research. Increased investment along with government support, integrated efforts of engineers, biologists, and environmentalists to build a skilled workforce is prudential. Current research efforts focus on optimizing operating parameters, selecting and enriching biocatalysts within bioreactor ecosystems. Advances are directed at identifying robust microorganisms adaptable to various substrates, emphasizing economic and environmental sustainability (Vinayak et al., 2023) by employing pretreatment strategies (Mohan, 2008). Operating conditions such as substrate availability (Cheng et al., 2011), hydraulic retention time (Babu et al., 2009), pH, and intermediate metabolites have a significant impact on process efficiency. There is no one-size-fits-all bioreactor design; the ideal design is determined by the microorganism species and the desired products. Recent studies have incorporated innovative techniques involving membranes, nanomaterials, bio-nanotechnology, and physics. Integrating and upgrading bioreactor design holds promise for improving operating economics, scalability, and sophistication. Precise genomic analysis and genetic engineering are key to developing strains resistant to oxygen sensitivity, a major hindrance during bio-photolysis and photo-fermentation. These engineered strains should be tested within efficient biosystems, incorporating nutritional regulation (C:N) under hybrid systems, employing machine learning and design of experiments. These experiments will be conducted in suitable bioreactors engineered with nanomaterials and metal additives, harnessing bio-nanotechnology to unlock the full potential of this technology.
2. **Upstream Prospect**: Herein, the consumption of the energy, performance of the plant, its production rate, along with purity and storage, are the key indicators of the performance, which needs to be transparent and attractive to gain the interest of the investors. Besides a skilled workforce, technology acceleration is critical. Our progressive Industrial Revolution 4.0 is an evolution era of artificial intelligence; photobiological H_2 production technology also needs to couple its hands with the machine learning algorithms to handle the multivariate data complexity, to anticipate the nonlinear relationships for precise control over the complex dynamics (Pandey et al., 2022). Artificial intelligence and Internet of Things are needed to optimize the automation of complete systems via intensified data administration and analysis (Pandey et al., 2022). Recent use of machine learning as a tool for biological mass conversion processes has quietly been increased, but it is still not methodically investigated in hydrogen technologies, especially for photobiological routes.

Support vector machine, artificial neural network, and Random Forest are some of the machine learning algorithms that should be explored more intensively as a tool to model the nonlinear complex relationships of the parameters to develop the technology to its best possibilities (Pandey et al., 2022). These machine algorithms could calculate and depict the results at initial stages, saving time and expenses by utilizing digital twins and multiple model scenarios (Sharma et al., 2022). Herein, artificial intelligence provides the platform for Internet of Things (IoT) that could expedite the revolution of green hydrogen (Sharma et al., 2022). IoT could process real-time monitoring and amplify control over key performance indicators and can visualize the data. IoT can also establish the platform for bio foundry in hydrogen technology by providing the interconnections between the researchers for collaborative and updated research (Centurion Plus, 2021).

3. **Downstream Prospect**: There is a rising concern, particularly regarding the sustainability of the biomass-to-energy conversion. Life cycle assessment (LCA) is a quantitative methodology that assesses the utilization of resources and the potential environmental impacts of any product, encompassing aspects like global warming, eutrophication, acidification, and threats to human health, among others. This evaluation process takes into account the entire life cycle of a product, starting from the procurement of raw materials through production, utilization, and eventual disposal or recycling. The LCA sustainability comparison for biomass gasification and conventional processes of hydrogen as a sustainable substitute showed the concerns about the commercial and social aspects (Valente et al., 2019). Romagnoli et al. (2011) introduced an initial endeavor to establish a scientifically-grounded, decision-oriented framework aimed at evaluating and pinpointing viable biohydrogen production routes utilizing LCA. This approach encompassed a wide array of sustainability factors, including considerations of availability, accessibility, and acceptability. The analysis outcomes indicate that employing biohydrogen for electricity generation yields superior environmental advantages when compared to relying on fossil fuel sources. The analysis specifically quantified the reduction in CO_2 emissions achieved by switching to a cycling photobiological hydrogen production method using green algae (*C. reinhardtii*) under forced sulfur deprivation conditions. This reduction can be maximally estimated at approximately 25.5 metric tons of CO_2 per year when coal is replaced as the primary energy source for electricity generation (Romagnoli et al., 2011). The further work should aim to deliver a comprehensive LCA encompassing all aspects of the industrial-scale photobiological hydrogen production processes.

Therefore, industrial scale-up needs further standardization and optimization for a comprehensive environmental implementation to be an effective share of energy mix in the coming years. Moreover, with the exploration of the full potential of green hydrogen, there is a need to realize the circular economy of hydrogen. The organizational boundaries of multiple domains need to be crossed to support the circular framework of hydrogen, and the key is to share data effectively and securely. The circular economy could provide an additional edge to the technology over the economics. Leveraging models of analysis could provide closed-loop design for the investors to take tactical and strategic decisions to optimize the returns from the technology.

7.8 CONCLUSION

In this chapter, we looked at three different photobiological H_2 production processes, each with its own set of advantages and disadvantages on the road to large-scale deployment. As we conclude, it is clear that these techniques have considerable potential for long-term H_2 production, but they also face unique feasibility challenges. Photo-fermentation processes include the recycling of biomass; these processes provide an admirable level of environmental sustainability. However, they face constraints like low sunlight-to-H_2 conversion efficiency and low yields. Future developments must prioritize efficiency and yield while addressing issues such as light and oxygen sensitivity. Modulating redox potentials in electron transfer systems and developing multicomponent-integrated systems are two promising strategies. Concurrently, optimization of bioreactor materials, designs, and biofouling prevention are critical factors that must be addressed. While bio-photolysis requires less energy than photoelectrochemical processes, it currently confronts challenges in producing higher H_2 yields, which often dip below 15%. The focus of research is on overcoming scientific hurdles such as sensitivity to oxygen, light, and competing routes. The search for higher light-to-hydrogen conversion efficiency is crucial. Advances in this subject have the potential to transform sustainable H_2 generation, potentially competing with biodiesel and natural gas-derived H_2. Finally, photobiological H_2 production offers a possible path toward greener, more sustainable energy options. Overcoming the aforementioned obstacles is critical to reaching the full potential of these strategies. Photobiological H_2 production may emerge as a compelling alternative in our pursuit of a cleaner, H_2-powered future as our scientific understanding and engineering skills progress.

ACKNOWLEDGMENT

The Department of Science and Technology (DST), Government of India, funded the research with a grant (DST/TMD-EWO/AHFC-2021/2021/32). PK acknowledges the Ministry of Social Justice & Empowerment for providing the UGC-NFSC fellowship. The authors wish to thank the Director, CSIR-IICT, for support and encouragement to carry out this work (Manuscript No: IICT/Pubs./2023/346).

REFERENCES

Ahmed, Shams Forruque, Nazifa Rafa, M. Mofijur, Irfan Anjum Badruddin, Abrar Inayat, MdSawkat Ali, Omar Farrok, and T. M. Yunus Khan. "Biohydrogen production from biomass sources: metabolic pathways and economic analysis." *Frontiers in Energy Research* 9 (2021): 753878.

Akca, Mehmet Sadik, Okan Bostancı, Aslı Kirectepe Aydin, Ismail Koyuncu, and Mahmut Altinbas. "BioH2 production from food waste by anaerobic membrane bioreactor." *International Journal of Hydrogen Energy* 46, no. 55 (2021): 27941–27955.

Allen, John F. "Origin of oxygenic photosynthesis from anoxygenic type I and Type II reaction centers." In *The Biophysics of Photosynthesis*, pp. 433–450. Springer New York, 2014.

Babu, V. Lalit, S. Venkata Mohan, and P. N. Sarma. "Influence of reactor configuration on fermentative hydrogen production during wastewater treatment." *International Journal of Hydrogen Energy* 34, no. 8 (2009): 3305–3312.

Baeyens, Jan, Huili Zhang, Jiapei Nie, Lise Appels, Raf Dewil, Renaud Ansart, and Yimin Deng. "Reviewing the potential of bio-hydrogen production by fermentation." *Renewable and Sustainable Energy Reviews* 131 (2020): 110023.

Biswal, Trinath, Krushna Prasad Shadangi, and Prakash Kumar Sarangi. "Application of nanotechnology in the production of biohydrogen: a review." *Chemical Engineering & Technology* 46, no. 2 (2023): 218–233.

Blankenship, Robert E. "Origin and early evolution of photosynthesis." *Photosynthesis Research* 33 (1992): 91–111.

Blankenship, Robert E. *Molecular Mechanisms of Photosynthesis*. John Wiley & Sons, 2021.

Boichenko, A., Elisas Greenbaum, and Michael Seibert. "Hydrogen production by photosynthetic microorganisms." *Photoconversion of Solar Energy, Molecular to Global Photosynthesis* 2 (2004): 397–452.

Brown, Katherine A., Derek F. Harris, Molly B. Wilker, Andrew Rasmussen, Nimesh Khadka, Hayden Hamby, Stephen Keable et al. "Light-driven dinitrogen reduction catalyzed by a CdS: nitrogenase MoFe protein biohybrid." *Science* 352, no. 6284 (2016): 448–450.

Centurion Plus. *Growing The Green Hydrogen Economy: What AI, IoT and ML Can Deliver*, 2021. https://centurionlgplus.com/growing-the-green-hydrogen-economy-what-ai-iot-and-ml-can-deliver/

Chandra, Rashmi, Hafiz M.N. Iqbal, Garima Vishal, Hyung-Sool Lee, and Sunil Nagra. "Algal biorefinery: a sustainable approach to valorize algal-based biomass towards multiple product recovery." *Bioresource Technology* 278 (2019): 346–359.

Chandra, Rashmi, G. N. Nikhil, and S. Venkata Mohan. "Single-stage operation of hybrid dark-photo fermentation to enhance biohydrogen production through regulation of system redox condition: evaluation with real-field wastewater." *International Journal of Molecular Sciences* 16, no. 5 (2015): 9540–9556.

Chandra, Rashmi, and S. Venkata Mohan. "Microalgal community and their growth conditions influence biohydrogen production during integration of dark-fermentation and photo-fermentation processes." *International Journal of Hydrogen Energy* 36, no. 19 (2011): 12211–12219.

Chandra, Rashmi, and S. Venkata Mohan. "Enhanced bio-hydrogenesis by co-culturing photosynthetic bacteria with acidogenic process: Augmented dark-photo fermentative hybrid system to regulate volatile fatty acid inhibition." *International Journal of Hydrogen Energy* 39, no. 14 (2014): 7604–7615.

Cheng, J., H. Su, J. Zhou, W. Song, and K. Cen. "Hydrogen production by mixed bacteria through dark and photo fermentation." *International Journal of Hydrogen Energy* 36, no. 1 (2011), 450–457.

Chiranjeevi, P., and S. Venkata Mohan. "Critical parametric influence on microalgae cultivation towards maximizing biomass growth with simultaneous lipid productivity." *Renewable Energy* 98 (2016): 64–71.

Dahiya, Shikha, Sulogna Chatterjee, Omprakash Sarkar, and S. Venkata Mohan. "Renewable hydrogen production by dark-fermentation: Current status, challenges and perspectives." *Bioresource Technology* 321 (2021): 124354.

Dahiya, Shikha, Ranaprathap Katakojwala, Seeram Ramakrishna, and S. Venkata Mohan. "Biobased products and life cycle assessment in the context of circular economy and sustainability." *Materials Circular Economy* 2 (2020b): 1–28.

Dahiya, Shikha, Sushmitha Lakshminarayanan, and S. Venkata Mohan. "Steering acidogenesis towards selective propionic acid production using co-factors and evaluating environmental sustainability." *Chemical Engineering Journal* 379 (2020a): 122135.

Dahiya, Shikha, Yaswanth Lingam, and S. Venkata Mohan. "Understanding acidogenesis towards green hydrogen and volatile fatty acid production–critical analysis and circular economy perspective." *Chemical Engineering Journal* 464 (2023): 141550.

Dalena, Francesco, Alessandro Senatore, Antonio Tursi and Angelo Basile. "Bioenergy production from second- and third-generation feedstocks." In *Bioenergy Systems for the Future*, edited by Raffaele Barretta, Angelo Basile, T. Nejat Veziroğlu, and Fabrizio Dalena, pp. 559–599. Woodhead Publishing, 2017.

Einsle, Oliver, and Douglas C. Rees. "Structural enzymology of nitrogenase enzymes." *Chemical Reviews* 120, no. 12 (2020): 4969–5004.

Eroglu, Ela, and Anastasios Melis. "Photobiological hydrogen production: recent advances and state of the art." *Bioresource Technology* 102, no. 18 (2011): 8403–8413.

Giri, Deen Dayal, Himanshu Dwivedi, Abdulmohsen Khalaf D. Alsukaibi, Dan Bahadur Pal, Ahmed Al Otaibi, Mohammed Y. Areeshi, Shafiul Haque, and Vijai Kumar Gupta. "Sustainable production of algae-bacteria granular consortia based biological hydrogen: New insights." *Bioresource Technology* 352 (2022): 127036.

Iragavarapu, Gayathri Priya, Syed Shahed Imam, Omprakash Sarkar, Srinivasula Venkata Mohan, Young-Cheol Chang, Motakatla Venkateswar Reddy, Sang-Hyoun Kim, and Naresh Kumar Amradi. "Bioprocessing of waste for renewable chemicals and fuels to promote bioeconomy." *Energies* 16, no. 9 (2023): 3873.

Kamshybayeva, Gulzhanay K., Bekzhan D. Kossalbayev, Asemgul K. Sadvakasova, Bolatkhan K. Zayadan, Ayshat M. Bozieva, Dmitry Dunikov, Saleh Alwasel, and Suleyman I. Allakhverdiev. "Strategies and economic feasibilities in cyanobacterial hydrogen production." *International Journal of Hydrogen Energy* 47, no. 69 (2022): 29661–29684.

Katakojwala, Ranaprathap, and S. Venkata Mohan. "A critical view on the environmental sustainability of biorefinery systems." *Current Opinion in Green and Sustainable Chemistry* 27 (2021): 100392.

Kiran, Boda Ravi, and S. Venkata Mohan. "Phycoremediation potential of Tetradesmus sp. SVMIICT4 in treating dairy wastewater using Flat-Panel photobioreactor." *Bioresource Technology* 345 (2022): 126446.

Kose, Ayse, and Suphi S. Oncel. "Biohydrogen production." In A. Singh & D. Rathore (Eds.), *Sustainability of Current Technology and Future Perspective* (p. 181), 2016, Springer.

Kosourov, Sergey, Maximilian Böhm, Moritz Senger, Gustav Berggren, Karin Stensjö, Fikret Mamedov, Peter Lindblad, and Yagut Allahverdiyeva. "Photosynthetic hydrogen production: Novel protocols, promising engineering approaches and application of semi-synthetic hydrogenases." *Physiologia Plantarum* 173, no. 2 (2021): 555–567.

Kosourov, Sergey, Valéria Nagy, Dmitry Shevela, Martina Jokel, Johannes Messinger, and Yagut Allahverdiyeva. "Water oxidation by photosystem II is the primary source of electrons for sustained H2 photoproduction in nutrient-replete green algae." *Proceedings of the National Academy of Sciences* 117, no. 47 (2020): 29629–29636.

Kumari, Poonam, Boda Ravi Kiran, and S. Venkata Mohan. "Polyhydroxybutyrate production by Chlorella sorokiniana SVMIICT8 under nutrient-deprived mixotrophy." *Bioresource Technology* 354 (2022): 127135.

Kuravi, Sri Divya, and S. Venkata Mohan. "Mixotrophic cultivation of Monoraphidium sp. In dairy wastewater using Flat-Panel photobioreactor and photosynthetic performance." *Bioresource Technology* 348 (2022): 126671.

Lau, Nyok-Sean, Minami Matsui, and Amirul Al-Ashraf Abdullah. "Cyanobacteria: photoautotrophic microbial factories for the sustainable synthesis of industrial products." *BioMed Research International* 2015 (2015): 754934.

Li, Shengnan, Fanghua Li, Xun Zhu, Qiang Liao, Jo-Shu Chang, and Shih-Hsin Ho. "Biohydrogen production from microalgae for environmental sustainability." *Chemosphere* 291 (2022): 132717.

Madigan, Michael T., John M. Martinko, and Jack Parker. *Brock biology of microorganisms*. Vol. 11. Prentice Hall, 1997.

Melis, Anastasios. "Green alga hydrogen production: progress, challenges and prospects." *International Journal of Hydrogen Energy* 27, no. 11 (2002): 1217–1228.

Melis, Anastasios. "Photosynthetic H2 metabolism in *Chlamydomonas reinhardtii* (unicellular green algae)." *Planta* 226, no. 5 (2007): 1075–1086.

Melis, Anastasios. "Photosynthesis-to-fuels: from sunlight to hydrogen, isoprene, and botryococcene production." *Energy & Environmental Science* 5, no. 2 (2012): 5531–5539.

Mohan, S. Venkata. "Fermentative hydrogen production with simultaneous wastewater treatment: influence of pretreatment and system operating conditions." *Journal of Scientific and Industrial Research 67*, no. 11 (2008): 950.

Mohan, S. Venkata, V. Lalit Babu, and P. N. Sarma. "Anaerobic biohydrogen production from dairy wastewater treatment in sequencing batch reactor (AnSBR): effect of organic loading rate." *Enzyme and Microbial Technology* 41, no. 4 (2007b): 506–515.

Mohan, S. Venkata, P. Chiranjeevi, K. Chandrasekhar, P. Suresh Babu, and Omprakash Sarkar. "Acidogenic biohydrogen production from wastewater." In *Biohydrogen*, pp. 279–320. Elsevier, 2019.

Mohan, S. Venkata, Manupati Hemalatha, Debkumar Chakraborty, Sulogna Chatterjee, Palle Ranadheer, and Rajesh Kona. "Algal biorefinery models with self-sustainable closed loop approach: Trends and prospective for blue-bioeconomy." *Bioresource Technology* 295 (2020): 122128.

Mohan, S. Venkata, G. Mohanakrishna, S. Veer Raghavulu, and P. N. Sarma. "Enhancing biohydrogen production from chemical wastewater treatment in anaerobic sequencing batch biofilm reactor (AnSBBR) by bioaugmenting with selectively enriched kanamycin resistant anaerobic mixed consortia." *International Journal of Hydrogen Energy* 32, no. 15 (2007a): 3284–3292.

Mohan, S. Venkata, G. N. Nikhil, P. Chiranjeevi, C. Nagendranatha Reddy, M. V. Rohit, A. Naresh Kumar, and Omprakash Sarkar. "Waste biorefinery models towards sustainable circular bioeconomy: critical review and future perspectives." *Bioresource Technology* 215 (2016): 2–12.

Mohan, S. Venkata, N. Chandrashekara Rao, K. Krishna Prasad, B. T. V. Madhavi, and P. N. Sharma. "Treatment of complex chemical wastewater in a sequencing batch reactor (SBR) with an aerobic suspended growth configuration." *Process Biochemistry* 40, no. 5 (2005): 1501–1508.

Mohan, S. Venkata, M. Venkateswar Reddy, G. Venkata Subhash, and P. N. Sarma. "Fermentative effluents from hydrogen producing bioreactor as substrate for poly (β-OH) butyrate production with simultaneous treatment: An integrated approach." *Bioresource Technology* 101, no. 23 (2010): 9382–9386.

Mohan, S. Venkata, M. V. Rohit, P. Chiranjeevi, Rashmi Chandra, and B. Navaneeth. "Heterotrophic microalgae cultivation to synergize biodiesel production with waste remediation: progress and perspectives." *Bioresource Technology* 184 (2015): 169–178.

Mohanakrishna, G., and S. Venkata Mohan. "Multiple process integrations for broad perspective analysis of fermentative H2 production from wastewater treatment: technical and environmental considerations." *Applied Energy* 107 (2013): 244–254.

Mosebach, Laura, Claudia Heilmann, Risa Mutoh, Philipp Gäbelein, Janina Steinbeck, Thomas Happe, Takahisa Ikegami, Guy Hanke, Genji Kurisu, and Michael Hippler. "Association of Ferredoxin: NADP+ oxidoreductase with the photosynthetic apparatus modulates electron transfer in *Chlamydomonas reinhardtii*." *Photosynthesis Research* 134 (2017): 291–306.

Mukherjee, Nandini, and Rohit Srivastava. "Photobiological hydrogen production: Introduction and fundamental concept." In *Solar-Driven Green Hydrogen Generation and Storage*, pp. 193–222. Elsevier, 2023.

Mulder, David W., Eric M. Shepard, Jonathan E. Meuser, Neelambari Joshi, Paul W. King, Matthew C. Posewitz, Joan B. Broderick, and John W. Peters. "Insights into [FeFe]-hydrogenase structure, mechanism, and maturation." *Structure* 19, no. 8 (2011): 1038–1052.

Nagy, Valéria, André Vidal-Meireles, Anna Podmaniczki, Klára Szentmihályi, Gábor Rákhely, Laura Zsigmond, László Kovács, and Szilvia Z. Tóth. "The mechanism of photosystem-II inactivation during sulphur deprivation-induced H2 production in *Chlamydomonas reinhardtii*." *The Plant Journal* 94, no. 3 (2018): 548–561.

Nikhil, G. N., Omprakash Sarkar, and S. Venkata Mohan. "Biohydrogen production: an outlook of fermentative processes and integration strategies." In *Optimization and Applicability of Bioprocesses*, pp. 249–265. Springer Singapore, 2017.

Oey, Melanie, Anne Sawyer, Ian L. Ross and Ben Hankamer. "Challenges and opportunities for hydrogen production from microalgae." *Plant Biotechnology Journal* 14 (2016): 1487–1499.

Ogbonna, James C., Toshihiko Soejima, and Hideo Tanaka. "An integrated solar and artificial light system for internal illumination of photobioreactors." In *Progress in Industrial Microbiology*, vol. 35, pp. 289–297. Elsevier, 1999.

Oh, You-Kwan, Subramanian Mohan Raj, Gyoo Yeol Jung, and Sunghoon Park. "Metabolic engineering of microorganisms for biohydrogen production." In *Biohydrogen*, pp. 45–65. Elsevier, 2013.

Oncel, Suphi S., Ayse Kose, and Cecilia Faraloni. "Genetic optimization of microalgae for biohydrogen production." In *Handbook of Marine Microalgae*, pp. 383–404. Academic Press, 2015.

Oyserman, Ben O., Daniel R. Noguera, Tijana Glavina del Rio, Susannah G. Tringe, and Katherine D. McMahon. "Metatranscriptomic insights on gene expression and regulatory controls in *Candidatus Accumuli bacterphosphatis*." *The ISME Journal* 10, no. 4 (2016): 810–822.

Pandey, Ashok, S. Venkata Mohan, Jo-Shu Chang, Patrick C. Hallenbeck, and Christian Larroche, eds. *Biomass, Biofuels, Biochemicals: Biohydrogen*. Elsevier, 2019.

Pandey, Ashutosh Kumar, Jungsu Park, Jeun Ko, Hwan-Hong Joo, Tirath Raj, Lalit Kumar Singh, Noopur Singh, and Sang-Hyoun Kim. "Machine learning in fermentative biohydrogen production: Advantages, challenges, and applications." *Bioresource Technology* 370 (2022): 128502.

Pathy, Abhijeet, Krishnamoorthy Nageshwari, Ramesh Prabu Ramaraj, Gaanty Pragas Maniam, Natanamurugaraj Govindan, and Paramasivan Balasubramanian. "Biohydrogen production using algae: Potentiality, economics and challenges." *Bioresource Technology* 360 (2022): 127514.

Pilon, Laurent, and Halil Berberoğlu. *Photobiological Hydrogen Production*. CRC Press, 2014.

Raleiras, Patrícia, Namita Khanna, Hélder Miranda, Lívia S. Mészáros, Henning Krassen, Felix Ho, Natalia Battchikova et al. "Turning around the electron flow in an uptake hydrogenase. EPR spectroscopy and in vivo activity of a designed mutant in HupSL from *Nostoc punctiforme*." *Energy & Environmental Science* 9, no. 2 (2016): 581–594.

Ramprakash, Balasubramani, Peter Lindblad, Julian J. Eaton-Rye, and Aran Incharoensakdi. "Current strategies and future perspectives in biological hydrogen production: A review." *Renewable and Sustainable Energy Reviews* 168 (2022): 112773.

Romagnoli, Francesco, Dagnija Blumberga, and Iluta Pilicka. "Life cycle assessment of biohydrogen production in photosynthetic processes." *International Journal of Hydrogen Energy* 36, no. 13 (2011): 7866–7871.

Sarkar, O., J. Santosh, S.V. Mohan, Y.C. Chang. Waste-derived biohydrogen enriched CNG/biohythane: Research trend and utilities. In *Biofuels*, pp. 333–353. CRC Press, 2023 April 18.

Sarkar, Omprakash, Ranaprathap Katakojwala, and S. Venkata Mohan. "Low carbon hydrogen production from a waste-based biorefinery system and environmental sustainability assessment." *Green Chemistry* 23, no. 1 (2021): 561–574.

Savasturk, Dilan, Emine Kayahan, and Harun Koku. "Photofermentative hydrogen production from molasses: scale-up and outdoor operation at low carbon-to-nitrogen ratio." *International Journal of Hydrogen Energy* 43, no. 26 (2018): 11676–11687.

Schlegel, Hans Günter, and Joseph Barnea, eds. *Microbial Energy Conversion: The Proceedings of a Seminar Sponsored by the UN Institute for Training and Research (UNITAR) and the Ministry for Research and Technology of the Federal Republic of Germany Held in Göttingen, October 1976*. Elsevier, 2013.

Sharma, Amit Kumar, Praveen Kumar Ghodke, Nishu Goyal, S. Nethaji, and Wei-Hsin Chen. "Machine learning technology in biohydrogen production from agriculture waste: Recent advances and future perspectives." *Bioresource Technology* 364 (2022): 128076.

S. A. Sherif, D. Y. Goswami, E. K. L. Stefanakos, A. Steinfeld. *Handbook of Hydrogen Energy*. CRC Press. 2014

Srivastava, Neha, Manish Srivastava, P.K. Mishra, Mohd Adnan Kausar, Mohd Saeed, Vijai K. Gupta, Rajeev Singh, and P.W. Ramteke. "Advances in nanomaterials induced biohydrogen production using waste biomass." *Bioresource Technology* 307 (2020): 123094.

Suresh, G., Poonam Kumari, and S. Venkata Mohan. "Light-dependent biohydrogen production: Progress and perspectives." *Bioresource Technology* 380 (2023): 129007.

Suresh, P. V., Kudre, T. G., & Johny, L. C. (2017). Sustainable Valorization of Seafood Processing By-Product/Discard. In R. R. Singhania, R. A. Agarwal, R. P. Kumar, & R. K. Sukumaran (Eds.), *Waste to Wealth* (pp. 111–139). Springer Singapore

Valente, Antonio, Diego Iribarren, José-Luis Gálvez-Martos, and Javier Dufour. "Robust eco-efficiency assessment of hydrogen from biomass gasification as an alternative to conventional hydrogen: A life-cycle study with and without external costs." *Science of the Total Environment* 650 (2019): 1465–1475.

Verne, Jules. *The Mysterious Island*. Penguin, 2004.

Vignais, Paulette M., and Bernard Billoud. Occurrence, classification, and biological function of hydrogenases: an overview. *Chemical Reviews* 107, no. 10 (2007 October): 4206–4272. doi: 10.1021/cr050196r PMID: 17927159.

Vinayak, Vandana, Vandana Sirotiya, Priyanka Khandelwal, Anshuman Rai, Dipak A. Jadhav, Arivalagan Pugazhendhi, Benoit Schoefs, Justine Marchand, and Kyu-Jung Chae. "Recent trends in engineering algae for biohydrogen production: State of art strategies." *Fuel* 348 (2023): 128636.

Wang, Yue, Huanling Yang, Xinna Zhang, Fei Han, Wenfeng Tu, and Wenqiang Yang. "Microalgal hydrogen production." *Small Methods* 4, no. 3 (2020): 1900514.

8 Sustainable Processing of Biomass as an Eco-Friendly Biosorbent

A Green Material for Biosorption of Dyes from Wastewater

Kasturi Bidkar, Akshay Prakash, and Jerold Manuel

8.1 INTRODUCTION

The use of water has significantly increased with the rapid growth of industrial activities in both developed and developing countries. This increase has led to the deterioration of the aquatic ecosystem due to the release of various pollutants into water bodies, including dyes, toxic metals, organic compounds such as pesticides, phenolic compounds, detergents, coloring agents, humic substances, and other persistent organic substances. Among these pollutants, dyes are a significant contributor to water pollution as they have the ability to chemically interact with the substrate they are used in, leading to heavy contamination of water bodies.

The effluents from the textile or dye-using industries are highly complex in nature compared to other wastewaters due to the existence of diverse dye varieties, derivatives, and additives that vary with seasons, making it challenging to determine an effective and viable treatment method. Industries currently use treatment processes developed for other types of contaminants to remediate complex dye-containing effluents, resulting in poor efficiency in dye remediation (Geetha and Belagali, n.d.). Therefore, more efforts are needed to develop remediation technologies specifically for each dye class. The treatment techniques available to remove dyes from wastewaters are generally categorized under chemical, biological, and physical methods.

Several physical methods have been used in wastewater treatment for the separation of synthetic dyes, including ion-exchange, irradiation, adsorption, electrokinetic coagulation, and membrane filtration (Figure 8.1). Among these methods, adsorption is a widely used method for separating dyes from solution in both laboratory and industrial scales. Adsorption is a rapid process where soluble dyes are transferred from a contaminated solution to the surface of a highly porous solid material (sorbent), producing high-quality water (Bhattacharyya & Sharma, 2004). Adsorption is a surface phenomenon that describes the enhancement in the concentration of a particular solute at the interface between two phases. The extent of adsorption is determined by the surface chemistry of the sorbent and the properties of the dye. Adsorbent materials can be classified as inorganic or organic. Inorganic adsorbents include silica gel, fly ash, clays, zeolites, and metal oxides, while organic adsorbents include woodchips and peats, among others. Activated carbon is a popular adsorbent widely used in separating dye molecules from wastewater, but its high cost of preparation and problems during rejuvenation hinder its use in real situations.

Chemical techniques used to remediate dye wastewaters include oxidation (H_2O_2 and sodium hypochlorite), ozonation, and photochemical methods. In recent years, the biosorption process has been widely studied for dye removal. Biosorption is a phenomenon that describes the passive uptake

DOI: 10.1201/9781003732907-8

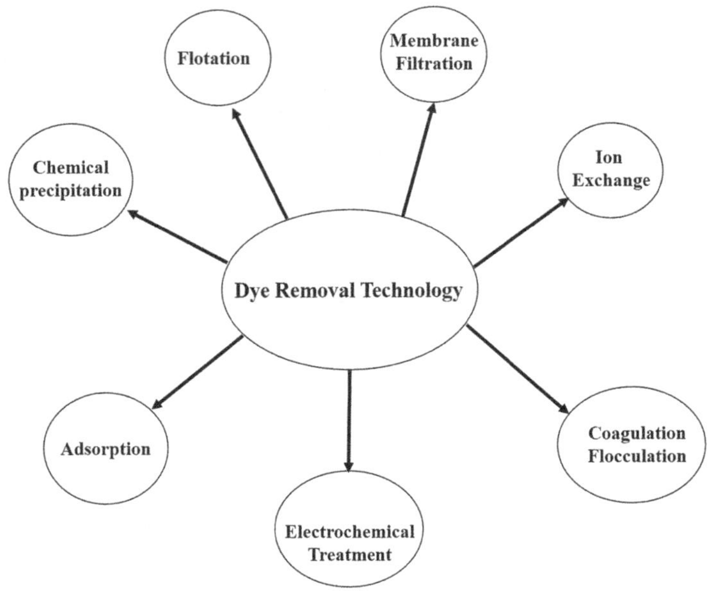

FIGURE 8.1 Dye removal technologies.

of toxic pollutants by dead or inactive biological matter or biological material derivatives. The mechanisms for contaminant remediation by biosorption usually involve physical adsorption, complexation, ion exchange, micro-precipitation, and chelation processes. Biosorbents for dyes are generally derived from fungi, bacteria, agricultural wastes, industrial wastes, algae, and other polysaccharide sorbents. The performance of these biosorbents varies with the type of dye, co-ions, and experimental parameters, including dye concentration, temperature, and solution pH. Many researchers have investigated the application of low-cost biosorbents for the bioremediation of dyes, such as Malachite green, Rhodamine B, Acid blue 25, Brilliant green, Methylene blue, Methyl violet 2B, Crystal violet (Ali etal., 2014) and Congo red (Faria et al., 2004; Osagie et al., 2021).

In conclusion, the growing release of dyes into water bodies poses a significant threat to the environment and human health. To mitigate this issue, the use of biomass-derived biosorbents has emerged as a promising approach for the bioremediation of dye-contaminated wastewater. The following sections will provide an overview of the latest developments in this field, including the characterization of different biomass-derived biosorbents and their performance in removing various dyes from contaminated water. Additionally, the challenges and opportunities associated with the use of biomass-derived biosorbents for dye removal will also be discussed.

8.2 BIOSORPTION

Biosorption is the passive uptake of toxic pollutants by biological materials or their derivatives. In the context of dye remediation, it involves the use of biomass-derived biosorbents for the removal of dyes from wastewater. These biosorbents can be derived from various sources such as fungi, bacteria, agricultural wastes, industrial wastes, and algae, among others. The performance of these biosorbents is influenced by several factors, such as the type of dye, co-ions, and experimental parameters, including dye concentration, temperature, and solution pH (Papirio et al., 2017).

Numerous studies have been conducted to investigate the potential of biomass-derived biosorbents for the bioremediation of different types of dyes. One such biosorbent that has received significant attention is biochar. Biochar is a highly porous carbon-rich material produced from biomass through a process called pyrolysis. (Chen etal., 2016). It has been shown to have excellent adsorption properties and can be used to remove a wide range of dyes from wastewater (Ngah & Hanafiah, 2008).

Research investigations have also been performed utilizing different adsorbents for the remediation of real effluents bearing dye particles. The development of efficient and cost-effective biosorbents for dye remediation is crucial in mitigating the negative impacts of dye pollution on the environment.

8.2.1 Mechanism of Biosorption

Biosorption is a widely researched technology for the removal of pollutants from wastewater. This process involves the use of biological materials as sorbents to remove pollutants from water. Physical and chemical interactions occur between the sorbent and pollutant to enable the binding of pollutants on the surface of the sorbent. Various factors such as pH, temperature, concentration, and contact time influence the biosorption process, and optimization of these factors can improve the efficiency of pollutant removal. The use of biosorbents has shown promising results for the removal of heavy metals, dyes, and other organic compounds from wastewater (Lata, n.d.).

On the surface of an adsorbent, adsorption is a technique for separating some of the components from the aqueous phase (Figure 8.2). The adsorbent has both internal and exterior surfaces, but the internal surface is more significant than the external surface because it has a greater impact on the adsorption process. Pollutants are spread out on the adsorbent's exterior surface during the first stage of adsorption. If given enough time in the second step, adsorbed molecules diffuse internally, penetrating the adsorbent's depth and moving contaminated mass to the inner surfaces. The adsorption procedure then proceeds until a layer is formed on the surface of the active pores, at which point it is complete. Adsorption is a cycle that repeats itself until balance is reached. When removing contaminants with

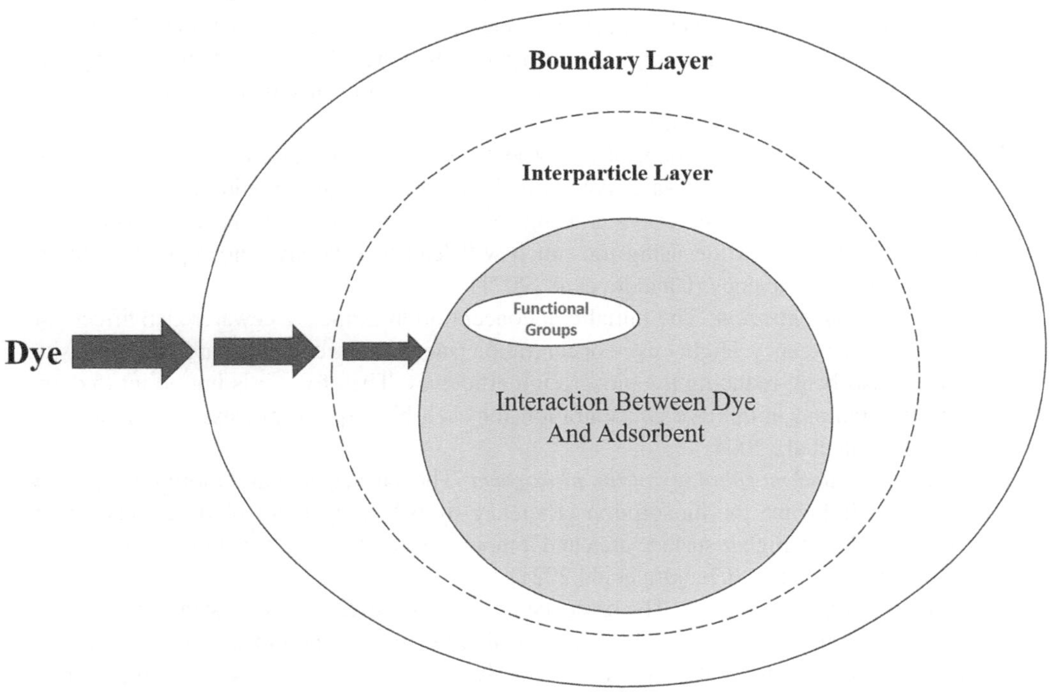

FIGURE 8.2 Biosorption mechanism.

adsorbents, one of the two primary adsorption mechanisms—physical or chemical—is employed. Target molecules bind to pores in the walls of high-level adsorbents through physical adsorption using van der Waals forces, which have a minimal adsorption heat. In chemical adsorption, the target pollutant is subjected to a chemical covalent reaction that attaches to specific sites in the adsorbent; its heat is significantly greater than adsorption and is roughly equivalent to the heat of reaction (Nayeri & Mousavi, 2020).

Several studies have been conducted to understand the mechanism of biosorption and optimize the process for efficient pollutant removal. Dhillon et al. (2013) highlighted the biosorption mechanism of heavy metals and dyes using low-cost adsorbents. Gupta and Suhas (2009) provided an extensive review of the application of low-cost adsorbents for dye removal. Islam et al. (2018) discussed the mechanism, methods, and applications of biosorption in their review article.

Biosorption is a promising technology that offers various advantages, including low cost, high efficiency, and environmental friendliness. However, the optimization of several factors is necessary to achieve effective biosorption of pollutants from wastewater.

8.2.2 Factors Affecting Biosorption

1. *pH*: The pH of the solution plays a crucial role in the biosorption of dyes. The optimal pH for biosorption depends on the type of dye and the biosorbent used. Generally, a pH range of 2–9 is suitable for most biosorption processes, and pH values outside this range may lead to a decrease in biosorption efficiency (Chandra et al., 2021).
2. *Temperature*: Temperature affects the rate of biosorption by altering the physical and chemical properties of the dye molecules and the biosorbent. Generally, higher temperatures increase the rate of biosorption due to increased molecular motion and better accessibility of the binding sites. However, excessively high temperatures can also denature the biosorbent and reduce the biosorption efficiency (Chandra et al., 2021).
3. *Contact time*: Contact time refers to the duration of time that the biosorbent is in contact with the dye solution. Longer contact times increase the chances of dye molecules binding to the biosorbent, leading to higher biosorption efficiency. However, there is an optimal contact time beyond which further contact may not significantly increase the biosorption efficiency (Chandra et al., 2021).
4. *Biosorbent dosage*: The amount of biosorbent used for biosorption affects the amount of dye removed from the wastewater. An optimal biosorbent dosage should be used to achieve maximum biosorption efficiency, as using too little biosorbent will result in incomplete removal of the dye, while using too much will lead to excessive biosorption, reducing the biosorption efficiency (Chandra et al., 2021).
5. *Initial dye concentration*: The initial dye concentration in the wastewater also affects the biosorption efficiency. Higher dye concentrations increase the competition for binding sites on the biosorbent, reducing the biosorption efficiency. Therefore, it is important to determine the optimal initial dye concentration for each biosorption process (Chandra et al., 2021) (Gupta et al., 2004)
6. *Surface area and morphology of the biosorbent*: The surface area and morphology of the biosorbent influence the biosorption efficiency by providing more binding sites for the dye molecules. A higher surface area and a more porous structure generally lead to higher biosorption efficiency (Chandra et al., 2021).
7. *Presence of other pollutants*: The presence of other pollutants in the wastewater can affect the biosorption efficiency by competing for binding sites on the biosorbent or by altering the chemical and physical properties of the dye molecules. Therefore, it is important to consider the composition of the wastewater when designing a biosorption process (Chandra et al., 2021).

8.3 BIOSORBENT

Biomass-derived biosorbents are materials that are made from renewable organic sources such as agricultural waste, plant residues, or other natural materials. These materials are abundant and low-cost, and can be processed to produce biosorbents with high surface areas and functional groups that can selectively bind to specific contaminants in wastewater (Pereira et al., 2015).

These types of biosorbents have been widely studied for their ability to remove various pollutants from wastewater, including dyes, heavy metals, and organic compounds. They are considered to be a promising alternative to traditional chemical methods of wastewater treatment, which can be expensive and environmentally damaging (Jeguirim et al., 2016).

Activated carbon is one of the most commonly used biomass-derived biosorbents. It can be produced from various sources such as coconut shells, sawdust, and agricultural waste, and has been shown to effectively remove a wide range of pollutants from wastewater (Foo and Hameed, 2012). Chitosan is another widely used biomass-derived biosorbent, which is produced from shrimp and crab shells. It has been shown to be effective in removing heavy metals and dyes from wastewater (Ngah and Hanafiah, 2008). Lignocellulosic biomass, such as rice husks and corn stalks, is another potential source of biosorbents that have been shown to effectively remove dyes and heavy metals from wastewater (Li et al., 2014).

Biomass-derived biosorbents can be used in a variety of applications such as water treatment, air purification, and soil remediation (Jeguirim et al., 2016). They offer a sustainable and low-cost solution to the problem of wastewater treatment and have the potential to make a significant contribution to environmental sustainability.

8.3.1 Activated Carbon

Activated carbon is a commonly used biosorbent for the removal of dyes from wastewater. Activated carbon has a high surface area and a porous structure, which allows it to adsorb a large amount of dye molecules (Sun et al., 2018). The biosorption process involves the binding of dye molecules onto the surface of the activated carbon through various mechanisms such as hydrogen bonding, electrostatic interactions, and π-π interactions (Rajabzadeh et al., 2015).

The efficiency of activated carbon as a biosorbent for dye removal is influenced by several factors, including the pH of the solution, the initial concentration of dye, the contact time between the biosorbent and the dye solution, and the temperature (Dawood and Sen, 2012). The pH of the solution plays a critical role in the biosorption process since it affects the surface charge of the activated carbon and the charge of the dye molecules. The optimal pH for dye removal varies depending on the type of dye and the properties of the activated carbon used (Sun et al., 2018).

The initial concentration of dye in the solution also affects the biosorption process. At low initial dye concentrations, the amount of dye adsorbed onto the activated carbon increases with increasing concentration, while at high initial dye concentrations, the amount of dye adsorbed decreases due to saturation of the active sites on the activated carbon surface (Dawood and Sen, 2012).

The contact time between the activated carbon and the dye solution is another important factor in the biosorption process. Initially, the rate of dye adsorption is high, but it decreases over time as the active sites on the activated carbon surface become saturated with dye molecules (Rajabzadeh et al., 2015).

Temperature also affects the biosorption process, with higher temperatures generally leading to higher rates of dye adsorption. This is because higher temperatures increase the kinetic energy of the dye molecules, which makes them more likely to interact with the activated carbon surface (Sun et al., 2018).

Overall, activated carbon is a promising biosorbent for the removal of dyes from wastewater due to its high adsorption capacity, low cost, and wide availability. However, the efficiency of the process

TABLE 8.1
Activated Carbon Biosorbent for Dye Removal

Activated Carbon Biosorbent Type	Dye Removed	Optimized Parameters	Sorption Percentage	Reference
Coconut shell activated carbon	Congo Red	pH 4, 50 mg/L, 0.2 g	96.5%	(Foo & Hameed, 2012)
Pineapple peel activated carbon	Methylene Blue	pH 7, 50 mg/L, 0.2 g	90.6%	(Ashtaputrey and Ashtaputrey, 2016)
Bamboo activated carbon	Malachite Green	pH 6, 50 mg/L, 0.1 g	98.4%	(Órfão et al., 2006)
Olive stones activated carbon	Reactive Red 195	pH 2, 50 mg/L, 0.1 g	98.9%	(Nayeri & Mousavi, 2020)
Palm kernel shell activated carbon	Acid Blue 25	pH 2, 100 mg/L, 0.2 g	98.6%	(Zainal et al., 2018)

is affected by several factors, and optimization of these factors is necessary for effective dye removal using activated carbon biosorbents.

Table 8.1 presents a summary of some of the activated carbon biosorbents that have been studied for dye removal, along with the specific dye removed, optimized parameters such as pH, dye concentration, and mass of biosorbent, and the sorption percentage achieved. The data presented in Table 8.1 can serve as a helpful reference for researchers and practitioners working in the field of wastewater treatment and dye removal.

8.3.2 Agricultural Wastes

The utilization of agricultural solid wastes as biosorbents has garnered attention in recent years due to their low cost, abundance, and renewable nature. The use of these biosorbents for the removal of various pollutants, including dyes, from wastewater has shown promising results. Agricultural solid wastes, such as rice husk, sugarcane bagasse, coconut shell, and corn cob, have been investigated for their potential as biosorbents (Moharm et al., 2022).

The high surface area and abundant functional groups on the surface of these biosorbents allow for effective binding of dye molecules through various mechanisms, such as hydrogen bonding, electrostatic interactions, and π-π interactions. The efficiency of these biosorbents for dye removal is influenced by several factors, including the pH of the solution, the initial dye concentration, the contact time between the biosorbent and the dye solution, and the temperature. The optimal pH for dye removal varies depending on the type of dye and the properties of the biosorbent used (Dawood & Sen, 2012).

The use of agricultural solid wastes as biosorbents for dye removal offers a sustainable and cost-effective solution for wastewater treatment. However, the optimization of several factors is necessary to achieve effective dye removal. These factors include pH, initial dye concentration, contact time, and temperature.

The utilization of agricultural waste materials as biosorbents for dye removal has received considerable attention due to their abundance, low cost, and potential to mitigate environmental pollution. Agricultural waste materials such as rice husk, corn cob, sugarcane bagasse, and coconut coir have been studied for their efficacy as biosorbents for dye removal (Foo & Hameed, 2012; Sackey et al., 2021; Zhang et al., 2018). Table 8.2. summarizes some of the agricultural waste materials that have been used as biosorbents for dye removal, the specific dyes removed, the optimized parameters such as pH, dye concentration, and mass of biosorbent, and the sorption percentage achieved. The data presented in the table can be used as a reference for researchers and practitioners working in the field of sustainable waste management and the utilization of agricultural waste materials.

TABLE 8.2
Agricultural Waste Biosorbent for Dye Removal

Agricultural Waste Biosorbent	Dye Removed	Optimized Parameters	Sorption Percentage	Reference
Rice husk	Methyl Orange	pH 2, 50 mg/L, 0.2 g	95%	(Sackey et al., 2021)
Corn cob	Crystal Violet	pH 6, 50 mg/L, 0.1 g	98.6%	(Zhang et al., 2018)
Sugarcane bagasse	Congo Red	pH 6, 50 mg/L, 0.2 g	90.4%	(Zhang et al., 2018)
Coconut coir	Direct Red 81	pH 4, 50 mg/L, 0.2 g	98.7%	(Foo & Hameed, 2012)
Wheat straw	Acid Red 18	pH 2, 50 mg/L, 0.5 g	94.6%	(Moharm et al., 2022)

8.3.3 Biological Materials

Biological materials have been extensively studied for their potential as biosorbents for dye removal from wastewater. These materials are derived from natural sources, such as bacteria, fungi, algae, and plants, and possess unique physicochemical properties that allow them to effectively bind with various types of dyes.

Bacteria have been shown to produce extracellular polymeric substances (EPS) that contain functional groups capable of adsorbing dyes. The EPS can be obtained from bacterial cultures and used as a biosorbent for dye removal. Fungi are another type of biological material that can be used as biosorbents due to their high surface area and the presence of chitin and chitosan, which can bind with dyes through electrostatic interactions. Algae, such as seaweed, have also been studied as biosorbents for dye removal due to their high surface area, porosity, and the presence of functional groups, such as carboxyl and amino groups, that can adsorb dyes (Jayasantha Kumari et al., 2017).

Plants are another source of biological materials that have been explored as biosorbents for dye removal. These materials can be obtained from various plant parts, such as leaves, stems, and roots, and can be used either in their natural form or after undergoing various treatments, such as drying, grinding, and chemical modification. The effectiveness of plant-based biosorbents is dependent on the type of plant, as well as the degree of processing and chemical modification (Lata, n.d.).

Studies have demonstrated the potential of biological materials as effective biosorbents for dye removal from wastewater (Table 8.3). For example, EPS produced by *Bacillus subtilis* has been shown to effectively remove cationic dyes, while chitosan derived from fungi has been shown to remove anionic dyes. Algae-based biosorbents have also been shown to remove various types of dyes, including acid dyes and reactive dyes. Finally, plant-based biosorbents, such as banana peels and tea waste, have been demonstrated to be effective in removing dyes from wastewater.

TABLE 8.3
Biological Material Biosorbent for Dye Removal

Biological Material Biosorbent	Dye Removed	Optimized Parameters	Sorption Percentage	Reference
Chitosan	Methylene Blue	pH 7, 2 g/L biosorbent, 120 min	94.8%	(Jayasantha Kumari et al., 2017)
Wheat Bran	Crystal Violet	pH 9, 1 g/L biosorbent, 120 min	95.8%	(Ali et al., 2014,&Wu et al., 2020)
Pomegranate Peel	Congo Red	pH 2, 5 g/L biosorbent, 120 min	91.5%	(Al-Mahmoud, 2020)
Banana Peel	Malachite Green	pH 6, 3 g/L biosorbent, 120 min	97.2%	(Lata, n.d.)
Sawdust	Acid Red 18	pH 3, 3 g/L biosorbent, 60 min	92.5%	(Shirmardi et al., n.d.)

8.3.4 Biochar

Biochar, a carbon-rich material produced through the pyrolysis of biomass, has shown great potential as a biosorbent for dye removal from wastewater. Biochar has a high surface area, porosity, and specific surface chemistry, which make it an effective adsorbent for various types of dyes (Tan et al., 2016).

Biochar can be produced from various types of biomass, including wood, crop residues, and animal manure. The properties of biochar can be modified through various processing techniques, such as activation, surface modification, and impregnation, to improve its adsorption properties. The effectiveness of biochar as a biosorbent is influenced by several factors, such as the type of biomass used, the degree of pyrolysis, the activation method used, and the pH of the solution (Eldeeb et al., 2022).

Several studies have demonstrated the potential of biochar as an effective biosorbent for dye removal from wastewater. For example, biochar produced from sugarcane bagasse has been shown to effectively remove dyes such as methylene blue and crystal violet from wastewater due to its high surface area and porosity. Similarly, biochar produced from rice straw has been shown to effectively remove dyes such as methyl orange and acid red from wastewater due to its high adsorption capacity and surface chemistry (Sackey et al., 2021).

- Biochar-based biosorbents have also been shown to have several advantages over conventional treatment methods, such as their low cost, easy availability, and potential for reuse. Additionally, biochar has been shown to be effective (Table 8.4) in removing other pollutants from wastewater, such as heavy metals and organic compounds.

8.3.5 Fungal Biosorbent

Fungal biomass-derived biosorbents have shown great potential for the removal of dyes from wastewater through a process called bioremediation (Table 8.5). Bioremediation is the use of living organisms to remove or detoxify pollutants from contaminated environments. In this case, fungal biomass is used as the biosorbent to bind and remove dyes from wastewater.

Several types of fungal biomass-derived biosorbents have been tested for their effectiveness in dye removal, including *Aspergillus niger*, *Trametes versicolor*, and *Pleurotus ostreatus*, among

TABLE 8.4
Biochar Biosorbent for Dye Removal

Biochar Biosorbent	Dye Removed	Optimized Parameters	Sorption Percentage	Reference
Pineapple peel biochar	Congo red	pH 5, contact time 120 min, dosage 1 g/L	99.8%	(Ashtaputrey and Ashtaputrey, 2016)
Sawdust biochar	Rhodamine B	pH 6.8, contact time 120 min, dosage 2 g/L	97.4%	(Eldeeb et al., 2022)
Rice straw biochar	Methylene blue	pH 6, contact time 60 min, dosage 2 g/L	91.3%	(Sackey et al., 2021)
Coconut shell biochar	Acid Orange 7	pH 4, contact time 180 min, dosage 1 g/L	91.1%	(Muralikrishnan & Jodhi, 2020)
Maize cob biochar	Reactive Black 5	pH 2, contact time 240 min, dosage 2 g/L	84.5%	(Zhang et al., 2018)

TABLE 8.5
Fungal Biosorbent for Dye Removal

Fungal Biosorbent	Dye Removed	Optimized Parameters	Sorption Percentage	Reference
Pleurotus ostreatus	Methylene blue	pH 7, 120 min, 50 mg/L, 2 g/L	99.48%	(Maurya et al., 2018)
Aspergillus niger	Malachite green	pH 4, 150 min, 50 mg/L, 2.5 g/L	97%	(Srivastava and Thakur, 2016)
Trametes versicolor	Congo red	pH 5.5, 4 h, 100 mg/L, 0.5 g/L	97.6%	(Guo et al., 2015)

others. The effectiveness of these biosorbents is influenced by various parameters such as pH, contact time, initial dye concentration, temperature, and agitation speed.

A study by Maurya et al. (2018) investigated the use of *Pleurotus ostreatus* biomass for the removal of methylene blue dye. The researchers optimized the process parameters and achieved a sorption percentage of 99.48% at pH 7, a contact time of 120 minutes, an initial dye concentration of 50 mg/L, and a biosorbent dose of 2 g/L.

Another study by Srivastava and Thakur (2016) used *Aspergillus niger* biomass to remove malachite green dye from wastewater. The researchers optimized the process parameters and achieved a sorption percentage of 97% at pH 4, a contact time of 150 minutes, an initial dye concentration of 50 mg/L, and a biosorbent dose of 2.5 g/L.

8.3.6 Bacterial Biosorbent

Bacterial biomass-derived biosorbents have also shown great potential for the removal of dyes from wastewater through bioremediation (Table 8.6). Bacterial biosorbents work similarly to fungal biosorbents by binding and removing dyes from wastewater.

Various bacterial species have been used as biosorbents for dye removal, including *Pseudomonas aeruginosa*, *Bacillus subtilis*, and *Escherichia coli*, among others. The effectiveness of these biosorbents is influenced by parameters such as pH, contact time, initial dye concentration, temperature, and agitation speed.

A study by Chatterjee and Chandra (2011) investigated the use of *Bacillus subtilis* biomass for the removal of methyl red dye. The researchers optimized the process parameters and achieved a sorption percentage of 96% at pH 6, a contact time of 60 minutes, an initial dye concentration of 50 mg/L, and a biosorbent dose of 0.1 g/L.

Another study by Li et al. (2017) used *Pseudomonas aeruginosa* biomass to remove reactive blue 19 dye from wastewater. The researchers optimized the process parameters and achieved a sorption percentage of 98% at pH 7, a contact time of 120 minutes, an initial dye concentration of 100 mg/L, and a biosorbent dose of 4 g/L.

TABLE 8.6
Bacterial Biosorbent for Dye Removal

Bacterial Biosorbent	Dye Removed	Optimized Parameters	Sorption Percentage	Reference
Bacillus subtilis	Methyl red	pH 6, 60 min, 50 mg/L, 0.1 g/L	96%	(Chatterjee and Chandra, 2011)
Pseudomonas aeruginosa	Reactive blue 19	pH 7, 120 min, 100 mg/L, 4 g/L	98%	(Li et al., 2017)
Escherichia coli	Acid orange 7	pH 5, 60 min, 150 mg/L, 1 g/L	85.9%	(Asgher et al., 2010)

8.3.7 Conclusion

In conclusion, the use of biomass-derived biosorbents for bioremediation of dyes offers a promising approach for addressing the challenges associated with the discharge of these pollutants into the environment. The advantages of these biosorbents include their low cost, high availability, and environmentally friendly nature. The performance of these biosorbents can be optimized through the modification of their surface properties and the use of appropriate pretreatment techniques.

However, there are still some challenges that need to be addressed to ensure the widespread adoption of these biosorbents for dye removal. These include the need for further research to optimize the biosorption process and the development of efficient and cost-effective methods for the recovery and reuse of the biosorbents.

Overall, the use of biomass-derived biosorbents holds great promise for the bioremediation of dyes, and further research in this area is needed to fully harness their potential. It is essential to explore more efficient and sustainable solutions for the remediation of dye-contaminated environments to ensure a safe and healthy environment for all.

REFERENCES

Ali, I., Al-Othman, Z. A., Alwarthan, A., & Asim, M. (2014). Removal of crystal violet dye from waste water by wheat bran. *Journal of Molecular Liquids*, 190, 41–47.

Al-Mahmoud, S. M. (2020). Kinetic and thermodynamic studies for the efficient removal of methylene blue using hordeum murinum as a new biosorbent. *Egyptian Journal of Chemistry*, 63(9), 3381–3390.

Asgher, M., et al. (2010). Decolorization potential of Acid Orange 7 by bacterial consortium AR3 immobilized in polyurethane foam cubes. *Journal of Hazardous Materials*, 176(1–3), 766–772.

Ashtaputrey, P. D., & Ashtaputrey, S. D. (2016). Removal of dye from aqueous solution by pineapple peel activated carbon as an adsorbent. *International Journal of Advanced Research*, 4(10), 1513–1518. https://doi.org/10.21474/IJAR01/1957

Bhattacharyya, K., & Sharma, A. (2004). Adsorption of Pb(II) from aqueous solution by Azadirachta indica (Neem) leaf powder. *Journal of Hazardous Materials*, 113, 97–109. https://doi.org/10.1016/j.jhazmat.2004.05.034

Chandra, R., Yadav, S., & Singh, J. (2021). Advances in biosorption of dyes from wastewater: A critical review. *Chemosphere*, 274, 1297.

Chatterjee, S., and Chandra, G. (2011). Biosorption of a few acid dyes by a bacterium isolated from tannery effluent: equilibrium and kinetic studies. *Journal of Environmental Management*, 92(1), 11–18.

Chen, Y., Jiang, W., Wang, J., Zheng, H., & Wang, Z. (2016). Biochar-based materials: An emerging adsorbent for the removal of organic contaminants from water. *Chemosphere*, 157, 137–146.

Dawood, S., & Sen, T. K. (2012). Removal of anionic dye Congo red from aqueous solution by raw pine and acid-treated pine cone powder as adsorbent: equilibrium, thermodynamic, kinetics, mechanism and process design. *Water Research*, 46(6), 1933–1946.

Dhillon, G. S., Kaur, S., & Brar, S. K. (2013). Biosorption mechanism of heavy metals and dye using low cost adsorbents. *Water, Air, & Soil Pollution*, 224(8), 1–23.

Eldeeb, T. M., Aigbe, U. O., Ukhurebor, K. E., Onyancha, R. B., El-Nemr, M. A., Hassaan, M. A., Ragab, S., Osibote, O. A., & El Nemr, A. (2022). Adsorption of methylene blue (MB) dye on ozone, purified and sonicated sawdust biochars. *Biomass Conversion and Biorefinery*. https://doi.org/10.1007/s13399-022-03015-w

Faria, P. C. C., Órfão, J. J. M., & Pereira, M. F. R. (2004). Adsorption of anionic and cationic dyes on activated carbons with different surface chemistries. *Water Research*, 38(8), 2043–2052. https://doi.org/10.1016/J.WATRES.2004.01.034

Foo, K. Y., & Hameed, B. H. (2012). Coconut husk derived activated carbon via microwave induced activation: Effects of activation agents, preparation parameters and adsorption performance. *Chemical Engineering Journal*, 184, 57–65. https://doi.org/10.1016/j.cej.2011.12.084

Geetha K. S., & Belagali S. L. (n.d.). Removal of heavy metals and dyes using low cost adsorbents from aqueous medium-, a review. *IOSR Journal of Environmental Science*, 4(3), 56–68. https://www.Iosrjournals.Org

Guo, J., et al. (2015). Bioremediation of Congo red dye in aqueous solution by Trametes versicolor. *Journal of Chemistry*, 2015, 1–8.

Gupta, V. K., Mittal, A., Krishnan, L., Gajbe, V., & Adsul, M. (2004). Removal of a basic dye from aqueous solution using Azadirachta indica (neem) leaf powder as a biosorbent. *Journal of Hazardous Materials*, 113(1–3), 97–109.

Gupta, V. K., & Suhas (2009). Application of low-cost adsorbents for dye removal—A review. *Journal of Environmental Management*, *90*(8), 2313–2342.

Islam, M. A., Patel, R., & Kumar, A. (2018). Biosorption: Mechanisms, methods and applications. In S. K. Gupta and S. J. Su (eds.) *Comprehensive Water Quality and Purification* (pp. 408–428). Amsterdam: Elsevier

Jayasantha Kumari, H., Krishnamoorthy, P., Arumugam, T. K., Radhakrishnan, S., & Vasudevan, D. (2017). An efficient removal of crystal violet dye from waste water by adsorption onto TLAC/Chitosan composite: A novel low cost adsorbent. *International Journal of Biological Macromolecules*, 96, 324–333. https://doi.org/10.1016/J.IJBIOMAC.2016.11.077

Jeguirim, M., Delavari, A., Pereira, M. F. R., & Harel, S. (2016). Biochar from lignocellulosic biomass: Innovative products for environmental management. *Bioresource Technology*, 215, 57–65.

Lata, S. (n.d.). Studies on removal of malachite green dye from aqueous solution using plant based biosorbents, Master of Technology thesis, National Institute of Technology Rourkela]. E-Thesis @ NIT Rourkela.

Li, Y., Du, Q., Liu, T., & Sun, J. (2014). Preparation of lignocellulosic waste and its application in removal of dyes from wastewater. *Journal of Industrial and Engineering Chemistry*, 20(3), 897–903.

Li, Y., et al. (2017). Bacterial biomass-based biosorbent for the removal of reactive blue 19 from aqueous solution. *Ecotoxicology and Environmental Safety*, 138, 268–274.

Maurya, N. S., et al. (2018). Biosorption of methylene blue dye from wastewater using Pleurotus ostreatus biomass: equilibrium, kinetic and thermodynamic studies. *Desalination and Water Treatment*, 126, 129–136.

Moharm, A. E., El Naeem, G. A., Soliman, H. M. A., Abd-Elhamid, A. I., El-Bardan, A. A., Kassem, T. S., Nayl, A. A., & Bräse, S. (2022). Fabrication and characterization of effective biochar biosorbent derived from agricultural waste to remove cationic dyes from wastewater. *Polymers*, 14(13). https://doi.org/10.3390/polym14132587

Muralikrishnan, R., & Jodhi, C. (2020). Biodecolorization of reactive dyes using biochar derived from coconut shell: batch, isotherm, kinetic and desorption studies. *ChemistrySelect*, 5(26), 7734–7742. https://doi.org/10.1002/slct.202001454

Nayeri, D., & Mousavi, S. A. (2020). Dye removal from water and wastewater by nanosized metal oxides-modified activated carbon: a review on recent researches. *Journal of Environmental Health Science and Engineering*, 18(2), 1671–1689. https://doi.org/10.1007/s40201-020-00566-w

Ngah, W. S. W., & Hanafiah, M. A. K. M. (2008). Removal of heavy metal ions from wastewater by chemically modified plant wastes as adsorbents: A review. *Bioresource Technology*, 99(10), 3935–3948.

Órfão, J. J. M., Silva, A. I. M., Pereira, J. C. V., Barata, S. A., Fonseca, I. M., Faria, P. C. C., & Pereira, M. F. R. (2006). Adsorption of a reactive dye on chemically modified activated carbons—Influence of pH. *Journal of Colloid and Interface Science*, 296(2), 480–489. https://doi.org/10.1016/J.JCIS.2005.09.063

Osagie, C., Othmani, A., Ghosh, S., Malloum, A., Kashitarash Esfahani, Z., & Ahmadi, S. (2021). Dyes adsorption from aqueous media through the nanotechnology: A review. *Journal of Materials Research and Technology*, 14, 2195–2218. https://doi.org/10.1016/J.JMRT.2021.07.085

Papirio, S., Frunzo, L., Mattei, M. R., Ferraro, A., Race, M., D'Acunto, B., Pirozzi, F., & Esposito, G. (2017). Heavy metal removal from wastewaters by biosorption: Mechanisms and modeling. In *Sustainable Heavy Metal Remediation: Volume 1: Principles and Processes* (pp. 25–63). Springer International Publishing. https://doi.org/10.1007/978-3-319-58622-9_2

Pereira, M. F. R., Soares, O. S. G. P., Órfão, J. J. M., & Figueiredo, J. L. (2015). Adsorption of dyes on activated carbons: Influence of surface chemical groups. *Carbon*, 41(5), 811–821.

Rajabzadeh, S., Alemzadeh, I., Kaghazchi, T., & Soleimani, M. (2015). Adsorption of anionic dyes on activated carbon: Influence of surface chemical groups. *Chemical Engineering Journal*, 262, 965–974.

Sackey, E. A., Song, Y., Yu, Y., & Zhuang, H. (July 7, 2021). Biochars derived from bamboo and rice straw for sorption of basic red dyes. *PLoS ONE*, 16. https://doi.org/10.1371/journal.pone.0254637

Shirmardi, M., Mesdaghinia, A., Mahvi, A. H., Nasseri, S., & Nabizadeh, R. (n.d.). Kinetics and equilibrium studies on adsorption of acid red 18 (Azo-Dye) using multiwall carbon nanotubes (MWCNTs) from aqueous solution (Vol. 2012, Issue 4). http://www.ejchem

Srivastava, S., and Thakur, I. S. (2016). Bioremediation of malachite green dye from aqueous solution by Aspergillus niger. *Journal of Environmental Chemical Engineering*, 4(3), 3295–3302.

Sun, Y., Li, Q., Li, Y., Li, A., Li, R., & Ahmad, Z. (2018). Adsorption of dyes from aqueous solutions or suspensions with clay-based and related materials: A review. *Journal of Hazardous Materials*, 344, 179–195.

Tan, X. F., Liu, Y. G., Gu, Y. L., Xu, Y., Zeng, G. M., Hu, X. J., Liu, S. B., Wang, X., Liu, S.-M., & Li, J. (2016). Biochar-based nano-composites for the decontamination of wastewater: A review. *Bioresource Technology*, 212, 318–333. https://doi.org/10.1016/J.BIORTECH.2016.04.093

Wu, K., Pan, X., Zhang, J., Zhang, X., Salah Zene, A., & Tian, Y. (2020). Biosorption of congo red from aqueous solutions based on self-immobilized mycelial pellets: Kinetics, isotherms, and thermodynamic studies. *ACS Omega*, 5(38), 24601–24612. https://doi.org/10.1021/acsomega.0c03114

Zainal, N. H., Aziz, A. A., Ibrahim, M. F., Idris, J., Hassan, M. A., Bahrin, E. K., Jalani, N. F., Wafti, N. S. A., & Abd-Aziz, S. (2018). Carbonisation-activation of oil palm kernel shell to produce activated carbon and methylene blue adsorption kinetics. *Journal of Oil Palm Research*, 30(3), 495–502. https://doi.org/10.21894/jopr.2018.0039

Zhang, L., Tong, L., Zhu, P., Huang, P., Tan, Z., Qin, F., Shi, W., Wang, M., Nie, H., Yan, G., & Huang, H. (2018). Adsorption of chlortetracycline onto biochar derived from corn cob and sugarcane bagasse. *Water Science and Technology*, 78(6), 1336–1347. https://doi.org/10.2166/wst.2018.407

9 Life Cycle Assessment Informed Process Development for Sustainable Biocement Bricks Reinforced with Plant-Based Natural Fibers and Supplemented with Biosilica from Renewable Biomass

Yogesh Kumar Murugesan, V. Sivasubramanian, and Nithya V. Pillai

9.1 INTRODUCTION

Rising energy prices and global urgency to limit carbon emissions are driving the search for sustainable manufacturing technologies that can mitigate the environmental impact of energy-intensive processes. Currently, the cement industry contributes to 8% of anthropogenic carbon emissions [1], and the Global Cement and Concrete Association has committed to cutting CO_2 emissions by a quarter by 2030 and achieving net carbon neutrality by 2050. The demand for cement-based construction materials is projected to grow in the coming years, driven by rapid economic development and urbanization in emerging economies and the need for building climate change-resilient infrastructure globally [2]. There is enormous incentive to develop novel technologies and materials for the construction industry that are cleaner, sustainable, and scalable. In this chapter, three strategies for addressing the sustainability issues associated with conventional cement manufacturing through biomass processing are discussed. Section 9.2 presents microbially induced calcite precipitation (MICP) based synthesis of biocements and bioconcretes that employs microbial cells and enzymes synthesized by them to catalyze precipitation of calcite through biomineralization that binds together sand particles. Application of life cycle assessment to identify sustainability issues arising from biocement manufacturing is emphasized. Using pretreated cellulose fibers derived from agricultural waste, a renewable resource, as reinforcement in cement-based composites to increase the lifetime of construction materials, is discussed in Section 9.3. Admixtures for biocement from silicon-rich plant residues that exhibit pozzolanic activity for partial replacement of cement are presented in Section 9.4. Section 9.5 highlights the sustainability potential and summarizes the challenges and opportunities of biomass processing technologies for biocement production.

DOI: 10.1201/9781003732907-9

9.2 MICROBIALLY INDUCED CALCITE PRECIPITATION (MICP)

MICP, a bottom-up approach for the fabrication of masonry from sand with the aid of bacteria/algae, has been developed and successfully used to manufacture biocement, as a sustainable alternative to conventional cement manufacturing [3]. By replacing the high-temperature, CO_2 emission-intensive calcination process of conventional cement manufacturing with ambient temperature microbial process, MICP greatly reduces the greenhouse gas (GHG) emissions. Globally, only two biocement products are available in the market to date: BioMason bricks employing ureolytic bacteria and Basilisk – a crack-healing bacterial concrete mix and Prometheus photosynthetic algal bricks are in the final stage of commercialization. Such eco-friendly alternatives for environmentally conscious buyers are lacking in the global market, and the sustainability of using such alternatives is justified only if they are fabricated locally to reduce the carbon footprint associated with the transporting of raw materials and finished products. This indicates the need for developing indigenous technology based on locally abundant minerals, bioresources, and microbial strains that optimally perform biomineralization at local environmental and weather conditions. Additionally, MICP also has the potential to develop processes that can sequester CO_2 from the atmosphere [4] and store it in built environments in the form of calcium carbonate in masonry blocks and construction materials with self-healing capabilities. This section summarizes various mechanisms of MICP and their sustainability, technology merits, and challenges.

9.2.1 Mechanisms of MICP

In nature, biomineralization of calcite by bacteria [3], cyanobacteria [5], and microalgae [1] occurs in limestone caves and soil and in marine environments. In these native environments, nutrient and calcium ion concentrations are very low, and the sedimentation process occurs over much slower geological timescales. This calcite precipitation occurs as a byproduct of various metabolic activities such as aerobic heterotrophic ureolysis and hydration of CO_2, autotrophic photosynthesis and methane oxidation and anoxic heterotrophic denitrification and sulfate reduction performed by a variety of microorganisms under varied environments [1]. Some of these processes have been engineered to bind together sand through calcite precipitation under a controlled nutrient-rich environment for optimal conditions for growth of microorganisms and calcite precipitation. MICP depicted in Figure 9.1 binds together sand particles in the presence of four essential components: a bacteria-producing enzyme that can catalyze the biochemical synthesis of carbonate, a source of carbon, a

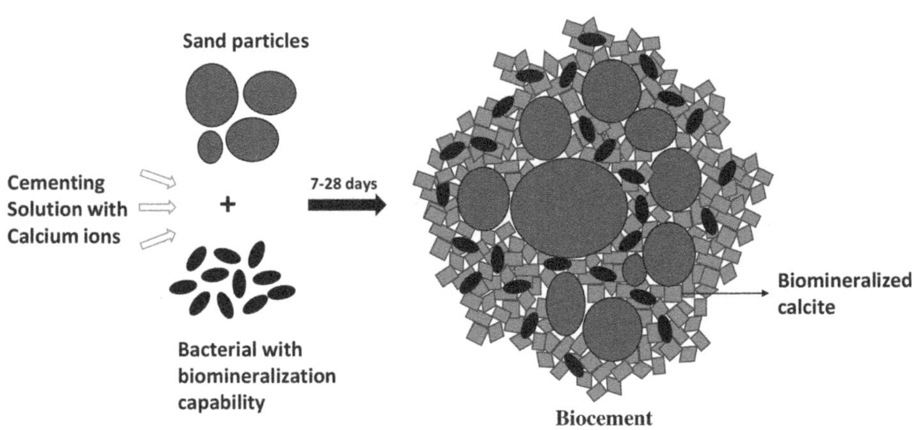

FIGURE 9.1 MICP-based biocement.

cementing solution that supplies calcium ions, and a pH-regulating component that maintains the pH in the alkaline range to facilitate formation of calcium carbonate.

This biochemical mineralization process has been applied for geotechnical soil improvement [6], restoration of degraded marble structures [7], production of biocement construction blocks [8], crack healing in concrete structures [9], and reduction of water permeability [10] through concrete. The improvements in mechanical properties and reduction in water permeability due to microbially induced calcite precipitation are listed in Table 9.1.

Owing to the mild process conditions of these biochemical processes, these methods have the potential for reducing carbon emissions associated with conventional cement manufacturing. In this section, various biochemical pathways for calcite precipitation are presented and their application to sustainable construction material production is discussed.

TABLE 9.1
Improvement in Mechanical Properties and Reduction of Water Permeability of Concrete Due to MICP

Sl No.	Pathway	Microorganisms Involved	Medium	Mechanical Property of Concrete	Decrement of Water Absorption	Ref.
1	Ureolysis	*Sporosarcina pasteurii*	Corn steep liquor, soy bean peptone, urea, sodium dihydrogen phosphate	21.3% increase in compressive strength	-	[11]
		Enterobacter Sp	Lentil seeds, Sugar, urea, beef extract	23% increase in flexural strength	15.40%	[1]
		Bacillus cereus	Tofu wastewater, Nutrient broth	27.8% increase in compressive strength	-	[12]
2	Denitrification	*Pseudomonas aeruginosa*	Urea, Calcium formate, Calcium nitrate, Yeast extract	-	42%	[1]
		Diaphorobacter nitroreducens	Urea, Calcium formate, Calcium nitrate, Yeast extract	-	47%	
3	Sulfate Reduction	Sulfate-reducing bacteria (SRB)	Nutrient Broth	15% increase in compressive strength, 14% increase in tensile strength	-	[13]
4	Hydration of CO_2	*Bacillus mucilaginous*	Powdered bacteria in steel slag and slaked lime mixture	For low slacked lime content of 0.1–0.3 slacked lime/steel slag, Compressive and flexural strength increased with the content of slaked lime.		[14]
5	Photosynthetic MICP	Cyanobacteria *Gloeocapsa* PCC73106		7.7% higher compressive strength	10%	[15]
		Synechocystis pevalekii BDHKU 35101		25.54% enhancement in compressive strength	Water absorption decreased by 1.5 times	[16]

9.2.1.1 MICP by Denitrification

Denitrifying bacteria such as *Pseudomonas aeruginosa*, *Diaphorobacter nitroreducens*, *Pseudomonas denitrificans*, *Spirillum Micrococcus*, *Thiobacillus Alcaligenes*, and *Denitro bacillus* are commonly involved in denitrification MICP [17]. These bacteria are anaerobic microorganisms that utilize nitrate (NO_3^-) as an alternative electron acceptor in the absence of oxygen during their metabolic processes, facilitated by the enzyme nitrate reductase. Denitrifying bacteria reduce nitrate (NO_3^-) to nitrite (NO_2^-) catalyzed by which is further reduced to nitric oxide (NO) and nitrous oxide (N_2O), eventually to nitrogen gas (N_2). These bacteria utilize organic carbon as a carbon and energy source and release CO_2 as a metabolic byproduct. This CO_2 equilibrates with water and forms bicarbonate ions, which precipitate as calcite when combined with Ca^{2+} ions. The overall reaction can be represented by:

$$\left(CH_3COOH\right)_2 Ca + NO_3^- \rightarrow CaCO_3 + 4/5\,N_2 + 3CO_2 + 3H_2O + OH^-$$

The resultant precipitation of calcium carbonate helps in strengthening soil, filling pore spaces, and providing additional stability to the structure. This process is advantageous in various applications, such as soil reinforcement [18], repair of concrete in cracks [19], and bioremediation by immobilizing heavy metals or pollutants through the formation of insoluble compounds [20]. Erşan et al. demonstrated the self-healing ability of *Diaphorobacter nitroreducens* and *Pseudomonas aeruginosa* by applying the microbe-containing mortar to cracks in concrete. The microbial mortar, along with calcium nitrate and calcium formate, healed cracks of width 370 µm in 28 days and 480 µm in 56 days [21]. Though the final byproduct is nitrogen, which is not harmful, the intermediate nitrous oxide is a potent greenhouse gas that can negatively impact the sustainability of this metabolic pathway [1], undermining the development of MICP based on denitrification as an environmentally friendly alternative to the conventional cement manufacturing process.

9.2.1.2 MICP by Sulfate Reduction

Sulfate-reducing bacteria of *Desulfovibrio* sp. in anaerobic environments rich in organic matter induce calcite precipitation using sulfate ions as an electron acceptor [1]. This is a form of anaerobic respiration that reduces sulfate ions to hydrogen sulfate under the enzymatic action of sulfate adenylyltransferase (EC 2.7.7.4), adenylyl-sulfate reductase (EC 1.8.99.2), and sulfite reductases (EC 1.8.99.1). The overall reaction, assuming gypsum as a source of sulfate and calcium, is given by:

$$6CaSO_4 + 4H_2O + 6CO_2 \rightarrow 6CaCO_3 + 4H_2S + 2S + 11O_2$$

However, the environmental conditions that modulate the pH determine the effectiveness of calcite precipitation and determine the suitability of employing the mechanism for biocement production. For instance, oxidation of the metabolic byproduct hydrogen sulfide to sulfur by anoxygenic sulfide phototropic bacteria that can use hydrogen sulfide as electron donor instead of water without generating oxygen can increase the pH of the system promoting calcite precipitation, while autotropic aerobic sulfide oxidizing bacteria in oxygen-rich environment could oxidize sulfide to sulfate, producing sulfuric acid that decreases the pH and inhibits calcite precipitation [17]. Challenges in maintaining anaerobic conditions throughout the biomineralization process limit the development of MICP through sulfite reduction as a viable route for commercial biocement production. Also, the metabolic byproduct hydrogen sulfide is toxic, odorous, and flammable that leads to acute and chronic health effects and negatively impacts the environment. Braissant et al. reported sulfate-reducing bacteria *Desulfovibrio* sp that reduce sulfate ions and consume organic acids to increase alkalinity, which favors calcite precipitation. It was observed that the extracellular polymeric substance produced by the cyanobacteria modulates the acid-base properties of the surrounding environment and the calcium-binding capacity and promotes calcium carbonate precipitation [22].

9.2.1.3 MICP Through Photosynthesis

In aquatic environments, certain cyanobacteria and microalgae actively transport bicarbonate ions (HCO_3^-) from their surroundings into the cell, and the enzyme carbonic anhydrase (CA) synthesized by these microorganisms catalyzes the dissociation of the bicarbonate into CO_2 and OH^-. The generation of OH^- ions increases the pH, which precipitates calcite in the presence of Ca^{2+} ions. The reaction mechanism of photosynthetic mineralization can be summarized by following the equations:

$$CO_2 + H_2O \xrightarrow{CA} H_2CO_3$$

$$H_2CO_3 \rightarrow HCO_3^- + H^+$$

$$Ca^{2+} + HCO_3^- \rightarrow CaCO_3 + H^+$$

Calcification by live and UV-killed photosynthetic biofilm-forming cyanobacteria *Gloeocapsa* PCC73106 has been investigated for their ability to improve the strength and durability in mortars [15]. While the amount of carbonate precipitated is the highest when the cells are alive and illuminated, the UV-killed cells improved the compressive strength and reduced water permeability and porosity of the mortar. When blends of cyanobacteria *Synechocystis pevalekii* BDHKU 35101 and cement mortar of weight-to-volume ratio of 0.47 were achieved, it was reported that the compressive strength increased by 25.5% and water absorption decreased by 1.5 times [16]. Recently, Prometheus Materials, based in Colorado, USA, has developed masonry blocks from microalgae grown in photobioreactors using sunlight, seawater, and CO_2. It is claimed that the material has mechanical properties comparable to Portland cement while having 90% lesser carbon footprint. This MICP pathway has the potential to sequester CO_2 during cell growth and the biomineralization process, and hence has the potential for process development for achieving carbon-neutral cement manufacturing.

9.2.1.4 Biomineralization Through Methane Oxidation

Methane-oxidizing bacteria *Methylocystis Parvus* OBBP, under aerobic conditions, precipitate calcite through autotrophic carbon fixation [23]. In this process, methane is converted to methanol by the activity of the enzyme mono-oxygenase in the presence of oxygen. The bacterial cells use methanol as a carbon source and produce formate through the action of enzymes methanol oxidase and formaldehyde dehydrogenase. Subsequently, formate is oxidized to CO_2 by formate dehydrogenase activity, converted to CO_2^{-3} and precipitated as calcium carbonate in the presence of calcium ions around the cells. The reactions involved in biomineralization through methane oxidation can be summarized as follows:

$$CH_4 + O_2 \rightarrow CH_3OH + H_2O$$

$$CH_3OH \rightarrow CHOH$$

$$CHOH + H_2O \rightarrow HCOO^- + H^+$$

$$HCOO^- + H_2O \rightarrow HCOOH + OH^-$$

$$HCOOH \rightarrow CO2$$

$$Ca^{2+} + CO_2 + 2OH^- \rightarrow CaCO_3 + H_2O$$

This MICP pathway has the potential to mitigate the global warming impact of methane released into the atmosphere due to natural processes and human activity. This pathway has been investigated for biogenic concrete protection through alteration of material porosity by filling of micropores in concrete samples that reduced water absorption by up to 2.92 ±0.91 kg/m^2. Recently, Caesar et al. (2019) showed the potential of MICP in the mitigation of methane release into the atmosphere due to anaerobic methane oxidation. However, there are no studies on the use of this mechanism in biocement production yet.

9.2.1.5 Ureolytic Calcite Precipitation

Currently, the calcite precipitation through the Ureolytic pathway has been the most investigated route because of its efficacy and faster rate, and has been successfully demonstrated at a commercial scale. *Sporosarcina pasteurii*, *Bacillus sphaericus*, *Bacillus licheniformis*, and *Bacillus megaterium* SS3 are some of the urease-producing bacteria investigated for their biocementation potential [5]. In this pathway, urea is hydrolyzed to form carbamic acid and ammonia following the reaction:

$$CO(NH_2)_2 + H_2O \rightarrow NH_2COOH + NH_3$$

The carbamic acid spontaneously decomposes to form carbonic acid and one mole of ammonia (NH3), which further hydrolyze to generate ammonium and hydroxide ions:

$$NH_2COOH + H_2O \rightarrow NH_3 + H_2CO_3$$

$$2NH_3 + 2H_2O \rightarrow 2NH_4^+ + 2OH^-$$

Due to hydroxyl ions, the pH of the medium will increase, which encourages the production of bicarbonate and carbonate ions.

$$H_2CO_3 \rightarrow HCO_3^- + H^+$$

$$HCO_3^- + H^+ + 2OH^- \rightarrow CO_3^{2-} + 2H_2O$$

Calcium carbonate precipitation occurs during the saturated state of both dissolved calcium ions (Ca^{2+}) and carbonate ions ($CO3^{2-}$). Calcium is supplemented as calcium chloride to accomplish the $CaCO_3$ precipitation.

$$Ca^{2+} + CO_3^{2-} \rightarrow CaCO_3(s)$$

The overall reactions showed that urease-producing bacteria catalyze the hydrolysis of urea into ammonium ions and raise the pH level with subsequent precipitation of carbonate. *Sporosarcina pasteurii* has been extensively used as the model urease-producing organism in ureolysis-driven $CaCO_3$ precipitation studies, due to its high ureolytic activity [5]. There has been tremendous research on this process, and biomasonry blocks have been developed and commercialized based on this mechanism. A response surface methodology-based optimization of the ureolytic biocementation determined 98% quartz sand of particle size between 125 and 500 µm when inoculated with 7.47 ml of bacterial solution and fed with urea and calcium chloride solutions of concentrations 1492 and 1391 mM, respectively, achieved a compressive strength of 1877 kPa after three treatment cycles with 24 hour inoculation time, each followed by 72 hours of drying [24]. However, sustainability issues that arise from the use of urea, as it is known to produce nitrogenous byproducts, especially ammonium, which has the potential to pollute water bodies and lead to eutrophication, have been identified [8, 25].

9.2.1.6 MICP Through Hydration of CO_2 Catalyzed by Carbonic Anhydrase

MICP based on carbonic anhydrase (CA) enzyme can convert and sequester CO_2 in biocement sustainably by using sodium bicarbonate as a source of CO_2 [26]. *Bacillus megaterium*-based MICP has been employed to bind together calcareous sand and recycled concrete aggregates by blending with urea and feeding it with a cementing solution rich in calcium ions following the ureolytic pathway [5]. In this species, urease and carbonic anhydrase act synergistically to precipitate calcite [27]. A recent study on biocalcifying marine bacteria has shown carbonic anhydrase (CA) positive, urease-negative bacterial species to mineralize calcite as efficiently as ureolytic bacteria [28]. The enzyme CA can catalyze the reversible conversion of carbon dioxide and water into bicarbonate. The biomineralization process involves calcite precipitation using carbon dioxide in the atmosphere by the carbonic anhydrase enzyme, thus paving the way for CO_2 sequestration. The bicarbonate molecule thus formed can induce calcite precipitation in the presence of Ca^{2+} ions [3] following the reaction mechanism:

$$CO_2 + H_2O \leftrightarrow H_2CO_3 \leftrightarrow H^+ + HCO_3^-$$

$$Ca^{2+} + 2HCO^-_3 \rightarrow CaCO_3 + HCO_3^- + H^+ \rightarrow CaCO_3 + CO_2 + H_2O$$

Hence, CA-producing bacteria can be an effective alternative to urease-producing bacteria in the production of sustainable biocement. During the MICP process, the bacterial cells function as a source of CA enzyme and also as a nucleation site for carbonate mineralization. Currently, biomineralization using CA-MICP shows a slower setting time compared to urease-based MICP. This can be improved through optimizing the feeding trajectory of sodium bicarbonate and calcium chloride cementing solution during the MICP process to improve the setting time and maximize the mechanical properties of the biocement block. Another advantage of using carbonic anhydrase-based MICP is that the bacteria can sequester CO_2 from the atmosphere or industrial exhaust gas from kilns and furnaces and change it to HCO_3^- and store it in the form of $CaCO_3$. Research focusing on MICP driven predominantly by CA enzyme is scarce, and there is a potential for improving the sustainability of biocement blocks based on MICP by adapting this enzyme pathway [25].

9.2.2 Factors Influencing the Mechanical Properties of Ureolytic MICP-Based Biocement

The mechanical property of biocement manufactured by ureolytic MICP is strongly influenced by bacterial strain and concentration, calcium source and concentration, urea concentration, pH and environmental conditions, treatment duration and curing conditions, soil characteristics, mixing and distribution [6]. The choice of bacterial strain used in MICP significantly affects the mechanical properties of the biocement. Different bacteria have varying abilities to precipitate calcium carbonate. Additionally, the concentration of bacteria in the solution can impact the efficiency of calcium carbonate precipitation. The source and concentration of calcium ions play a crucial role. Calcium chloride ($CaCl_2$) or calcium acetate ($Ca(CH_3COO)_2$) are commonly used as calcium sources. The concentration of the calcium solution influences the rate and amount of calcium carbonate precipitation, thus affecting the mechanical properties. Urea serves as a substrate for bacteria, providing the necessary carbon dioxide for the precipitation of calcium carbonate. The concentration of urea affects the metabolic activity of bacteria and influences the rate and amount of carbonate precipitation. The pH level influences the solubility of calcium ions and affects bacterial activity. Different bacterial strains have specific pH ranges for optimal calcium carbonate precipitation. Typically, carbonate precipitation is favored in alkaline pH. Environmental factors such as temperature, moisture content, and nutrient availability also impact bacterial activity and the mechanical properties of biocement. Urease activity is optimal at the temperature and pH of 35°C and 9, respectively [5]. The duration of treatment and curing conditions, such as temperature, humidity, and curing time, affect the rate and extent of calcium carbonate precipitation, thereby impacting the final mechanical properties of the biocement. The type of soil or substrate used can influence the effectiveness of MICP-based biocement. Different soils have varying compositions and properties that can affect the uptake and distribution of calcium carbonate, ultimately affecting the mechanical strength and durability of the treated material. Silica sand, sandy soil, recycled concrete aggregates, and red brick have been evaluated for their potential for biocementation by urease-based MICP. Proper mixing and distribution of the bacterial solution, calcium source, and urea throughout the soil or construction material are critical to achieving uniform calcium carbonate precipitation and consistent mechanical properties.

9.2.3 Sustainability Issues in Ureolytic Biocement

Life Cycle Sustainability Assessment (LCSA) of the successfully commercialized ureolytic pathway employed by BioMason to grow biocement bricks indicates higher eutrophication risk than the conventional cement manufacturing process due to its use of ammonium chloride as pH moderating ingredient, and urea and soy bean meal as carbon and nitrogen sources. Considerable negative impact on respiratory health due to particulate matter emission and increased carcinogenic risk arising from the use of urea [8, 25] has also been estimated by LCSA. Hence, there is tremendous scope to improve the sustainability of biocement manufacturing by adopting an alternative pathway that eliminates the use of urea and ammonium chloride and replacing them with low- or negative-carbon-footprint chemicals in the MICP process. LCSA of seven possible pathways for MICP concludes that biomineralization using the carbonic anhydrase (CA) pathway is the most sustainable, representing the lowest GHG emissions and eutrophication risk among the known biochemical mechanisms at a cost comparable to the current calcium carbonate manufacturing process [25]. Thus, MICP by CA-producing bacteria (CA-MICP) has the potential for economically and sustainably producing biocement blocks.

9.2.4 Challenges in MICP

MICP has several potential advantages due to its eco-friendly and sustainable nature. However, there are also some disadvantages and challenges associated with this technology. The cost and scalability of this technology is a concern. The merits and key technological and sustainability issues associated with various calcite precipitation pathways are listed in Table 9.2. Scaling up the production of these microbial biocements for large-scale applications in construction may pose technical and economic challenges, making it less cost-effective compared to conventional cement in some scenarios. While being excellent catalysts, microbial enzymes are sensitive to environmental conditions such as pH, temperature, and chemical interference. Maintaining enzyme stability during the construction process and ensuring their effectiveness in different environmental conditions can be challenging. Due to the variability in performance of microorganisms, the efficiency of MICP biocement may vary due to factors like soil composition, bacterial activity, and the distribution of the enzyme within the treated material. Achieving consistent and predictable mechanical properties across different conditions can be challenging. Biocement formation through MICP is generally slower compared to conventional cement setting times. The precipitation of calcium carbonate by bacterial action might require longer curing times, impacting project timelines. The long-term durability and compatibility of biocement in various environmental conditions need further study.

Factors such as freeze–thaw cycles, chemical exposure, and structural integrity over time may affect its performance. Maintaining optimal bacterial activity and controlling the growth of bacteria after the biocement formation process to prevent further carbonate precipitation might be challenging, especially in field applications. Compared to traditional cement, which has been extensively researched and refined, biocement technology is still in the early stages of development. Further research is needed to overcome these limitations and optimize the technology for practical

TABLE 9.2
Merits and Demerits of Various MICP Pathways [29]

Sl No	Pathway	Microorganisms involved	Enzymes involved	Merits	Demerits
1.	Urea Hydrolysis	*Bacillus megaterium, Bacillus subtilis, P. vulgaris, B. sphaericus, B. thuringiensis, Sporosarcina pasterurii, S. ginsengisoli, Kocuria flavia, Sporolactobacillus, Bacillus sphaericus, Bacillus mucilaginous, Enterobacter sp, Bacillus cereus, Sporosarcina pasterurii ATCC 11859, ACRN*	Urease	• Quick process • Easily controllable • Commonly applied and developed	• Generates ammonium byproducts that are undesirable from an environmental standpoint • When ammonium is oxidized, it may provide acidic conditions that dissolve the CaCO3 precipitate over time. High carbon footprint and inefficient process
2.	Hydration of CO_2	CA producing bacteria	Carbonic anhydrase	• Reduced carbon footprint due to CO_2 sequestration • Energy-saving method • No harmful byproducts	• Instability and inefficient usage of CA
3.	Denitrification	*Pseudomonas aeruginosa, Diaphrobacter nitroreducens, Pseudomonas denitrificans, Spirillium, Micrococcus, Thiobacillus, Alcaligens, Denitro bacillus,*		• Useful in anoxic and anaerobic environments • Healing of cracks. • Employable at depth/low oxygen level • Denitrification can be generated in most soils by stimulating indigenous microorganisms.	• Incomplete reactions may result in the formation of nitrous oxide, a greenhouse gas, as well as nitrite and nitric oxide byproducts, which are potentially hazardous and damaging to the environment and also hinder the MICP process. • Because of the nitrogen gas production, it has a significant eutrophication potential. Intermediate products are harmful to aquatic ecosystems, agriculture, and the environment

(Continued)

TABLE9.2 (Continued)

Sl No	Pathway	Microorganisms involved	Enzymes involved	Merits	Demerits
4.	Methane Oxidation	Methane oxidizing bacteria, *Methylocystis parvus* OBBP, Anaerobic methanotrophic bacteria.	Methane mono-oxygenase Formate dehydrogenase	• Ability to absorb and store methane, a well-known greenhouse gas, in biocement/ building materials. Methane is rendered harmless, and the process may help to mitigate climate change.	• Hydrogen sulphide emission.
5.	Photosynthesis	Cyanobacteria and microalgae	Carbonic anhydrase	• Cheap and simple	• The necessity for sunlight limits its applicability, particularly for geotechnical applications at depth. • Likely significant carbon footprint due to CO_2 generation
6.	Sulphate Reduction	Desulfovibrio sp. Cyanobacteria Anoxygenic sulfide phototrophic bacteria Sulfate-reducing bacteria Autotrophic aerobic sulphide oxidising bacteria		• Generate biocement under anaerobic conditions, as well as the usage of organic matter, which may be obtained easily from waste sources such as food waste	• Slow process • Not viable for engineering applications since maintaining anaerobic conditions indefinitely is challenging under real-world situation • Evolution of highly toxic hydrogen sulphide
7.	Ammonification of Aminoacids	*Myxococcus xanthus, Alcanivorax borkumensis*		• Uranium is precipitated as meta-autunite, which can safeguard concrete structures exposed to radioactive waste.	• Ammonia gas is produced and it is toxic to health

applications. Additionally, by using sustainable bio-silica and plant fibers derived from renewable bio-resources as bioadditives, the strength, durability, workability, and setting characteristics of the resulting composite can be improved while the carbon footprint associated with conventional cement manufacturing can be reduced.

9.3 PLANT FIBERS AS STRUCTURAL REINFORCEMENT

The sustainability of the process is further improved through the incorporation of locally available plant fibers as reinforcement. Fiber reinforcement has been used since antiquity to improve the mechanical properties of construction materials. In nature, biological fibrous composites such as bone, fish scale, insect, and crab exoskeleton combine a fibrous component and a cementitious matrix to synthesize materials with excellent mechanical properties. Incorporating plant-derived fibers into cementitious materials offers various advantages, enhancing the performance and properties of the resulting composite, such as improved tensile strength, crack resistance and durability, flexibility and ductility, reduced shrinkage and dimensional stability, improved workability and mix homogeneity, thermal and acoustic insulation, weight reduction, and cost effectiveness. Adding discontinuous fibers to brittle materials has been known to improve the engineering properties such as tensile strength, compressive strength, flexural strength, ductility, and toughness [30]. Inclusion of secondary fibrous phase in cement has been known to increase the crack propagation resistance of the composite material [31]. Adding plant fibers from agricultural waste to biocement will decrease raw material cost and improve the sustainability and carbon economy of the material manufacturing process. Inclusion of naturally abundant and locally sourced fibers like coir, banana pseudo stem, rice and millet straw, bagasse from sugarcane as reinforcement in biocement can improve the mechanical properties, sustainability, and carbon neutrality of the construction material.

In order to improve the adhesion in the fibers with the mortar and interaction of the fibers with the bacteria and enzymes that require nucleation sites for biomineralization, modification of plant fibers with pretreatment is required. Fiber surface modification can increase interfacial adhesion while also increasing the fiber–matrix interfacial bonding, roughness, and wettability [32]. Physical treatment, such as heat treatment and fiber beating, can modify the mechanical and thermal properties of the fibers [33]. Heat treatment is considered to raise the fiber's crystallinity index and increase fiber strength [34]. Fiber beating increases fiber surface area and improves fiber bonding [35]. Chemical treatment methods, such as alkali treatment and acetylation, boost the fiber–matrix interface adhesion through chemical bonding or mechanical interlocking at the interface and lower the water absorption capacity of the fibers [36]. Adding alkali can increase the cellulose content, mechanical strength, and roughness of the fibers [37]. Acetic acid enhances fiber swelling by increasing the accessibility of the fiber's hydroxyl groups, which improves the interfacial bonding [38]. Incorporation of synthetic [39] and natural fibers [40, 41] in biocemented sand has been reported to improve the compressive strength, flexural resistance, and tensile strength while reducing the water permeability and brittleness of the resulting material. Biocemented sand with 0.75% of jute fibers as reinforcement has been reported to improve the compressive strength by six times. The improvement in mechanical properties was attributed to stimulation of bacterial activity by the fibers and hence calcite precipitation for an extended period of the process without the need for intermittent addition of bacteria [40]. Adding 0.1% by weight of wool with loose sand biocemented by *Sporosarcina pasteurii* enhanced the compressive strength by 3.34 folds and significantly decreased the porosity of the material by providing many nucleation sites for calcite precipitation and uniformly distributing fiber-binding microstructure [41]. Up to 5% of the mass of the cement can be replaced with pretreated plant fibers from agricultural residue such as rice straw, wheat straw, sugarcane bagasse, cotton stalk, coir and banana fibers to reinforce the composites that form CO_2-rich environment where steel cannot be used owing to CO_2 corrosion that could potentially lead to failure of the material.

9.4 BIOSILICA AS ADDITIVES IN CEMENT

Biosilica, a form of silica produced by living organisms, can be sustainably obtained through various methods that harness natural processes or utilize agricultural waste [42]. Utilizing agricultural residues such as rice husks, sugarcane bagasse, wheat straw, and other plant materials rich in silica content is an environmentally friendly approach. Plants naturally accumulate silica from the environment. Through controlled processes like combustion or chemical extraction, silica can be obtained from these residues without additional environmental burden. Some industries, like the rice milling industry, generate significant quantities of silica-rich byproducts, which can be processed to extract biosilica. Repurposing industrial waste streams for biosilica extraction reduces the environmental impact associated with waste disposal. Certain bacteria and microorganisms can also produce silica as part of their natural metabolic processes. Utilizing these microorganisms or genetically engineering bacteria to produce biosilica in controlled environments could be a sustainable method. Employing eco-friendly extraction and purification methods that use non-toxic solvents or environmentally benign processes can enhance the sustainability of biosilica production.

Biosilica, which is derived from natural sources like rice husks, sugarcane bagasse, or other plant materials, can be used as an additive in cement due to its pozzolanic properties. Pozzolans are materials that, when combined with calcium hydroxide, present in hydrated cement, create additional cementitious materials. This enhances the strength and durability of concrete. Biosilica contains amorphous silica, which reacts with calcium hydroxide released during cement hydration to form additional calcium silicate hydrate (CSH) gel. CSH gel is the main binding material in concrete and contributes to its strength and durability. The pozzolanic reaction between amorphous bio-silica and calcium hydroxide must be optimized to produce biocement with good mechanical and physical properties. It is optimized when an amorphous (non-crystalline) silica form with high surface area (particle size less than 45 mm) is combined in the appropriate proportion with Al_2O_3 and Fe_2O_3 (at least 70% by weight) (ASTM, 2003). For example, corn cob ash improved the properties of cement materials, based on their pozzolanic properties [43]. The corn cob ash contained more than 70% (mass basis) of SiO_2, Al_2O_3, and Fe_2O_3. Ash from plant byproducts may contain 26–94% SiO_2 content. Good pozzolanic properties are obtained when the combined mass of SiO_2, Al_2O_3, and Fe_2O_3 exceeds 70%. By this standard, plant byproducts such as sawdust, rice husk, corn cob, and bamboo leaf are excellent candidates for biocement. The addition of biosilica can improve various properties of concrete, including compressive strength, durability, resistance to chemical attacks, and reduction in permeability. Biosilica can typically replace a portion of the cement in concrete mixtures. The optimum dosage may vary depending on the specific material and the desired properties of the concrete. Investigation on the effect of replacement of cement with ash from invasive plant species *Prosopis juliflora* indicates an increase in compressive strength, and impact resistance and flexural behavior of the resulting biocement due to pozzolanic activity of the ash. Up to 20% of concrete can be replaced by biosilica from *Prosopis juliflora*, paving the way for efficient control of invasive species while producing a sustainable biocement additive. Thus, the use of biosilica in concrete can have environmental benefits by reducing the dependency on traditional cement production, which is a significant source of carbon dioxide emissions.

9.5 CONCLUSION

In conclusion, microbial biomass that precipitates calcite, plant-derived fibers, and biosilica from agricultural residue have the potential to reduce the colossal carbon footprint associated with conventional cement manufacturing and pursue carbon neutrality. These materials can be designed to be sustainable, and they demonstrate the ability to self-heal cracks, and the manufacturing process can sequester CO_2 and store it in built environments. Life cycle assessment informed material and process design is a vital tool to identify pathways for manufacturing these materials and bioresources that can be incorporated to improve the mechanical properties of these materials and the sustainability of the process that builds them.

Despite the proven and perceived sustainability merits of manufacturing sustainable construction material from bacteria performing biomineralization that binds together sand particles, there are challenges in the scalability of production and increasing the productivity to meet the demand. To address these concerns, biocement is manufactured as a dry powder that can be shipped to the construction site, and blending biocement with concrete to improve its strength and self-healing properties is being explored. While the addition of plant fibers and biosilica to biocement can potentially improve the mechanical properties and durability performance of the resulting composite material, the effect of blending these additives on interfacial stability and workability of the cement mortar is to be evaluated. While an enormous body of research on MICP-based biocements and commercialized technology focuses on the ureolytic pathway, sustainability assessment studies indicate that this mechanism of biomineralization presents eutrophication risks and pollution threats to waterbodies. Alternative mechanisms for MICP are yet to be optimized and commercialized. The nutrients used for the growth of microorganisms that perform biomineralization using non-food feedstock, with their impact on the environment, are to be identified based on a circular economy approach. In conclusion, biocement manufacturing through biomineralization has the potential for producing sustainable, carbon-neutral construction material and opens up research opportunities for technological innovation and novel product development.

ACKNOWLEDGMENT

We authors are thankful to the Chemical Engineering Department, NIT Calicut, and the Directorate of Technical Education, Kerala, for the support of the research. The NIT Calicut provided financial support through for the Faculty Research Grant (FRG), File No: NITC/(R&C)/2023-2024/FRG/PHASE III/ (75).

REFERENCES

[1] Castro-Alonso MJ, Montañez-Hernandez LE, Sanchez-Muñoz MA, Macias Franco MR, Narayanasamy R, Balagurusamy N. Microbially induced calcium carbonate precipitation (MICP) and its potential in bioconcrete: Microbiological and molecular concepts. *Front Mater* 2019;6:1–15. https://doi.org/10.3389/fmats.2019.00126

[2] York IN, Europe I. Concrete needs to lose its colossal carbon footprint. *Nature* 2021;597:593–4. https://doi.org/10.1038/d41586-021-02612-5

[3] Dhami NK, Alsubhi WR, Watkin E, Mukherjee A. Bacterial community dynamics and biocement formation during stimulation and augmentation: Implications for soil consolidation. *Front Microbiol* 2017;8. https://doi.org/10.3389/fmicb.2017.01267

[4] Reddy MS, Joshi S. *Carbon dioxide sequestration on biocement-based composites*. Elsevier Ltd; 2018. https://doi.org/10.1016/B978-0-08-102444-7.00010-1

[5] Chuo SC, Mohamed SF, Setapar SHM, Ahmad A, Jawaid M, Wani WA, et al. Insights into the current trends in the utilization of bacteria for microbially induced calcium carbonate precipitation. *Materials (Basel)* 2020;13:1–28. https://doi.org/10.3390/ma13214993

[6] Al Qabany A, Soga K, Santamarina C. Factors affecting efficiency of microbially induced calcite precipitation. *J Geotech Geoenvironmental Eng* 2012;138:992–1001. https://doi.org/10.1061/(asce)gt.1943-5606.0000666

[7] Minto JM, Tan Q, Lunn RJ, El Mountassir G, Guo H, Cheng X. 'Microbial mortar'-restoration of degraded marble structures with microbially induced carbonate precipitation. *Constr Build Mater* 2018;180:44–54. https://doi.org/10.1016/j.conbuildmat.2018.05.200

[8] Iezzi B, Brady R, Sardag S, Eu B, Skerlos S. Growing bricks: Assessing biocement for lower embodied carbon structures. *Procedia CIRP* 2019;80:470–5. https://doi.org/10.1016/j.procir.2019.01.061

[9] Seifan M, Samani AK, Berenjian A. Bioconcrete: next generation of self-healing concrete. *Appl Microbiol Biotechnol* 2016;100:2591–602. https://doi.org/10.1007/s00253-016-7316-z

[10] Noshi CI, Schubert JJ. Self-healing biocement and its potential applications in cementing and sand-consolidation jobs: A review targeted at the oil and gas industry. *Soc Pet Eng - SPE Liq Basins Conf - North Am* 2018, LRBC 2018 2018:1–19. https://doi.org/10.2118/191778-ms

[11] Chen L, Song Y, Fang H, Feng Q, Lai C, Song X. Systematic optimization of a novel, cost-effective fermentation medium of Sporosarcina pasteurii for microbially induced calcite precipitation (MICP). *Constr Build Mater* 2022;348:128632. https://doi.org/10.1016/j.conbuildmat.2022.128632

[12] Fang C, He J, Achal V, Plaza G. Tofu wastewater as efficient nutritional source in biocementation for improved mechanical strength of cement mortars. *Geomicrobiol J* 2019;36:515–21. https://doi.org/10.1080/01490451.2019.1576804

[13] Teddy T, Irwan JM, Othman N. The influence of Sulphate reduction bacteria on the durability of concrete in seawater condition. *Pertanika J Sci Technol* 2017;25:117–22.

[14] Wang K, Qian C, Wang R. The properties and mechanism of microbial mineralized steel slag bricks. *Constr Build Mater* 2016;113:815–23. https://doi.org/10.1016/j.conbuildmat.2016.03.122

[15] Zhu T, Lu X, Dittrich M. Calcification on mortar by live and UV-killed biofilm-forming cyanobacterial Gloeocapsa PCC73106. *Constr Build Mater* 2017;146:43–53. https://doi.org/10.1016/j.conbuildmat.2017.04.026

[16] Sidhu N, Goyal S, Reddy MS. Biomineralization of cyanobacteria Synechocystis pevalekii improves the durability properties of cement mortar. *AMB Express* 2022;12. https://doi.org/10.1186/s13568-022-01403-z

[17] Jain S, Fang C, Achal V. A critical review on microbial carbonate precipitation via denitrification process in building materials. *Bioengineered* 2021;12:7529–51. https://doi.org/10.1080/21655979.2021.1979862

[18] van Paassen LA, Daza CM, Staal M, Sorokin DY, van der Zon W, van Loosdrecht MCM. Potential soil reinforcement by biological denitrification. *Ecol Eng* 2010;36:168–75. https://doi.org/10.1016/j.ecoleng.2009.03.026

[19] Ersan Y, Gruyaert E, Louis G, Lors C, De Belie N, Boon N. Self-protected nitrate reducing culture for intrinsic repair of concrete cracks. *Front Microbiol* 2015;6. https://doi.org/10.3389/fmicb.2015.01228

[20] Eltarahony M, Zaki S, Abd-El-Haleem D. Aerobic and anaerobic removal of lead and mercury via calcium carbonate precipitation mediated by statistically optimized nitrate reductases. *Sci Rep* 2020;10:1–20. https://doi.org/10.1038/s41598-020-60951-1

[21] Erşan YÇ, de Belie N, Boon N. Microbially induced CaCO3 precipitation through denitrification: An optimization study in minimal nutrient environment. *Biochem Eng J* 2015;101:108–18. https://doi.org/10.1016/j.bej.2015.05.006

[22] Braissant O, Decho AW, Dupraz C, Glunk C, Przekop KM, Visscher PT. Exopolymeric substances of sulfate-reducing bacteria: Interactions with calcium at alkaline pH and implication for formation of carbonate minerals. *Geobiology* 2007;5:401–11. https://doi.org/10.1111/j.1472-4669.2007.00117.x

[23] Ganendra G, De Muynck WD, Ho A, Arvaniti EC, Hosseinkhani B, Ramos JA, et al. Formate oxidation-driven calcium carbonate precipitation by Methylocystis parvus OBBP. *Appl Environ Microbiol* 2014;80:4659–67. https://doi.org/10.1128/AEM.01349-14

[24] Erdmann N, de Payrebrune KM, Ulber R, Strieth D. Optimizing compressive strength of sand treated with MICP using response surface methodology. *SN Appl Sci* 2022;4. https://doi.org/10.1007/s42452-022-05169-8

[25] Porter H, Mukherjee A, Tuladhar R, Dhami NK. Life cycle assessment of biocement: An emerging sustainable solution? *Sustain* 2021;13:1–14. https://doi.org/10.3390/su132413878

[26] Li W, Liu L, Chen W, Yu L, Li W, Yu H. Calcium carbonate precipitation and crystal morphology induced by microbial carbonic anhydrase and other biological factors. *Process Biochem* 2010;45:1017–21. https://doi.org/10.1016/j.procbio.2010.03.004

[27] Dhami NK, Reddy MS, Mukherjee A. Synergistic role of bacterial urease and carbonic anhydrase in carbonate mineralization. *Appl Biochem Biotechnol* 2014;172:2552–61. https://doi.org/10.1007/s12010-013-0694-0

[28] Vincent J, Colin B, Lanneluc I, Sabot R, Sopéna V, Turcry P, et al. New biocalcifying marine bacterial strains isolated from calcareous deposits and immediate surroundings. *Microorganisms* 2022;10:1–22. https://doi.org/10.3390/microorganisms10010076

[29] Mwandira W, Mavroulidou M, Gunn MJ, Purchase D, Garelick H, Garelick J. Concurrent carbon capture and biocementation through the carbonic anhydrase (CA) activity of microorganisms-a review and outlook. *Environ Process* 2023;10:56. https://doi.org/10.1007/s40710-023-00\867-2

[30] Al Imran M, Gowthaman S, Nakashima K, Kawasaki S. The influence of the addition of plant-based natural fibers (Jute) on biocemented sand using MICP method. *Materials (Basel)* 2020;13. https://doi.org/10.3390/MA13184198

[31] Banthia N, Nandakumar N. Crack growth resistance of hybrid fiber reinforced cement composites. *Cem Concr Compos* 2003;25:3–9. https://doi.org/10.1016/S0958-9465(01)00043-9

[32] Taib MNAM, Julkapli NM. *Dimensional stability of natural fiber-based and hybrid composites*. Elsevier Ltd; 2018. https://doi.org/10.1016/B978-0-08-102292-4.00004-7

[33] Venkatachalam N, Navaneethakrishnan P, Rajsekar R, Shankar S. Effect of pretreatment methods on properties of natural fiber composites: A review. *Polym Polym Compos* 2016;24:555–66. https://doi.org/10.1177/096739111602400715

[34] Ariawan D, Salim MS, Taib RM, Thirmizir MZA, Ishak ZAM. Interfacial characterisation and mechanical properties of heat treated non-woven kenaf fibre and its reinforced composites. *Compos Interfaces* 2018;25:187–203. https://doi.org/10.1080/09276440.2017.1354562

[35] Ferreira SR, Silva FDA, Lima PRL, Toledo Filho RD. Effect of fiber treatments on the sisal fiber properties and fiber-matrix bond in cement based systems. *Constr Build Mater* 2015;101:730–40. https://doi.org/10.1016/j.conbuildmat.2015.10.120

[36] Zhou Y, Fan M, Chen L. *Interface and bonding mechanisms of plant fibre composites: An overview*. vol. 101. Elsevier Ltd; 2016. https://doi.org/10.1016/j.compositesb.2016.06.055

[37] Chand N, Fahim M. 1 - Natural fibers and their composites. In: Chand N, Fahim M, editors. *Tribology of natural fiber polymer composites* (Second Edition, Woodhead Publishing; 2021, p. 1–59). https://doi.org/10.1016/B978-0-12-818983-2.00001-3

[38] Sawpan MA, Pickering KL, Fernyhough A. Effect of various chemical treatments on the fibre structure and tensile properties of industrial hemp fibres. *Compos Part A Appl Sci Manuf* 2011;42:888–95. https://doi.org/10.1016/j.compositesa.2011.03.008

[39] Choi SG, Wang K, Chu J. Properties of biocemented, fiber reinforced sand. *Constr Build Mater* 2016;120:623–9. https://doi.org/10.1016/j.conbuildmat.2016.05.124

[40] Spencer CA, van Paassen L, Sass H. Effect of jute fibres on the process of MICP and properties of biocemented sand. *Materials (Basel)* 2020;13:1–24. https://doi.org/10.3390/ma13235429

[41] Yao D, Wu J, Wang G, Wang P, Zheng JJ, Yan J, et al. Effect of wool fiber addition on the reinforcement of loose sands by microbially induced carbonate precipitation (MICP): mechanical property and underlying mechanism. *Acta Geotech* 2021;16:1401–16. https://doi.org/10.1007/s11440-020-01112-6

[42] Alishah Aratboni H, Rafiei N, Garcia-Granados R, Alemzadeh A, Morones-Ramírez JR. Biomass and lipid induction strategies in microalgae for biofuel production and other applications. *Microb Cell Fact* 2019;18:1–17. https://doi.org/10.1186/s12934-019-1228-4

[43] Adesanya DA, Raheem AA. Development of corn cob ash blended cement. *Constr Build Mater* 2009;23:347–52. https://doi.org/10.1016/j.conbuildmat.2007.11.013

10 Turning Biomass into Bioplastic

A Sustainable Approach for Management of Agrowaste

R. Divya and Jerold Manuel

10.1 INTRODUCTION

Petroleum-derived plastics such as Polyvinyl Chloride (PVC), Polyethylene (PE), Polypropylene (PP), Polystyrene (PS), Poly ethyl terephthalate (PET), and Nylons are the most widely used polymers in day-to-day life due to their versatility, light weight, and excellent thermal and rheological properties. The ecosystem has been severely impacted by the overuse of plastics, and it is estimated that humans manufacture roughly 34 million tons of plastic annually. Only 7% of that is recycled, while the other 93% ends up in landfills, oceans, and seas. Plastic burning releases a variety of hazardous emissions, including carbon dioxide and methane, which are greenhouse gases that have an impact on the global climate. The incineration or burning of plastic is poisonous and cancer-causing due to the chemical additives used in its production. In order to replace traditional plastics, industries have started using natural, renewable materials like starch and vegetable oils in the manufacture of bioplastics. However, greater production costs for bioplastics when compared to conventional plastics hinder their development. They are two to three times more expensive than typical polymers, and their manufacture is plagued by low yields and high costs. Some bioplastics are less durable than oil-based plastics due to their weaker mechanical properties, such as more water vapor permeability than traditional plastic, the tendency to tear easily like tissue paper, extreme brittleness, and poor mechanical and barrier qualities. In order to make bioplastics sustainable and to improve their qualities, additive synthesis, such as the incorporation of polymer and composite combinations, is frequently investigated. Furthermore, nanomaterials such as nanoclay have been employed to improve the thermal stability of starch-based bioplastics. According to Harunsyah et al. (2018), a bioplastic made from cassava starch, clay nanoparticles, and the plasticizer glycerine exhibited intriguing mechanical features such as transparency, clarity, homogeneity, flexibility, and ease of handling.

10.2 CLASSIFICATION OF BIOPLASTICS

Plastics can be produced using fossil or bio-based ingredients, and they can either be biodegradable or non-biodegradable. Plastics can be further categorized into four classes based on their biodegradability and biobased content. Figure 10.1 illustrates the different properties of plastic materials according to their biodegradability and origin.

1. Bioplastics that are biobased or partially biobased and non-biodegradable, such as bio-PE, bio-PET, bio-PA, and bio-PTT.
2. Bioplastics that are biobased and biodegradable, such as PLA, PHA, PBS, and starch blends.

DOI: 10.1201/9781003732907-10

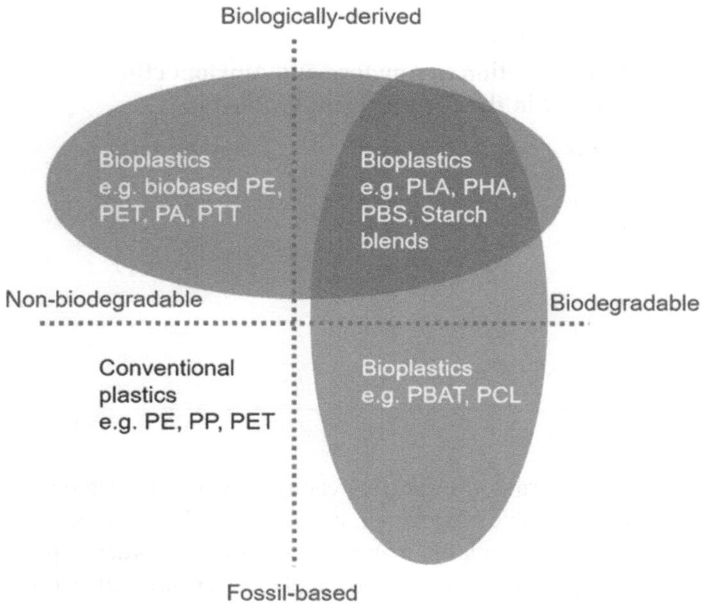

FIGURE 10.1 Classification of plastics based on their raw material and biodegradability.

(Source: Shah et al., 2021).

3. Plastics that are fossil-based and non-biodegradable, such as conventional PE, PP, and PET.
4. Bioplastics that are fossil-based and biodegradable, such as PBAT and PCL.

10.3 MATERIALS FOR BIOPLASTIC PRODUCTION

Nowadays, a variety of materials, including cellulose and its derivatives, starch, poly-beta-hydroxy-alkanoates (PHB), polylactide acid (PLA) plastics, and chitin/chitosan, can be used to make bioplastics.

10.3.1 Cellulose and Its Derivatives

Bioplastic made from cellulose is quite difficult to use as a packaging material because of its poor moisture barrier, hydrophilic nature, and highly crystalline structure. Due to this structure, the resulting packaging material is brittle with the least flexible and tensile strength (Hyon et al., 1998).

Cellulose derivatives, Hydroxypropyl cellulose, Hydroxypropyl methylcellulose, carboxymethyl cellulose, and methylcellulose, can be used as edible films or coatings. But these films also have properties like cellulose and are not recommended for packaging. One method employed for improving the quality of the moisture barrier is the addition of hydrophobic compounds such as fatty acids into the cellulose matrix to develop the films (Morillon et al., 2002).

10.3.2 Starch and Starch-Based Plastics

Several raw materials, including potato and its peel, corn, wheat, pea, etc., are used to generate starch and its derivatives. Beyond the food industry, it has a wide range of uses, including as a binding agent in the pharmaceutical industry, as well as in the textile and paper and board sectors (Laycock and Halley, 2014). Starch-based polymers are typically biodegradable polysaccharide polymers that are used in food processing, disposable tableware and cutlery, coffee machine

TABLE 10.1
Concentration of Amylose and Amylopectin Present in the Various Plant Products

Source	Amylose (in %)	Amylopectin (in %)
Banana	20.5	79.5
Potato	17.8	82.2
Corn	28	72
Rice	35	65
Wheat	20	80

Source: Shah et al., 2021.

capsules, and bottles as an alternative to polystyrene (PS) materials. Although starch is a readily available, affordable, and renewable biopolymer, hydrogen bonding and intermolecular forces prevent it from being used as a thermoplastic material. As a result, in order to produce thermoplastic starch (TPS), a deformable thermoplastic polymer, a plasticizer (urea, glycerol, or sorbitol) is also required (Khan et al. 2017). The concentration of amylose and amylopectin found in the various plant products has been noted in Table 10.1. It has been noted that the bioplastic's tensile properties increase with increasing amylose concentration (Ceseracciu et al., 2015). Additionally, it has been shown that the presence of amylose molecules, which release water, enhances the likelihood of biodegradation, and the plasticizer property increases due to the presence of the amylopectin molecules (Verma et al., 2012).

10.3.3 Poly-Beta-Hydroxyalkanoates (PHB)

PHB degrades in the presence of microorganisms that come in contact with the polymer; the microbes secrete enzymes and break down the polymer into smaller parts. Properties of PHB are 100% resistance to water, 100% biodegradability, and the ability to process thermoplastic (Kumar et al., 2014).

10.3.4 Polylactic Acid (PLA) Plastics

In terms of packaging, PLA is the most acceptable material because of its excellent biocompatibility, biodegradability, and processing capacity. It was the first biobased polymer used for commercial purposes that could be molded into different shapes and films (Rasal et al., 2010). It has replaced the packaging materials, including HDPE (High Density Polyethylene), LDPE (Low Density Polyethylene), PET (Polyethylene Terephthalate), and PS (Polystyrene).

10.3.5 Chitin/Chitosan

After cellulose, chitin is the second most prevalent natural polymer found in nature. Chitin is the source of chitosan. Chitosan is a byproduct of the fishing industry that is produced by the deacetylation process in the presence of alkali (Sánchez-González et al., 2010). It produces films without the use of any chemicals and has superior mechanical and antimicrobial properties that are resistant to fungi, yeast, and bacteria. It also has strong oxygen and carbon dioxide permeability, but its poor solubility in neutral solutions is a major negative. The required level of deacetylation must be between 80 and 85% in order to produce a soluble product. It forms transparent layers that improve the product's storage life and enhance its quality (Ribeiro et al., 2007).

10.4 ENZYMATIC CATALYST

Plasticizers are organic molecules that are added to polymers to reduce brittleness, reduce crystallinity, improve durability and toughness, and lower melting temperatures (De Groote et al., 2002). Different types of plasticizers, including polyols like glycol, glycerol, sorbitol, fructose, sucrose, and mannose, as well as fatty acids like palmitate or myristate, are utilized in the manufacturing of bioplastics. Because of its low cost, non-toxicity, and high boiling point (292°C), glycerol is utilized the most as a plasticizer (Forssell et al., 1997).

10.5 BIOPLASTIC REINFORCED WITH FIBER NANOFILLERS

Reinforcement by fiber has been widely used in developing polymeric composites and has successfully proven its value due to its excellent specific properties, e.g., high strength and stiffness and low weight. Starch reinforced by cellulose fiber is a typical example of natural polymer composites. Such composites have attracted great attention in the last two decades. In fact, improving mechanical properties is one of the major driving forces to reinforce starch-based materials using cellulose fibers. Wollerdorfer and Bader (1998) reported that the tensile properties of thermoplastic wheat starch were increased four times after reinforcement by cellulose fibers.

In recent years, the incorporation of nanoparticles (NPs) into composite materials has attracted a great deal of attention for their ability to enhance the thermal, mechanical, and gaseous barrier properties of polymers (Noshirvani et al., 2017). ZnO NPs have positive effects on the mechanical and barrier properties of the films, which is a useful property in food packaging applications. Zinc Oxide is an interesting agent in the food and pharmacy fields due to its antimicrobial properties, high stability, non-toxicity, and good UV absorbance properties. Additionally, it is recognized as safe (GRAS) from the FDA (182.8991 code) (Jebel and Almasi, 2016).

On the other hand, the Food and Drug Administration (FDA) has recognized it as typically safe for food and beverage packaging (Sufer et al., 2017). When the PLA is applied with the nanocomposite magnesium (MgO), approximately 44% of bacteria die after 24 h, and it is ideal for food preservation (Swaroop and Shukla, 2019). With the inclusion of 2% MgO nanocomposite, the plastic characteristics were increased by nearly 146%. Salvatore et al. (2016) reported that, however, adding PLA polymers of zinc oxide (ZnO) showed approximately a 99.99% decrease in *E. coli* after 24 h.

10.6 TECHNIQUES INVOLVED IN PROCESSING OF BIOPLASTICS

The selection of the technique and conditions used when processing bioplastics is also highly influential in their final properties. It should be noticed that Mass Flow Rate (MFR) is also of extreme importance when selecting a processing technique for bioplastics. Compression molding, extrusion molding, and injection molding are some of the different techniques for processing bioplastics.

Compression molding has two metal molds that apply a certain force on the sample to produce the bioplastic. As the sample flows into the mold, it acquires its shape (Ashter, 2016).

Extrusion is the most widely used technique for processing polymers. It is a continuous process, where the polymer is transported through a barrel where it is heated to enhance its flowability through a final, conveniently shaped orifice. The polymer cools as it exits and results in a constant section solid (Ashter, 2016).

Injection molding process can be divided into two stages: injection and compaction. In the first step, the substance to be molded is introduced into a tank that is heated to a certain temperature that allows the material to flow for subsequent injection molding, by means of a piston that exerts a certain pressure. In the second step, called compaction, the sample is already inside the cavities of the mold, which is heated to molding temperature. During this stage, the injected homogeneous mixture is subjected to a certain stable pressure by the piston until its extraction (Ashter, 2016). In the case of biopolymers, the molding temperature can be higher than that of the cylinder, so as to promote crosslinking and fix the structure.

10.7 APPLICATIONS OF BIOPLASTICS

Bioplastics are getting increased attention in a variety of industrial applications. This is due to the fact that industries and consumers both put a high value on a pollution-free environment, which may be accomplished with the help of bioplastics.

Currently, bioplastics are available in the following market sectors:

- Packaging
- Food-services
- Agriculture and horticulture
- Consumer electronics

Packaging: In the food industry, bioplastics in the nanocellulose application have shown rapid growth for the past several years and are expected to grow more in the near future. It is especially noticeable with the increased rate of research on food applications, especially in the food packaging sector. Some of the properties of bioplastics in nanocellulose applications, such as edibility, flexibility, biodegradability, and anti-microbial properties, are some of the promising factors that have attracted researchers. In their most recent study, Lu et al. (2020) employed nanocellulose as a colorimetric indicator for food freshness in intelligent food packaging. The freshness of the chicken used in the experiment will affect how the hydrogel's color changes. According to reports, the nanocellulose utilized in the experiment responds instantly to the rotting of the chicken, improving the value of intelligent packaging.

Food services: For the food and catering market, a wide range of bioplastics products is offered. There are both single-use and reusable options available for everything from plates and cutlery to cups, trays, and mugs. There are numerous conditions in which only single-use products are practical, particularly in closed systems with integrated waste management, such as on airplanes, at sporting events, or in penal facilities.

Agriculture and horticulture: Herb pots are a fantastic example of how biodegradable plastics offer marketing potential for potted plants. Everything, including the film, can be composted when the herbs have been picked. Alternatives include just planting the plant into its pot, which is very convenient for hobby gardeners. There are also flower bulbs that come prepackaged and can be planted right into the ground. The packaging disperses quickly, and then plant growth can begin. The Food and Agriculture Organization of the United Nations also openly acknowledges the benefits of agricultural plastics.

Consumer electronics: In the quickly evolving consumer electronics market, a wider variety of bioplastic items are being developed, including touch screen computer cases, loudspeakers, keyboard components, mobile cases, vacuum cleaners, and laptop mice. The world's first bioplastic touch screen computer, which was created in partnership with Taiwanese OEM/ODM Kuender, is the application being launched. This computer's high-gloss casing is manufactured of high-heat PLA (Akgul et al., 2022).

10.8 CHALLENGES OF BIOPLASTICS

Typically, bioplastics are marketed as a sustainable replacement for traditional plastics. The most difficult part of bioplastic synthesis, though, is the fact that it must be achieved without disturbing potential food sources. This situation can be mitigated by using non-food resources for the purpose

(Sidek et al., 2019). Additionally, some bioplastics made from the bacterial polymer PLA are only biodegradable under specific temperature and humidity conditions because their properties are only fixed under those conditions (Zulkafli, 2014). To ensure that bioplastics can decompose under any conditions in landfills, this constraint must be removed. In addition, other materials with hydrophilic properties, like starch and cellulose-based packaging materials, have a low water vapor barrier, which is the cause of their poor processability, brittleness, susceptibility to degradation, decreased long-term stability, and poor mechanical properties, like elongation and tensile strength (Cyras et al., 2007). The improved performance in natural fiber composites and green composites is required to provide more applications by industries (Misra et al., 2011).

10.9 CONCLUSION

The negative effects on the environment from the vast amount of non-biodegradable trash are encouraging research to create new biodegradable materials that can be produced from natural resources, including biomass, plants, and microorganisms. Future growth in the market for bioplastics might be attributed to their sustainability. In addition, the biotechnology of microorganisms offers the potential for the manufacture of bioplastics, which may find widespread use and commercialization in a range of industries, including agriculture, medicine, pharmaceuticals, veterinary medicine, etc.

Therefore, a new set of guidelines and standards for bioplastics should be developed for their manufacturing, use, and waste disposal globally. As a result, labeling regulations must be improved to take into account a product's raw material usage, energy consumption, and emissions during production and use.

However, if the world's oil supply becomes more and more scarce, bioplastics may eventually take the place of plastics produced from fossil fuels. Finally, the principles of sustainability, industrial ecology, eco performance, green chemistry, and engineering have all been included in the design of materials, products, and processes for future generations through the use of bio-based materials and natural fibers, bioplastics, and bio composites.

REFERENCES

Akgul, A., T. Palmeiro-Sanchez, H. Lange, D. Magalhaes, S. Moore, A. Paiva, … A. Trubetskaya. "Characterization of tars from recycling of PHA bioplastic and synthetic plastics using fast pyrolysis." *Journal of Hazardous Materials* 439 (2022): 129696.

Ashter, S. A. "Commercial applications of bioplastics." *Introduction to Bioplastics Engineering* (2016): 227–249.

Ceseracciu, L., J. A. Heredia-Guerrero, S. Dante, A. Athanassiou, & I. S. Bayer "Robust and biodegradable elastomers based on corn starch and polydimethylsiloxane (PDMS)." *ACS Applied Materials & Interfaces* 7.6 (2015): 3742–3753.

Cyras, V. P., et al. "Biodegradable double-layer films based on biological resources: Polyhydroxybutyrate and cellulose." *Journal of Applied Polymer Science* 106.2 (2007): 749–756.

De Groote, P. H., J. Devaux, and P. Godard. "Effect of benzenesulfonamide plasticizers on the glass-transition temperature of semicrystalline polydodecamide." *Journal of Polymer Science Part B: Polymer Physics* 40.19 (2002): 2208–2218.

Forssell, P. M., et al. "Phase and glass transition behaviour of concentrated barley starch-glycerol-water mixtures, a model for thermoplastic starch." *Carbohydrate Polymers* 34.4 (1997): 275–282.

Harunsyah, Sariadi, and Raudah. "The effect of clay nanoparticles as reinforcement on mechanical properties of bioplastic base on cassava starch." *Journal of Physics: Conference Series* 953 (2018, 1 January): 012021.

Hyon, S.-H., K. Jamshidi, and Y. Ikada. "Effects of residual monomer on the degradation of dl-lactide polymer." *Polymer International* 46.3 (1998): 196–202.

Jebel, F. S., and H. Almasi. "Morphological, physical, antimicrobial and release properties of ZnO nanoparticles-loaded bacterial cellulose films." *Carbohydrate Polymers* 149 (2016): 8–19.

Khan, B., et al. "Thermoplastic starch: A possible biodegradable food packaging material—A review." *Journal of Food Process Engineering* 40.3 (2017): e12447.

Kumar, Y., et al. "Bio-plastics. A perfect tool for eco-friendly food packaging: A review." *Journal of Food Product Development and Packaging* 1.1 (2014): 01–06.

Laycock, B. G., and P. J. Halley. "Starch applications: State of market and new trends." *Starch Polymers* (2014): 381–419.

Lu, P., Y. Yang, R. Liu, X. Liu, J. Ma, and M. Wu. "Preparation of sugarcane bagasse nanocellulose hydrogel as a colourimetric freshness indicator for intelligent food packaging." *Carbohydrate Polymers* 249 (2020): 116831.

Misra, M., et al. "Bioplastics and green composites from renewable resources: where we are and future directions." *18th International Conference on Composite Materials*, Jeju Island, Korea, August 21–26, 2011.

Morillon, V., et al. "Factors affecting the moisture permeability of lipid-based edible films: a review." *Critical Reviews in Food Science and Nutrition* 42.1 (2002): 67–89.

Noshirvani, N., et al. "Novel active packaging based on carboxymethyl cellulose-chitosan-ZnO NPs nanocomposite for increasing the shelf life of bread." *Food Packaging and Shelf Life* 11 (2017): 106–114.

Rasal, R. M., A. V. Janorkar, and D. E. Hirt. "Poly (lactic acid) modifications." *Progress in Polymer Science* 35.3 (2010): 338–356.

Ribeiro, C., et al. "Optimization of edible coating composition to retard strawberry fruit senescence." *Postharvest Biology and Technology* 44.1 (2007): 63–70.

Sánchez-González, L., et al. "Physical and antimicrobial properties of chitosan–tea tree essential oil composite films." *Journal of Food Engineering* 98.4 (2010): 443–452.

Shah, M., S. Rajhans, H. A. Pandya, & A. U. Mankad (2021). Bioplastic for future: A review then and now. *World Journal of Advanced Research and Reviews*, 9(2), 056–067. https://doi.org/10.30574/wjarr.2021.9.2.0054

Sidek, I. S., et al. "Current development on bioplastics and its future prospects: an introductory review." *INWASCON Technology Magazine* 1 (2019): 3–8.

Sufer, O., A. T. Oz, and Y. Çelebi Sezer. "Poly (lactic acid) films in food packaging systems." *Food Science & Nutrition Technology (FSNT)* 2.10 (2017): 23880.

Swaroop, C., and M. Shukla. "Development of blown polylactic acid-MgO nanocomposite films for food packaging." *Composites Part A: Applied Science and Manufacturing* 124 (2019): 105482.

Verma, D., P. C. Gope, M. K. Maheshwari, & R. K. Sharma. "Bagasse fiber composites-A review." *Journal of Materials and Environmental Science* 3.6 (2012), 1079–1092.

Wollerdorfer, M., and H. Bader. "Influence of natural fibres on the mechanical properties of biodegradable polymers." *Industrial Crops and Products* 8.2 (1998): 105–112.

Zulkafli, N. N. Production of Bioplastic from Agricultural Waste (Doctoral dissertation, UMP), 2014.

11 Valorization of Agro-Waste into Catalyst for Biodiesel Production

A Green Approach to Sustainable Waste Management

Akshay Prakash and Jerold Manuel

11.1 INTRODUCTION

The increasing global energy demand, coupled with the exponential expansion of urbanization, industrialization, and transportation, has emphasized the pivotal role of fuels in powering modern societies. This is evident because fuel accounts for approximately 81% of global energy consumption, and fossil fuels fulfill around 90% of it (Banković-Ilić et al., 2017; Muh et al., 2019; Nath et al., 2023). The transport sector alone, responsible for 54% of global energy consumption, is projected to drive a 57% increase in the world's energy consumption by the year 2050, with an estimated annual growth rate of 1.1% (Changmai et al., 2020; Nath et al., 2023). The world is leading toward an imminent energy crisis due to the overuse of fossil fuels, which also has several severe consequences, like the emission of greenhouse gases (GHGs) resulting in global warming and the rapid exponential depletion of petroleum reserves at a staggering rate of 104 times their natural formation (Banković-Ilić et al., 2017; Nath et al., 2023; Satyanarayana et al., 2011). If current trends continue, experts estimate that petroleum resources could become depleted by 2050, heightening the prospects of a critical energy crisis (Banković-Ilić et al., 2017; Nath et al., 2023; Seela et al., 2013). Moreover, the combustion of fossil fuels releases greenhouse gases, contributing significantly to environmental pollution and exacerbating global warming concerns (Banković-Ilić et al., 2017; Satyanarayana et al., 2011). As per an ongoing temperature study directed by researchers at NASA's Goddard Institute for Space Studies (GISS), the Earth's average global temperature has risen by a minimum of 1.1°C (1.9°F) since 1880. The most significant portion of this warming has occurred since 1975, with a rate of approximately 0.15–0.20°C per decade. In NASA's historical data, July 2023 stood out by being a warmer 0.43 degrees Fahrenheit (0.24°C) than any previous July. Moreover, it surpassed the average July temperature from 1951 to 1980 by a notable margin of 2.1 degrees Fahrenheit (1.18°C). The primary emphasis of the GISS examination revolves around examining temperature shifts across extended periods, spanning decades and even centuries. This approach employs a fixed base period, resulting in consistent anomalies over time. Temperature "normals" are typically established based on several decades or more, often encompassing 30 years (O'Shea, 2023).

To address these rapidly increasing energy demands and ecological challenges, the search for green, renewable, and environmentally friendly energy sources has become a crucial topic among researchers over the past decades (Algae, 2014). Researchers have explored renewable resources such as solar, wind, hydro, tidal, biomass, and geothermal energy. Still, amongst all the alternatives, biomass has demonstrated its potential to significantly mitigate the prevailing energy crisis by

DOI: 10.1201/9781003732907-11

producing fuel rather than electricity. In the realm of biomass, biodiesel stands out as a noteworthy candidate for the emerging need for sustainable green energy sources (Abdullah et al., 2017).

Biodiesel, a renewable and biodegradable fuel, has garnered substantial global attention as a promising alternative to conventional fossil fuels due to its comparable chemical and fuel properties to diesel fuel, as well as its capacity to yield similar engine performance (Muniyappa et al., 1996; Nath et al., 2023; Subramaniam et al., 2013). With attributes such as low viscosity, high flash point, and excellent lubricity, biodiesel outperforms conventional fossil fuels in terms of reduced emissions of particulate matter, carbon monoxide, carbon dioxide, and nitrogen oxides (Abdullah et al., 2017; Deka and Basumatary, 2011; Mendonça et al., 2019). Furthermore, biodiesel's high combustion efficiency contributes to enhanced engine longevity, thereby aligning with the goals of sustainability and environmental responsibility (Kumar et al., 2021). Biodiesel is a long-chain alkyl ester synthesized via a chemical reaction known as transesterification. This reaction combines triglycerides, often found in vegetable oils or animal fats, with an alcohol, typically methanol or ethanol, and an acidic or basic catalyst. During the transesterification reaction, the functional group on the triglyceride chain is cleaved and replaced by the alcohol's functional group, generally the methoxy (-OCH3¬) group. Transesterification is a chemical reaction for directly converting one type of ester into another, at least one step in this synthetic methodology of converting glycerides into Fatty Acid Methyl Ester (FAME) (Basumatry, 2013). In the biodiesel production process, the choice of catalyst plays a pivotal role in determining efficiency, selectivity, and overall feasibility (Nath et al., 2023). Various catalysts have been employed, and heterogeneous base catalysts, particularly those derived from plant and agricultural waste, have gathered attention due to their advantages over traditional homogeneous and enzyme catalysts. Developing economically viable and environmentally sustainable heterogeneous catalysts that utilize readily available and renewable resources is paramount for the large-scale production of biodiesel. As compared to petroleum-based diesel, biodiesel has several added advantages, including the absence of aromatic compounds, sulfur-free, non-toxicity, high flashpoint, making it easier to transport, and renewability. In addition, blending of 5% biodiesel to diesel significantly reduces emission of pollutants into the air (Qasemi et al., 2022). Triglyceride can be converted into FAME or biodiesel using different methods, like the transesterification process, which is the most commonly industrially accepted process as it requires low pressure, low temperature, and less time. Apart from transesterification, pyrolysis (thermal cracking) and micro-emulation are also well-researched methods for biodiesel production.

This chapter aims to bridge the existing gap in the literature by comprehensively evaluating the catalytic performance of diverse agricultural waste-based catalysts in biodiesel synthesis. Through an extensive survey of various catalysts, the study endeavors to elucidate the merits and demerits of each approach. Given the increasing need for eco-friendly energy sources and the importance of reducing waste, using agricultural waste as a catalyst for making biodiesel fits well with sustainable practices. This book chapter thoroughly examines different catalysts and uses innovative decision-making methods to suggest a more environmentally friendly way to produce biodiesel and manage waste. This chapter adds to the vision of a future in which clean energy and waste reduction collaborate harmoniously by integrating scientific expertise, technological advancements, and ecological consciousness.

11.2 ACID CHEMICAL CATALYST

Aranda et al. (2008) undertook a comparative evaluation of various homogeneous acid catalysts—sulfuric acid (98%), phosphoric acid (85%), trichloroacetic acid (98%), and methanesulfonic acid (95%)—for the biodiesel production from palm oil within a batch reactor. Zheng et al. (2006) conducted transesterification of waste frying soybean oil and canola oil employing excess methanol in the presence of sulfuric acid. Their study emphasized the direct correlation between the biodiesel yield and factors such as methanol-to-oil molar ratio, temperature, and catalyst load.(Akoh and Swanson 1988) The influence of water in the oil phase on the transesterification reaction was also noted. Through kinetic investigations, the optimal reaction conditions were ascertained, comprising a catalyst load of 1.5%, a methanol-to-oil-to-acid ratio of 1:245:3.8, and reaction temperatures of 70°C and 80°C. At 80°C, the overall yield reached 99.4%, while at 70°C, it attained 99.0%. Zheng et al. (2006) conducted transesterification of waste frying soybean oil and canola oil in a 5-L stainless steel jacketed reactor employing excess methanol in the presence of sulfuric acid. Chamola et al. (2019) conducted methanol-assisted transesterification of dry algae utilizing H_2SO_4 as the catalyst. Employing response surface methodology (RSM) to optimize reaction conditions, they achieved an overall yield of 89.58% at a reaction temperature of 50°C, a methanol-to-dry algae ratio of 8:1 (w/w), a catalyst load of 3.361% (w/w), and a reaction time of 1 hour. Notably, sulfuric acid (H_2SO_4) and hydrochloric acid (HCl) have emerged as focal points of investigation due to their resilience in the presence of free fatty acids (FFAs) and their ability to facilitate concurrent esterification and transesterification reactions. However, the acid-catalyzed method is associated with slower reaction kinetics, elevated reaction temperatures, higher alcohol-to-oil ratios, the need for catalyst separation, and environmental and corrosion concerns. The advantages and limitations of this method were elucidated in multiple studies (Jacobson et al., 2008; Kulkarni and Dalai, 2006; Lam et al., 2010; Wang et al., 2006; Zhang et al., 2003). For instance, Wang et al. (2006) highlighted that the yield of fatty acid methyl esters (FAME) increased with extended reaction durations, enhanced methanol-to-oil ratios, and higher catalyst loading. Freedman et al. (1984) achieved a 99% oil conversion rate using 1 mol% H_2SO_4 and a methanol-to-oil ratio of 30:1 over 69 hours. These findings underscore the stringent conditions, including H_2SO_4 extended reaction times, inherent to the acid-catalyzed process in comparison to the typical base-catalyzed approach. Owing to these limitations, the utilization of solid acid catalysts for the direct synthesis of biodiesel has not been subject to widespread exploration primarily due to impediments associated with the gradual progression of reaction kinetics and the plausible initiation of adverse parallel reactions.

11.3 ALKALINE CHEMICAL CATALYST

Leung and Guo (2006) investigated the transesterification process of waste frying oil and canola oil utilizing three alkaline or basic catalysts: sodium hydroxide (NaOH), potassium hydroxide (KOH), and sodium methoxide (CH_3NaO). In their study, the researchers assessed and compared the fuel properties of the resultant biodiesel, including ester content, kinematic viscosity, and acid value. Among the array of catalysts employed, NaOH was identified by Leung and Guo (2006) as the preeminent choice due to its intermediary catalytic activity and cost-effectiveness. The authors successfully attained optimized reaction conditions for used frying oil, considering specific parameters such as an acid value of 2 and a viscosity of 35 cSt. Leung and Guo (2006) achieved an overall yield of 88% at 80°C reaction temperature, 20 minutes reaction time, 1.1wt% NaOH, and 7:1 molar ratio of alcohol to oil. Notably, the authors observed that elevated reaction temperatures (>50°C) yielded adverse effects on the yield, while waste oils characterized by higher viscosities exhibited positive responses. Furthermore, the transesterification reaction revealed that the catalyst load and methanol-to-oil molar ratio exert more influence than other parameters.

In a similar vein, Demirbas (2009) conducted a parallel investigation utilizing KOH and supercritical methanol transesterification on waste cooking oil. Achieving an overall yield of 87% required a reaction temperature of 87°C, a methanol ratio of 9:1, and a reaction time of 2 hours. Demirbas (2009) underscored the challenge of reduced yields stemming from elevated levels of free fatty acids and water content in the oil. Additionally, advantageous features of supercritical methanol transesterification, including the absence of a catalyst requirement and insensitivity to water and free fatty acids, were highlighted. Kouzu et al. (2008) explored the employment of a solid base catalyst, calcium oxide (CaO), for the transesterification of waste cooking oil via refluxing methanol. The pinnacle yield attained was 66% for waste cooking oil, achieved after 1 hour of reaction time within the temperature range of 60–65°C, utilizing a 12:1 methanol-to-oil molar ratio and a 0.85% (w/w) catalyst load. Using homogeneous and heterogeneous base catalysts in biodiesel production presents advantages and disadvantages. Homogeneous base catalysts offer a notable advantage through their remarkably swift reaction kinetics, which can outpace acid-catalyzed transesterification by approximately 4000. Moreover, these catalysts enable reactions to occur under milder conditions, leading to reduced energy requirements. Notably, cost-effectiveness is a hallmark of catalysts such as sodium hydroxide (NaOH) and potassium hydroxide (KOH), both readily available and affordable. However, the oil feedstock's free fatty acid (FFA) content influences its efficacy. An elevated FFA content can instigate the formation of undesirable soap byproducts, particularly if the FFA content exceeds 2 wt%. This precipitates decreased biodiesel yield and complicates the product purification process, notably contributing to generating substantial wastewater volumes. Conversely, heterogeneous base catalysts offer their own set of merits and demerits. Though relatively faster than acid-catalyzed transesterification, the reaction rate aligns with the mild reaction conditions characteristic of both homogeneous and acid-catalyzed processes, thereby ensuring energy efficiency. The intrinsic advantage of facile catalyst separation from the product bolsters the feasibility of reusing and regenerating the catalyst, accentuating its sustainability. Nonetheless, challenges persist, including the susceptibility of the catalyst to ambient air-induced poisoning. Wang et al. (2006) presented a two-step catalytic strategy for biodiesel synthesis. This involved the transesterification of waste cooking oil with an acid value of 75 mg KOH/g. The process employed by Wang et al. (2006) involved a sequential application of 2% ferric sulfate followed by 1% potassium hydroxide as an alkali catalyst, culminating in an impressive overall yield of 97.22%. Reaction parameters included a reaction time of 4 hours, a methanol-to-oil ratio of 10:1, and a second step yielding 90% after 10 hours and a methanol-to-oil ratio of 20:1. This method was noted for its simpler management of byproducts such as glycerol and soap stock, rendering it more straightforward compared to the acid-catalyzed method.

Moreover, the FFA content in the oil, owing to the catalyst's inherent alkaline nature, can exert an adverse impact, potentially leading to soap formation upon surpassing the 2 wt% FFA thresholds. The harmful effects of excessive soap formation resonate with the scenarios witnessed in homogeneous catalysis, manifesting as a reduction in biodiesel yield and complicating the downstream purification phase. It is important to note that the heterogeneous framework has the potential for catalyst active site leaching, which could lead to undesirable product contamination. This intricate interplay of advantages and disadvantages underscores the necessity for prudent catalyst selection in base-catalyzed transesterification for biodiesel production.

11.4 ENZYMATIC CATALYST

According to Wu et al. (2017) and Hwang et al. (2014), lipases are the common enzymes employed in synthesizing biodiesel. These enzymes facilitate the conversion of triglycerides (TGs) into glycerol and fatty acids through hydrolysis, placing them within the category of hydrolases. Wu et al. (2017) have investigated a lipase-catalyzed transesterification reaction with crude *Chlorella* lipids. Under optimized conditions, such as a molar ratio of 1:12 of algae oil to methanol, a ratio of 1:1 (m/v) of algae to co-solvent, and a ratio of 20:1 (m/m) of algae oil to lipase, the study achieved an overall yield of 97% FAME after 12 hours of reaction time. Wu et al. (2017) conducted a study to optimize transesterification reactions involving recycled triglycerides sourced from restaurants. They achieved this optimization by employing Response Surface Methodology (RSM), utilizing 95% ethanol as the reactant and PS-30 lipase derived from *Pseudomonas cepacia*. Under optimal conditions of 38.4°C reaction temperature, a catalyst load of 13.7 wt%, and a molar ratio of ethanol to grease at 1:6.6, the study reported an overall yield of 85.4%. Authors have also reported more than 96% enhanced yield on adding lipase SP435 (from *Candida antarctica*) after 1 hour of reaction. Hwang et al. (2014) investigated transesterification reaction using Novozyme 435 (lipase from *C. antarctica*) using vegetable and waste cooking oil and reported an overall yield of 95.9% and 90.4%, respectively. Researchers investigated biodiesel production from waste edible oil using immobilized *Candida antarctica* lipase in a fixed-bed bioreactor. They employed a three-step methanolysis approach with sequential column reactions, achieving 90% waste oil conversion to methyl esters (ME). A one-step method also yielded 90% ME content in the eluate. A study by Guo et al. (2020) aimed to create an environmentally friendly method for producing biodiesel using frying palm oil (FPO) and *Candida antarctica* lipase A (CALA) as a catalyst. The optimal conditions for the highest FAME yield (94.6%) were determined as 16.6% water content, 5.5% enzyme load, a 7:1 molar ratio of methanol to FPO, at 30°C for 22 hours. The kinetics of the transesterification process were described by the Arrhenius equation, with an activation energy (Ea) of 32.96 kJ/mol. Some of the enzymatic catalysts, along with their feedstock and yield, are mentioned in Table 11.1.

Enzyme catalysts have garnered attention in transesterification reactions for biodiesel production due to their unique advantages, including substrate specificity, operation under mild reaction conditions, and alignment with sustainability principles. Economically, enzyme production and immobilization can incur substantial costs, which assume amplified significance when applied to expansive industrial scales. Compared to traditional chemical catalysts, the comparatively restrained reaction kinetics associated with enzymatic catalysis introduce prolonged reaction durations and potential reductions in production efficiency. One potential solution to the high cost of enzyme production and immobilization is the implementation of agro-waste-based catalysts. These catalysts are derived from agricultural waste and have demonstrated notable efficacy in catalyzing chemical reactions at a lower overall expense. Industries may minimize their ecological impact by integrating such sustainable and environmentally friendly catalysts while enhancing profitability. Enzymatic reactions are

TABLE 11.1
Enzymatic Catalysts (Yaashikaa et al., 2022)

Enzyme Catalyst	Preparation Methods	Synthesis Parameters	Feedstock	Yield (%)
Lipase–PDA–TiO_2 NPs	Hydrothermal	Heated at 180°C for 4 h	*Jatropha curcas* oil	92
SBA-15@oleate@lipase	Hydrothermal	Heated at 130°C for 24 h	*Synechococcus elongatus* (cyanobacterium)	85 ± 3.2
Fe_3O_4-poly(GMA-co-MAA) composite-immobilized lipase	Solvothermal	Heated at 200°C for 12 h	Soybean oil	92.8
P. mendocina magnetic whole-cell biocatalysts	Ultrasonication	Fe_3SO_4 mixed with cells using ultrasonic wave for 10 min	Soybean oil	87.32
Immobilized lipase on epoxy functionalized silica	Immobilization	Incubated at 25°C for 24 h	Canola oil	98
Immobilized CRL in chitosan beads	Encapsulation	1:6 mass ratio of lipase to chitosan for 120 min	Cooking oil	94.06
Immobilized fungal whole cells	Immobilization	Incubated at 30°C for 72 h	Aqueous plant oil emulsion	96.1
Rhodococcus opacus strain PD630	Microbial fermentation	Depolymerizing pretreatment	Fishery waste	61.87
Waste eggshells	Grinding mill	Grinded with rotating speed of 340 rpm for 8 min	Sunflower waste cooking oil	97.5
AOL@PDMS-ZIF-L	Immobilization	Stirred magnetically for 1 h at room temperature	Soybean oil	94
Magnetically recyclable immobilized lipase	Immobilization	Incubated at 30°C for 5 h	Soybean oil with methanol	92.8
Lipase from *Pseudomonas brenneri*	Fermentation	Fermented at 24 h at pH = 7.0	Olive oil	85 ± 5

inherently susceptible to variations in reaction parameters, encompassing temperature, pH, and solvent composition. Maintaining optimal conditions for sustained enzymatic activity is crucial. Any deviations from these ideal situations can result in compromised catalytic effectiveness. This set of challenges can be overcome by using an agro-waste-based catalyst, as no pH adjustment is required or a temperature limitation is related as these catalysts don't get degraded above 100°C. The phenomenon of product inhibition manifests as an additional intricacy in enzyme catalysis. Byproducts, notably glycerol, can hinder the enzyme's catalytic ability over time, necessitating countermeasures such as augmented enzyme loading or recurrent replenishment. These interventions engender augmented operational expenditures. Agro-waste-based catalyst has been utilized for multiple cycles since these are heterogenous and calcinated, which remain intact, providing a highly porous surface resulting in a greater surface area-to-volume ratio. Hence, there is a minor yield decrease after every transesterification cycle at a minimal cost.

Furthermore, the specific substrate preferences that enzymes exhibit confer advantages and limitations. While this specificity aligns with precise catalytic action, it constrains the breadth of applicable feedstocks, a pertinent aspect when contemplating utilizing diverse raw materials in biodiesel synthesis. Enzyme deactivation, consequent to various factors encompassing denaturation, microbial contamination, and aggregation, introduces an element of temporal instability in catalytic performance. Thus, recurrent enzyme supplementation may be requisite, further contributing to the economic considerations. The intricate process of catalyst recovery and reuse poses practical challenges. While immobilization strategies offer potential mitigation, they can concurrently introduce their complexities, potentially necessitating supplementary purification procedures. Environmental considerations pivot around the biodegradability of enzymes; however, their production routes may involve chemical processes with associated energy intensiveness, potentially attenuating the purported environmental benefits. The transition from laboratory-scale enzymatic processes to industrial magnitude presents various challenges. Factors encompassing mass transfer limitations, preservation of uniform reaction conditions, and maintenance of enzyme stability become notably intricate on an expanded scale.

In summary, enzyme catalysts offer promising advantages for biodiesel production. Still, their drawbacks, such as cost, reaction kinetics, sensitivity to parameters, and scaling issues, require careful consideration when evaluating their feasibility in commercial applications. On the other hand, agro-waste-based catalysts offer a sustainable alternative, addressing some challenges associated with traditional catalysts. Research into enzymatic and agro-waste-based catalysts contributes to the quest for greener and more efficient biodiesel production methods.

11.5 AGRO-WASTE-BASED HETEROGENEOUS CATALYST

In sustainable catalysis for transesterification reactions, the utilization of catalysts derived from agricultural waste has emerged as a noteworthy advancement. This innovative approach addresses pressing environmental concerns and aligns with the growing need for efficient biodiesel production. By harnessing agricultural waste, a readily available and underutilized resource, catalysts are synthesized using well-defined processes to render them active and selective for the transesterification reaction. This strategic amalgamation offers an eco-friendly alternative to conventional catalysts while contributing to the prudent management of agricultural byproducts. The catalyst demonstrates proficiency in converting triglycerides into biodiesel, underscoring its potential significance in propelling green chemistry practices within the agricultural and energy sectors. As ongoing research in this domain unfolds, a more comprehensive understanding of the catalyst's underlying mechanisms and optimal operating conditions is anticipated to pave the way for sustainable and impactful transformations in biodiesel production methodologies. Numerous studies have documented the preparation of several heterogeneous catalysts for biodiesel synthesis using agro-waste materials, including *Musa balbisiana* trunk (Deka and Basumatary, 2011), *Musa balbisiana* peel (Gohain et al., 2017), *Musa paradisiaca* peel (Betiku and Ajala, 2014), *Musa balbisiana*

TABLE 11.2
Agro-Waste-Based Ash Catalyst

Ash Catalyst	Raw Material	Catalyst wt. (%)	Molar Ratio	Time (mins)	Temperature (°C)	Yield (%)	Reference
Wheat Bran Ash	Waste cooking oil (WCO)	11.6	1.46:1	114.2	54.6	93.6	Gouran et al. (2021)
Rice husk ash	Palm oil	7	9:1	240	65	91.5	Chen et al. (2015)
Acai seeds	Refined soybean oil	12	18:1	60	100	98.5±0.21	Mares et al. (2021)
Walnut shell ash	Moringa oleifera oil	4	18:1	240	65	97.34	Foroutan et al. (2022)
Walnut shell ash	sunflower oil	5	12:1	96.5	60	98	Miladinović et al. (2020)

underground stem, banana peel (Betiku et al., 2017), *Musa acuminata* peel (Pathak et al., 2018), *Musa paradisiaca* peel (Etim et al., 2018), *Lemna perpusilla* (Chouhan and Sarma, 2013), coconut husk, rice husk (Vinukumar et al., 2018), cocoa pod husk (Betiku et al., 2017), and tucuma peels (Mendonça et al., 2019). The aforementioned raw materials used in catalyst preparation are essentially discarded agro-waste materials.

11.6 ASH-BASED CATALYST

The ashes sourced from agro-waste predominantly consist of alkali, alkaline earth metals, carbonate, and other metal oxides. These components exhibit notable catalytic activity while posing fewer leaching problems. Consequently, the overall reaction becomes more efficient in terms of recyclability, devoid of storage or disposal concerns (Basumatary et al., 2018). Thus, research is centered on the utilization of ash-based catalysts derived from agro-waste for biodiesel production (Table 11.2). Deka and Basumatary (2011) conducted the transesterification of yellow oleander (*Thevetia peruviana* schum) seed oil using a heterogeneous catalyst derived from the trunk of *Musa balbisiana* Colla. The authors achieved a 96% conversion to FAME (fatty acid methyl ester) under optimized conditions, which included utilizing a mixture of 20 mL of methanol per gram of oil, a 20 wt% catalyst load, a reaction temperature of 32°C, and a reaction time of 3 hours.

In 2017, Betiku et al. explored the utilization of calcined cocoa pod husk ash (CCPHA) as a catalyst for the transesterification of neem seed oil into FAME. The neem seed oil underwent pretreatment with iron (III) sulfate ($Fe_2(SO_4)_3$) to decrease the free fatty acid content through esterification. The research team optimized the reaction conditions using both the Box–Behnken design (BBD) and the central composite design (CCD) of the response surface methodology (RSM). They achieved an overall yield of 97% FAME by utilizing a methanol-to-oil molar ratio of 0.73 (v/v), a catalyst amount of 0.65 wt% and a reaction time of 57 minutes, while maintaining the reaction temperature at 65°C.

11.7 AGRO-WASTE-BASED NANO CATALYST

Abdelhady et al. (2020) investigate the potential of crushed nano-sized CaO-rich agro-industrial waste from sugar beets as a high-performing catalyst for efficient biodiesel production. The catalyst demonstrates excellent activity in transesterification, utilizing sunflower oil and methanol. The catalyst's physicochemical properties are examined through various analyses, including Thermogravimetric Analysis – Derivative Thermogravimetry (TGA-DTG), X-ray Fluorescence (XRF), X-ray Diffraction

(XRD), Fourier Transform Infrared spectroscopy (FTIR), Scanning Electron Microscopy (SEM), Brunauer–Emmett–Teller (BET), and CO_2-Temperature-programmed desorption (TPD). Under optimized conditions (1 wt% catalyst loading, 4.5:1 methanol/oil ratio, reflux at 75°C for 60 min), biodiesel conversion reaches around 93%, with rapid reaction kinetics highlighting its economic viability. The CaO-rich residue can be reused twice without losing catalytic potency. The produced biodiesel meets ASTM D6751 and EN-14214 standards. Sathish et al.'s (2022) investigation explores utilizing agricultural waste as a promising catalyst for efficient biodiesel synthesis. Azolla, a fast-growing macroalgae, is employed for a self-sustaining biorefinery model. The researchers treat tannery effluents using phytoremediation and utilize algal oil extracted from Azolla to produce biodiesel through transesterification. They assess the catalyst through N_2 sorption, SEM, FTIR, and X-ray diffraction analyses. Optimal conditions are determined for the conversion process, revealing that 2% weight catalyst loading and 1:9 molar ratio result in a high biodiesel yield of 92% using Graphene Oxide catalyst at 60–65°C. The graphene oxide (GO) catalyst proves to be economically viable and recyclable. Cholapandian et al. (2022) synthesized a calcium oxide (CaO) nanocatalyst from *Acalypha indica* for biodiesel production from waste cooking oil (WCO). The study optimized transesterification conditions, including catalyst concentration, methanol-to-oil molar ratio, reaction temperature, and time. Through response surface methodology (RSM), the optimal values were determined as 2.4 wt% catalyst concentration, 11.8:1 methanol-to-oil molar ratio, 63.7°C reaction temperature, and 70 minutes reaction time, resulting in the highest biodiesel yield of 94.74%. The synthesized catalyst's chemical bonds, size, and structure were characterized using FTIR, XRD, and SEM. The resulting biodiesel was analyzed using GC-MS, and its physicochemical properties were assessed. This research indicates the potential of utilizing a CaO nanocatalyst from *Acalypha indica* for promising biodiesel production from WCO.

11.8 AGRO-WASTE-BASED METAL CATALYST

Prashant et al. (2016) explore the effective use of agro-industrial residue combined with metal oxides as a promising heterogeneous catalyst for improving biodiesel production. They subjected the *Musa Balbisiana* Colla Underground Stem (MBCUS) catalyst to thermal treatment and mixed it with Strontium Oxide (SrO) in a 5:1 w/w ratio, regenerating active sites. MBCUS-SrO exhibited more distinct crystalline phases, verified by XRD and TEM analysis. BET, SEM, EDX, TGA, and XRF analyses characterized composition and morphology. Using the Taguchi method with orthogonal arrays, the study optimized biodiesel production from *Jatropha curcas* L. oil (JCO) with high free fatty acid content. The derived linear equation guided optimal conditions, yielding 96% FAME under milder conditions (200°C, lower internal vapor pressure) than using MBCUS alone. Reduced metal leaching and enhanced catalyst reusability were observed. Conversion efficacy appeared less influenced by particle size than by the presence of basic active sites (Table 11.3).

Dai et al. (2014) explore the application of Li_2CO_3-modified waste peanut husk ash as a solid base catalyst for biodiesel production. The waste peanut husk catalyst's physical and chemical properties are assessed via XRD, BET, SEM, and TGA. This study involves synthesizing biodiesel from soybean oil using the solid base catalyst through transesterification. Optimal conditions include a methanol/oil molar ratio of 12:1, a catalyst amount of 5% (wt/wt oil), and a reaction temperature of 65°C for 4 hours, achieving a notable 98.4% soybean oil to biodiesel conversion. Ofori-Boateng and Lee (2013) conducted a study to assess the potential of cocoa pod husk (CPH)-derived potash for the transesterification of soybean oil into biodiesel. Both supported (CPH/MgO) and unsupported (CPH ash) potash catalysts were evaluated as green heterogeneous catalysts for biodiesel production. Optimal conditions for transesterification using CPH/MgO catalyst (1 wt% of MgO-doped CPH ash, oil/methanol ratio 1:6, 60°C, 60 min) and CPH catalyst (1 wt% of CPH ash, oil/methanol ratio 1:6, 60°C, 120 min) yielded biodiesel samples with yields of 98.7% and 91.4%, respectively, meeting European biodiesel quality standards (EN 14112). Engine tests demonstrated superior performance for all fuel samples (B100 and B40), with B40 closely resembling petroleum diesel characteristics.

TABLE 11.3
Agro-Waste-Based Heterogeneous Metal Catalyst

		Parameters					
Metal Catalyst	Raw Material	Catalyst wt. (%)	Molar Ratio	Time (mins)	Temp (°C)	Yield (%)	Reference
Potato peels/mixed metal	Soybean waste cooking oil	3	9:1	120	60	97.6	Daimary et al. (2022)
Zinc oxide/banana corm	Fish waste lipid	2.5	6:1	240	70 ± 2	~90	Dutta et al. (2019)
Sesamum indicum/ mixed metal	Sunflower oil	7	12:1	40	65	94.2	Nath et al. (2020)
ZnO nano metal oxides/razma seeds powder	*Pongamiapinnata* oil	1.5	9:1	240	120	85.4	Reddy Yadav et al. (2018)

This pioneering report highlights the feasibility of using CPH as a catalyst for biodiesel, showcasing the potential of producing commercial green heterogeneous base catalysts from CPH for sustainable biodiesel production.

Takase et al. (2021) synthesized and examined Sodium molybdate (Na_2MoO_4) as a heterogeneous solid catalyst for *Camelina sativa* seed oil biodiesel production. Transesterification reactions were conducted at standard atmospheric conditions, yielding high efficiency with low temperature, short reaction time, and normal pressure. Optimal conditions for efficient transesterification (yield >95%) included a methanol-to-oil molar ratio of 17:1, 6% catalyst amount, 60°C reaction temperature, and 2.5-hour reaction time. The molybdate complex exhibited strong Lewis acidity, likely facilitating alcohol O–H bond interaction and forming nucleophilic species. The catalyst was easily recoverable and demonstrated recyclability for five cycles with consistent activity after washing. Odude et al. (2019) conducted a study to optimize the production of fatty acid methyl esters from esterified palm kernel oil, utilizing calcined banana peel ash (CBPA) and calcined cocoa pod husk ash (CCPHA) as catalysts. The researchers used response surface methodology (RSM) to explore the impact of variables such as methanol/oil ratio, catalyst weight, and reaction time. RSM indicated identical optimal conditions for both catalysts: 0.8 v/v methanol/oil ratio, 4 wt% catalyst weight, and a 65-minute reaction time at 65°C. These conditions resulted in observed yields of 99.5 wt% for CBPA and 99.3 wt% for CCPHA.

Deka and Basumatary (2011) investigated using agricultural waste as an effective catalyst for biodiesel synthesis. They study biodiesel production from yellow oleander (*Thevetia peruviana* Schum) seed oil using a heterogeneous catalyst derived from *Musa balbisiana* Colla trunk. Biodiesel conversion reaches 96 wt% at 32°C in 3 hours, yielding biodiesel with methyl oleate, methyl palmitate, methyl linoleate, methyl stearate, and methyl arachidate. The biodiesel meets ASTM D6751, EN 14214, BS II, and BS III standards, with a superior cetane number (61.5) and no sulfur content. Vadery et al. (2014) present an economical method for producing jatropha oil biodiesel at room temperature using a coconut husk-derived catalyst rich in potassium, achieved through controlled heating without chemicals. The catalyst operates at 45°C, enabling efficient transesterification with various methanol/oil ratios. Notably, the system's unique feature is its low-temperature requirement. The paper outlines the optimization of variables like molar ratio, reaction time, catalyst calcination temperature, and catalyst/oil weight percentage. Under 45°C reaction temperature conditions and catalyst calcination at 350°C, biodiesel's FAME content exceeds 97% within 30 minutes. Sharma et al. (2012) explore using agro-industrial residue as a promising heterogeneous catalyst to improve biodiesel production. The research evaluates wood ash's efficacy in catalyzing Jatropha oil transesterification. Thermal treatment (calcination) of wood ash at 500–1200°C yields calcined wood ash catalysts (CWC), while chemical activation with K_2CO_3 and $CaCO_3$ produces activated wood ash catalysts (AWC) via double carbonate solid-state reaction. Characterization encompasses surface morphology, crystalline phases, textural attributes, and alkalinity. Methyl ester conversion of Jatropha oil attains 97–99% using CWC and AWC catalysts, satisfying critical ASTM D-6751 biodiesel criteria. The study underscores wood ash's potential as a biodiesel synthesis heterogeneous catalyst, reducing costs through abundant wood ash utilization. Betiku and Ajala (2014) converted *Thevetia peruviana* oil into biodiesel via a two-step transesterification process. Esterification was modeled, optimized using RSM, and reduced the high FFA yellow oleander oil's acid value to 1.72 mgKOH g^{-1}. A novel solid plantain peel catalyst was developed and characterized. ANN and RSM optimized transesterification: 0.3 (v/v) methanol–oil ratio, 1.5 h reaction, 3.0% (w/v) calcinated plantain peels yielded 95.25% (w/w) yellow oleander oil biodiesel (validated 94.87% (w/w)). The acid value of yellow oleander oil (YOO) with high FFA was reduced to 1.72 mgKOH g^{-1} by esterification with a statistically significant quadratic model at the optimal condition of methanol–oil ratio 0.35 (v/v), H_2SO_4 0.78% (v/v), reaction time 60 min, and reaction temperature 55°C

11.9 AGRO-WASTE-BASED DUAL-FUNCTION CATALYST

Jamil et al. (2021) describe the development of a dual-functional heterogeneous catalyst to convert plant-based oils into biodiesel, addressing environmental concerns and energy demands. Pristine carbon, derived from waste date seeds powder before oil extraction, is a catalyst support. It is modified with alkaline earth metal oxides for effective transesterification and biodiesel production. The catalyst undergoes comprehensive pre-usage characterization using XRD, SEM, EDS, BET, and TPD. Characterization confirms catalyst suitability for transesterification. Optimization of the process occurs with the synthesized catalyst. Biodiesel yield optimization yields 94.27 wt%, achieved with parameters: 65°C temperature, 90 min time, methanol-to-oil molar ratio of 15, and 4 wt% catalyst loading. Catalyst commercialization is assessed through the use of oils from various sources. The kinetic analysis underscores the energy-efficient nature of non-edible oil transesterification with the synthesized heterogeneous catalyst for biodiesel production.

Atelge (2022) synthesized a heterogeneous catalyst using KOH as an activation agent and support derived from spent coffee grounds-derived activated carbon. The highest iodine number was achieved at a carbonization temperature of 600°C. The synthesized catalyst underwent characterization and was assessed for biodiesel and hydrogen production efficiency. Optimal parameters for transesterification were identified as 3 wt% catalyst loading, 9:1 methanol-to-waste cooking oil molar ratio, 90°C reaction temperature, and 120 min reaction time, yielding 91.57% biodiesel. Hydrogen production showed 100% conversion under all conditions. The fastest hydrogen production occurred with 0.25 g of the superior catalyst and 2.5% $NaBH_4$ concentration at 30°C. The catalyst demonstrated reusability for five cycles, maintaining performance. Activation energies were determined as 19.15 kJ/mol for transesterification and 15.48 kJ/mol for methanolysis. Thermodynamically, both reactions were endergonic and unspontaneous; however, reactant-to-product conversion was effective, suggesting a favorable entropy change.

11.10 BIOREACTOR FOR BIODIESEL PRODUCTION

There are basically three types of reactors used in the production of biodiesel: Batch reactors, semi-continuous-flow reactors, and continuous-flow reactors. The batch process offers cost advantages with minimal initial investment and infrastructure requirements. It offers flexibility in handling different feedstock types, compositions, and quantities (Farm-Energy, 2019). However, it suffers from drawbacks like low productivity, greater variability in product quality, and increased labor and energy demands. The semi-continuous process closely resembles the batch method but involves initiating the reaction in a smaller volume and gradually adding ingredients until the vessel is full.

This approach demands significant labor and is not widely utilized. In contrast, continuous transesterification processes are favored over batch methods for large-scale commercial production due to their consistent product quality and cost-effectiveness. The most prevalent type of continuous-flow reactor is the continuous stirred-tank reactor. Commercial applications also employ alternative continuous-flow reactors like ultrasonic and supercritical reactors, which can accelerate the reaction process (Malpartida et al., 2020).

Before continuous processes became prevalent, biodiesel synthesis primarily relied on batch production using Batch Reactors (BRs). BRs are essentially tanks equipped with various forms of agitating equipment. In a BR, the chamber or tank is consistently filled with reactants, including catalyst, oil, and alcohol, while agitators are operated for a specific duration. Reactants are introduced into the reactor at specific intervals. Over time, the chemical composition inside the reactor undergoes changes, leading to fluctuations in product concentration. To achieve this transformation, the reactor is sealed, and reaction conditions such as pressure, temperature, and agitation rates are adjusted to the desired levels. This adjustment helps alleviate the inhibitory effects of reaction byproducts. Once the predetermined reaction time has elapsed, the chemical substances in the reactor are separated and sent for further processing. An essential characteristic of BRs is their progression from unreacted substances at the outset, through the reaction process, to the eventual production of fully reacted substances after a certain time. Moreover, BRs offer advantages like efficient mixing capabilities and the relative ease of managing homogeneous catalysts in biodiesel transesterification reactions

Continuous Stirred-Tank Reactors (CSTRs) are designed for continuous pumping of reactants, ensuring efficient operation. While they may appear similar to Batch Reactors (BRs), CSTRs require internal controls. The agitation speed and in-out flow rate are key components, influencing residence time, mass transfer rate, and mixing efficiency. Adequate mixing is essential for product homogeneity but consumes more energy (Saratale et al., 2019). Efforts have been made to reduce energy consumption by exploring alternative mixing methods. CSTRs are widely used for large-scale biodiesel production due to their established technologies and operational understanding. Maintaining constant chemical concentrations throughout the reactor is ideal but often challenging, necessitating adjustments to ensure complete reactions. Two-stage CSTR processes are sometimes employed, with 80% of the catalyst and alcohol mixed with oil in the first stage and the remaining 20% added in the second stage. This approach promotes complete reactions and minimizes alcohol usage compared to single-phase processes. CSTRs offer benefits such as improved thermal and mass transfer, lower production costs, product uniformity, and scalability when compared to BRs. These advantages make transesterification via CSTR preferable, particularly when consistent product quality and cost considerations are important (Abubackar et al., 2018).

Ultrasonic biodiesel reactors are effective for enhancing the mixing of ingredients in biodiesel production. Ultrasonic waves create intense mixing by generating cavitation bubbles that, upon bursting, lead to a rapid contraction of the fluid, promoting thorough ingredient mixing. This high-energy process significantly accelerates the reaction, reducing the reaction time without requiring elevated temperatures, allowing for operation at or slightly above ambient temperature and energy savings. Research has demonstrated the effectiveness of ultrasound in biodiesel production, resulting in similar biodiesel yields with significantly reduced reaction times compared to conventional stirred-tank methods. Ultrasonic reactors are suitable for small-scale producers, typically producing up to 2 million gallons per year, requiring only one or two ultrasound probes per reactor vessel. However, employing ultrasound in large-scale production may pose challenges due to the need for numerous ultrasound probes to ensure uniform mixing throughout the reactant mixture

11.11 LIFE CYCLE ASSESSMENT

Biodiesel synthesis relies on diverse feedstock sources, each characterized by its unique composition and purity. The initial and pivotal step in biodiesel production entails carefully selecting feedstock guided by yield, composition, and economic feasibility considerations. This decision hinges on classifying biodiesel feedstock into distinct categories, namely edible, non-edible, and waste-based sources, predicated upon factors such as availability and source type. Additionally, the geographical location plays a significant role in determining the feedstock choice for biodiesel manufacturing, with national accessibility and economic conditions serving as primary considerations. An excess of feedstock options, encircling waste oils (e.g., waste edible oil or WEO, and waste cooking oil or WCO), animal fats, microalgae oil, and various vegetable oils, stands at the disposal of biodiesel producers. Vegetable oils, such as palm oil, canola oil, rapeseed oil, Moringa oleifera seed oil, jatropha oil, sunflower oil, soybean oil, and jojoba oil, offer distinct possibilities. Notably, palm oil, sunflower oil, jojoba oil, and soybean oil have found widespread application in the biodiesel generation process. Waste cooking oils, often stemming from both households and restaurant establishments, represent a valuable source of feedstock, particularly in the form of WCO. Furthermore, animal fats, including goat fat, mutton fat, chicken fat, poultry fat, and ostrich fat, constitute a separate category within the realm of biodiesel feedstock options. Biodiesel production from different feedstocks is represented in Table 11.4 (Yaashikaa et al., 2022).

TABLE 11.4
Biodiesel Production via Different Feedstock and Catalyst (Yaashikaa et al., 2022)

S. No	Feedstock	Catalyst	Reaction	Reaction Time	Alcohol-to-Oil Molar Ratio	Yield (wt%)	Reaction Temp (°C)	Catalyst Conc. (wt%)
1	Microalgal	Carbon-based solid	Esterification	6 h	10:01	94.23	80	5%
2	Waste cooking oil	Waste Carica papaya stem	Transesterification	3 h	09:01	95.23	60	2%
3	Waste cooking oil	Lipase	Transesterification	4 h	03:01	88–90	40	1.50%
4	Sunflower oil	lipase@Bio-MOF	Transesterification	4 h	08:01	>60	50	–
5	Palm fatty acid distillate	SCB-SO_3H	Esterification	1.5 h	10:01	98.6	60	2%
6	*Chlorella vulgaris* microalgae	ROL/MNP-AP-GA	Enzymatic	24 h	–	69.8	40	–
7	*Synechococcus elongatus*	SBA-15@oleate@ lipase	Transesterification	17 h	3.4:1	85	37	34%
8	*Pseudomonas brenneri*	Lipase	Transesterification	45 min	09:01	85	70	–
9	Soyabean oil	*Aspergillus oryzae*	Batch methanolysis	96 h	–	98	30	–
10	Spent coffee grounds	1,8-diazabicyclo[5.4.0] undec-7-ene	Transesterification	60 min	02:01	97.18	130	–
11	Soyabean oil	Pectin	Transesterification	4 h	07:01	99	65	1%
12	*Pongamia pinnata* oil	*Delonix regia*	Transesterification	90 min	06:01	99.39	60	4%
13	Sunflower oil	*Aspergillus niger*	Transesterification	72 h	–	85	30	–
14	Soyabean oil	*Aspergillus niger*	Transesterification	12 h	04:01	95.1	30	–

(Continued)

TABLE 11.4 (Continued)

S. No	Feedstock	Catalyst	Reaction	Reaction Time	Alcohol-to-Oil Molar Ratio	Yield (wt%)	Reaction Temp (°C)	Catalyst Conc. (wt%)
15	Spent coffee grounds	Lipase	Transesterification	6 h	06:01	91.6	60	–
16	Palm oil	Radiation-induced kenaf	Transesterification	8 h	50:01:00	99	25	–
17	Orange peels waste	Sulfonated carbon	Transesterification	274 min	19.95:1	91.68	65	5%
18	Eucalyptus oil	Waste eggshell-derived Zn-CaO	Transesterification	1.5 h	06:01	93.2	65	5%
19	Karanja and Castor oil	Bio support beads immobilized with lipase	Transesterification	24 h	06:01	82	50	10%
20	Edible oil	Activated carbon from corncob	Transesterification	20 min	06:01	88.7	75	20%

11.12 OIL CONTENT

The oil content in different biodiesel feedstock sources can vary significantly, influencing their suitability and efficiency for biodiesel production. Here is a brief overview of the oil content in some familiar feedstock sources: Soybean: Soybean is a widely used biodiesel feedstock and typically contains approximately 18–20% oil content (Yao et al., 2020), making it one of the more oil-rich options among vegetable oils. Canola: Canola oil, derived from rapeseed, contains around 38–43% oil content (Gao et al., 2010), making it an attractive choice for biodiesel production due to its relatively high oil content. Palm Oil: Palm oil is another popular feedstock, with an oil content of approximately 45–50% (Koushki et al., 2015), contributing to its widespread use in biodiesel production. Sunflower: Sunflower oil contains roughly 35–42% oil content (Guo et al., 2017), making it a viable source of biodiesel when considering oil yield. Waste Cooking Oil (WCO): Waste cooking oil, often collected from restaurants and households, can have varying oil content, typically 2–5% (Liu et al., 2012). Its collection and usage help reduce waste and environmental impact. Animal Fats: Animal fats like tallow and lard can contain between 2% and 20% oil content (Andreo-Martínez et al., 2022), depending on the specific type of fat. These fats can be used as feedstock for biodiesel when available. Biodiesel producers need to consider the oil content of their chosen feedstock, as it directly affects biodiesel production's yield and economic feasibility. The selection of an oil source for biodiesel production typically revolves around its regional availability. With many oil sources at our disposal, making the appropriate choice is often straightforward. In certain instances, a judicious blend of different oils, such as a combination of animal fat and plant-based oil, can be employed to create biodiesel. These oil sources mustn't encroach upon resources meant for human food consumption.

Furthermore, the compositional profile of fatty acids within the chosen oil source is pivotal in the decision-making process. Additionally, feedstock sources' sustainability and environmental impact are crucial factors to consider in the decision-making process. Different feedstock and their oil content are listed in Table 11.5.

TABLE 11.5
Oil Content in Various Vegetable Oils

Feedstock	Oil Content (%)	Feedstock	Oil Content (%)
Moringa oleifera seed	38–42	Rice bran	15–23
Neem	20–30	Castor	45–50
Mahua	30–40	Cottonseed	18–25
Jatropha	38	Sea mango	54
Coconut	63–65	Rubber seed	40–60
Palm	30–60	Pumpkin seed	31.5
Karanja	27–39	Tobacco	30.43
Jojoba	50	Rice bran (repeat)	15–23
Tung	30–40	Olive	45–70
Corn	24.44	Mustard seed	33
Hemp seed	23–38	Sunflower	25–35
Rapeseed	38–46	Safflower seed	35
Peanut	45–55		

11.13 ENVIRONMENT INDICATOR

Environmental indicators are indispensable tools when conducting a life cycle assessment (LCA) for biodiesel, as they are crucial for gauging the ecological consequences of biodiesel production and utilization. These metrics offer a comprehensive perspective on how biodiesel interacts with the environment across its entire life cycle, encompassing everything from feedstock cultivation to ultimate consumption. Here, we outline some critical environmental indicators that are routinely considered in biodiesel LCAs.

Greenhouse Gas Emissions: This stands as one of the most critical indicators, as it assesses the overall impact of biodiesel on global warming. It quantifies the emissions of greenhouse gases, including carbon dioxide (CO_2), methane (CH_4), and nitrous oxide (N_2O) linked to feedstock cultivation, processing, transportation, and combustion (Xu et al., 2022). Energy Balance: Evaluating the energy balance is pivotal in determining the energy efficiency of biodiesel production. It involves comparing the energy input required for feedstock cultivation, processing, and transportation with the energy output derived from biodiesel as a fuel (Karlsson et al., 2016). Land Use Change: This indicator considers land-use alterations, particularly instances of deforestation or the conversion of natural ecosystems into agricultural land for biodiesel feedstock cultivation. Such changes in land use can carry substantial environmental implications (Shrestha et al., 2019). Biodiversity Impact: Biodiesel feedstock cultivation can impact local ecosystems and biodiversity. Assessing this impact is essential to ensure that biodiesel production does not harm fragile ecosystems or jeopardize endangered species (Campbell and Doswald, 2009). Water Use: The assessment of the water footprint of biodiesel production is critical, as it examines the consumption of water resources during feedstock cultivation and processing, considering both the volume of water used and the potential impacts on water quality. Air Quality: Biodiesel combustion has consequences for local air quality. Assessing emissions of air pollutants, such as particulate matter, nitrogen oxides (NO_x), and sulfur dioxide (SO_2), is crucial for understanding their effects on human health and the environment (Shrestha et al., 2019; Xu et al., 2022). Chemical Use: This indicator evaluates the use of chemicals and fertilizers in feedstock cultivation, considering their potential for soil and water contamination (Campbell and Doswald, 2009). Waste Generation: Assessing waste generation during biodiesel production and determining disposal methods are integral in minimizing the environmental impact

(Campbell and Doswald, 2009). Ecotoxicity: This indicator considers the potential toxic effects of biodiesel and its byproducts on aquatic and terrestrial ecosystems. Transportation and Supply Chain Emissions: It encompasses emissions linked to feedstock transportation to processing facilities and biodiesel distribution to end-users (Sarkar et al., 2021).

Biodiesel LCA studies leverage these indicators to provide a comprehensive understanding of the environmental impact of biodiesel. This information is invaluable for guiding policymakers, producers, and consumers in making informed decisions that promote more sustainable and environmentally friendly biodiesel production and usage (Motevali et al., 2023).

11.14 CONCLUSION

In conclusion, the escalating global demand for energy, accompanied by the rapid growth of urbanization, industrialization, and transportation, has underscored the critical role of fuels in powering modern societies. With fossil fuels constituting around 90% of global energy consumption, the energy landscape faces challenges, including environmental concerns, greenhouse gas emissions, and the depletion of finite petroleum reserves. The transport sector, responsible for over half of global energy consumption, is projected to drive a substantial increase in energy demand by 2050. However, the overuse of fossil fuels has led to emissions of greenhouse gases, contributing to global warming and the depletion of petroleum reserves at an unsustainable rate. To address these challenges, researchers have been focusing on the development of green, renewable, and environmentally friendly energy sources. Among these, biodiesel derived from biomass has emerged as a promising alternative due to its comparable properties to diesel fuel and its potential to mitigate the energy crisis. Biodiesel offers advantages such as reduced emissions, improved combustion efficiency, and compatibility with existing engines. In the realm of biodiesel production, the choice of catalyst plays a crucial role in determining efficiency and feasibility. Various catalysts have been explored, including acid, base, and enzyme catalysts. Acid-catalyzed transesterification offers advantages such as faster reaction kinetics but faces challenges related to soap formation and energy-intensive conditions. Base-catalyzed transesterification using alkaline catalysts like sodium hydroxide (NaOH) and potassium hydroxide (KOH) provides rapid reactions and has the potential for catalyst regeneration. Although offering specificity and mild reaction conditions, enzyme catalysts are associated with higher costs and slower reaction kinetics. Recent research has delved into utilizing ash-based heterogeneous catalysts derived from agricultural waste, which present advantages such as reduced leaching and enhanced recyclability. These catalysts offer a promising avenue for sustainable biodiesel production by addressing issues associated with traditional catalysts. In light of the increasing need for eco-friendly energy sources and waste reduction, integrating agricultural waste as catalysts for biodiesel production aligns well with sustainable practices. This review has shed light on the different types of catalysts and their respective advantages and challenges. The development of economically viable and environmentally sustainable catalysts will be pivotal for the large-scale production of biodiesel, contributing to a future where clean energy and waste reduction coalesce through scientific innovation and ecological consciousness.

REFERENCES

Abdelhady, H.H., Elazab, H.A., Ewais, E.M., Saber, M. and El-Deab, M.S., 2020. Efficient catalytic production of biodiesel using nano-sized sugar beet agro-industrial waste. *Fuel*, 261, p.116481.

Abdullah, S.H.Y.S., Hanapi, N.H.M., Azid, A., Umar, R., Juahir, H., Khatoon, H. and Endut, A., 2017. A review of biomass-derived heterogeneous catalyst for a sustainable biodiesel production. *Renewable and Sustainable Energy Reviews*, 70, pp.1040–1051.

Abubackar, H. N., Veiga, M. C., & Kennes, C. 2018. Production of acids and alcohols from syngas in a two-stage continuous fermentation process. *Bioresource technology*, 253, 227–234.

Akoh, C.C. and Swanson, B.G., 1988. Base catalyzed transesterification of transester oils 1. *Journal of Food Processing and Preservation*, 12(2), pp.139–149.

Algae, G.M., 2014. A method for production and characterization of biodiesel from green micro algae. *International Journal of Bio-Science and Bio-Technology*, 69(5), pp.111–122.

Andreo-Martínez, P., Ortiz-Martínez, V.M., Salar-García, M.J., Veiga-del-Bano, J.M., Chica, A. and Quesada-Medina, J., 2022. Waste animal fats as feedstock for biodiesel production using non-catalytic supercritical alcohol transesterification: A perspective by the PRISMA methodology. *Energy for Sustainable Development*, 69, pp.150–163.

Aranda, D.A., Santos, R.T., Tapanes, N.C., Ramos, A.L.D. and Antunes, O.A.C., 2008. Acid-catalyzed homogeneous esterification reaction for biodiesel production from palm fatty acids. *Catalysis Letters*, 122, pp.20–25.

Atelge, M.R., 2022. Production of biodiesel and hydrogen by using a double-function heterogeneous catalyst derived from spent coffee grounds and its thermodynamic analysis. *Renewable Energy*, 198, pp.1–15.

Banković-Ilić, I. B., et al., 2017. Application of nano CaO–based catalysts in biodiesel synthesis. *Renewable and Sustainable Energy Reviews*, 72, pp.746–760.

Basumatary, K., Daimary, P., Das, S. K., Thapa, M., Singh, M., Mukherjee, A., & Kumar, S. 2018. Lagerstroemia speciosa fruit-mediated synthesis of silver nanoparticles and its application as filler in agar based nanocomposite films for antimicrobial food packaging. *Food packaging and shelf life*, *17*, 99–106.

Basumatary, S. 2013. Transesterification with heterogeneous catalyst in production of biodiesel: A Review. *Journal of Chemical and Pharmaceutical Research*, *5*(1), 1–7.

Betiku, E. and Ajala, S.O., 2014. Modeling and optimization of Thevetia peruviana (yellow oleander) oil biodiesel synthesis via Musa paradisiacal (plantain) peels as heterogeneous base catalyst: A case of artificial neural network vs. response surface methodology. *Industrial Crops and Products*, 53, pp.314–322.

Betiku, E., Etim, A.O., Pereao, O. and Ojumu, T.V., 2017. Two-step conversion of neem (Azadirachta indica) seed oil into fatty methyl esters using a heterogeneous biomass-based catalyst: An example of cocoa pod husk. *Energy & Fuels*, 31(6), pp.6182–6193.

Campbell, A. and Doswald, N., 2009. *The impacts of biofuel production on biodiversity: A review of the current literature.* UNEP-WCMC, pp.309–540.

Chamola, R., Khan, M.F., Raj, A., Verma, M. and Jain, S., 2019. Response surface methodology based optimization of in situ transesterification of dry algae with methanol, H2SO4 and NaOH. *Fuel*, 239, pp.511–520.

Changmai, B., Vanlalveni, C., Ingle, A.P., Bhagat, R. and Rokhum, S.L., 2020. Widely used catalysts in biodiesel production: a review. *RSC Advances*, 10(68), pp.41625–41679.

Chen, G.Y., Shan, R., Shi, J.F. and Yan, B.B., 2015. Transesterification of palm oil to biodiesel using rice husk ash-based catalysts. *Fuel Processing Technology*, 133, pp.8–13.

Cholapandian, K., Gurunathan, B. and Rajendran, N., 2022. Investigation of CaO nanocatalyst synthesized from Acalypha indica leaves and its application in biodiesel production using waste cooking oil. *Fuel*, 312, p.122958.

Chouhan, A. P. S., & Sarma, A. K. 2013. Biodiesel production from Jatropha curcas L. oil using Lemna perpusilla Torrey ash as heterogeneous catalyst. *Biomass and Bioenergy*, *55*, 386–389.

Dai, Y.M., Chen, K.T., Wang, Y.J. and Chen, C.C., 2014. Application of peanut husk ash as a low-cost solid catalyst for biodiesel production. *International Journal of Chemical Engineering and Applications*, 5(3), p.276.

Daimary, N., Eldiehy, K.S., Boruah, P., Deka, D., Bora, U. and Kakati, B.K., 2022. Potato peels as a sustainable source for biochar, bio-oil and a green heterogeneous catalyst for biodiesel production. *Journal of Environmental Chemical Engineering*, 10(1), p.107108.

Deka, D.C. and Basumatary, S., 2011. High quality biodiesel from yellow oleander (Thevetia peruviana) seed oil. *Biomass and Bioenergy*, 35(5), pp.1797–1803.

Demirbas, A., 2009. Biodiesel from waste cooking oil via base-catalytic and supercritical methanol transesterification. *Energy Conversion and Management*, 50(4), pp.923–927.

Dutta, S., Jaiswal, K.K., Verma, R., Basavaraju, D.M. and Ramaswamy, A.P., 2019. Green synthesis of zinc oxide catalyst under microwave irradiation using banana (Musa spp.) corm (rhizome) extract for biodiesel synthesis from fish waste lipid. *Biocatalysis and Agricultural Biotechnology*, 22, p.101390.

Etim, O., Ekpo, A., Bassey, U., & Akpan, A. 2018. Effect of aqueous and ethanol leaf extracts of musa paradisiaca on serum protein, liver and kidney function in albino wistar rats. *IOSR Journal of Biotechnology and Biochemistry*, *4*(6), 16–19.

Farm-Energy, 2019. Reactors for biodiesel production. *Farm Energy*. Available at: https://farm-energy.extension.org/reactors-for-biodiesel-production/ (Accessed: 08 November 2023).

Foroutan, R., Peighambardoust, S.J., Mohammadi, R., Peighambardoust, S.H. and Ramavandi, B., 2022. Application of walnut shell ash/ZnO/K2CO3 as a new composite catalyst for biodiesel generation from Moringa oleifera oil. *Fuel*, 311, p.122624.

Freedman, B., Pryde, E.H. and Mounts, T.L., 1984. Variables affecting the yields of fatty esters from transesterified vegetable oils. *Journal of the American Oil Chemists' Society*, 61(10), pp.1638–1643. Available at: https://doi.org/10.1007/bf02541649

Gao, J., Thelen, K.D., Min, D.H., Smith, S., Hao, X. and Gehl, R., 2010. Effects of manure and fertilizer applications on canola oil content and fatty acid composition. *Agronomy Journal*, 102(2), pp.790–797.

Gohain, M., Devi, A., & Deka, D. 2017. Musa balbisiana Colla peel as highly effective renewable heterogeneous base catalyst for biodiesel production. *Industrial Crops and Products*, *109*, 8–18.

Gouran, A., Aghel, B. and Nasirmanesh, F., 2021. Biodiesel production from waste cooking oil using wheat bran ash as a sustainable biomass. *Fuel*, 295, p.120542.

Guo, J., Sun, S. and Liu, J., 2020. Conversion of waste frying palm oil into biodiesel using free lipase A from Candida antarctica as a novel catalyst. *Fuel*, 267, 117323.

Guo, S., Ge, Y. and Na Jom, K., 2017. A review of phytochemistry, metabolite changes, and medicinal uses of the common sunflower seed and sprouts (Helianthus annuus L.). *Chemistry Central Journal*, 11, pp.1–10.

Hwang, H.T., Qi, F., Yuan, C., Zhao, X., Ramkrishna, D., Liu, D. and Varma, A., 2014. Lipase-catalyzed process for biodiesel production: Protein engineering and lipase production. *Biotechnology and Bioengineering*, 111(4), pp.639–653.

Jacobson, K., Gopinath, R., Meher, L. C., & Dalai, A. K. 2008. Solid acid catalyzed biodiesel production from waste cooking oil. *Applied Catalysis B: Environmental*, *85*(1-2), 86–91.

Jamil, F., Kumar, P.S.M., Al-Haj, L., Myint, M.T.Z. and Ala'a, H., 2021. Heterogeneous carbon-based catalyst modified by alkaline earth metal oxides for biodiesel production: Parametric and kinetic study. *Energy Conversion and Management: X*, 10, p.100047.

Karlsson, H., Ahlgren, S., Sandgren, M., Passoth, V., Wallberg, O. and Hansson, P.A., 2016. A systems analysis of biodiesel production from wheat straw using oleaginous yeast: process design, mass and energy balances. *Biotechnology for Biofuels*, 9(1), pp.1–13.

Koushki, M., Nahidi, M. and Cheraghali, F., 2015. Physico-chemical properties, fatty acid profile and nutrition in palm oil. *Archives of Advances in Biosciences*, 6(3), pp.117–134.

Kouzu, M., Kasuno, T., Tajika, M., Sugimoto, Y., Yamanaka, S. and Hidaka, J., 2008. Calcium oxide as a solid base catalyst for transesterification of soybean oil and its application to biodiesel production. *Fuel*, 87(12), pp.2798–2806.

Kulkarni, M.G. and Dalai, A.K., 2006. Waste cooking oil an economical source for biodiesel: a review. *Industrial & Engineering Chemistry Research*, 45(9), pp.2901–2913.

Kumar, S., Singhal, M.K. and Sharma, M.P., 2021. Utilization of mixed oils for biodiesel preparation: a review. *Energy Sources, Part A: Recovery, Utilization, and Environmental Effects*, 47(1), pp.4811–4844.

Lam, M.K., Lee, K.T. and Mohamed, A.R., 2010. Homogeneous, heterogeneous and enzymatic catalysis for transesterification of high free fatty acid oil (waste cooking oil) to biodiesel: a review. *Biotechnology Advances*, 28(4), pp.500–518.

Leung, D.Y.C. and Guo, Y., 2006. Transesterification of neat and used frying oil: optimization for biodiesel production. *Fuel Processing Technology*, 87(10), pp.883–890.

Liu, Y., Sotelo-Boyás, R., Murata, K., Minowa, T. and Sakanishi, K., 2012. Production of bio-hydrogenated diesel by hydrotreatment of high-acid-value waste cooking oil over ruthenium catalyst supported on Al-polyoxocation-pillared montmorillonite. *Catalysts*, 2(1), pp.171–190.

Malpartida, I., Maireles-Torres, P., Vereda, C., Rodríguez-Maroto, J.M., Halloumi, S., Lair, V., Thiel, J. and Lacoste, F., 2020. Semi-continuous mechanochemical process for biodiesel production under heterogeneous catalysis using calcium diglyceroxide. *Renewable Energy*, 159, pp.117–126.

Mares, E.K.L., Goncalves, M.A., da Luz, P.T.S., da Rocha Filho, G.N., Zamian, J.R. and da Conceição, L.R.V., 2021. Acai seed ash as a novel basic heterogeneous catalyst for biodiesel synthesis: Optimization of the biodiesel production process. *Fuel*, 299, p.120887.

Mendonça, I.M., Paes, O.A., Maia, P.J., Souza, M.P., Almeida, R.A., Silva, C.C., Duvoisin Jr, S. and de Freitas, F.A., 2019. New heterogeneous catalyst for biodiesel production from waste tucumã peels (Astrocaryum aculeatum Meyer): Parameters optimization study. *Renewable Energy*, 130, pp.103–110.

Miladinović, M.R., Zdujić, M.V., Veljović, D.N., Krstić, J.B., Banković-Ilić, I.B., Veljković, V.B. and Stamenković, O.S., 2020. Valorization of walnut shell ash as a catalyst for biodiesel production. *Renewable Energy*, 147, pp.1033–1043.

Motevali, A., Hooshmandzadeh, N., Fayyazi, E., Valipour, M. and Yue, J., 2023. Environmental impacts of biodiesel production cycle from farm to manufactory: an application of sustainable systems engineering. *Atmosphere*, 14(2), p.399.

Muh, E., Tabet, F. and Amara, S., 2019. Biomass conversion to fuels and value-added chemicals: a comprehensive review of the thermochemical processes. *Current Alternative Energy*, 4(1), pp.3–25.

Muniyappa, P.R., Brammer, S.C. and Noureddini, H., 1996. Improved conversion of plant oils and animal fats into biodiesel and co-product. *Bioresource Technology*, 56(1), pp.19–24.

Nath, B., Basumatary, B., Wary, N., Basumatary, U.R., Basumatary, J., Rokhum, S.L., Azam, M., Min, K. and Basumatary, S., 2023. Agricultural waste-based heterogeneous catalyst for the production of biodiesel: A ranking study via the VIKOR method. *International Journal of Energy Research*, 2023(1), p.7208754.

Nath, B., Kalita, P., Das, B. and Basumatary, S., 2020. Highly efficient renewable heterogeneous base catalyst derived from waste Sesamum indicum plant for synthesis of biodiesel. *Renewable Energy*, 151, pp.295–310.

O'Shea, C. (2023, August 14). NASA clocks July 2023 as hottest month on record ever since 1880. *NASA*. https://www.nasa.gov/press-release/nasa-clocks-july-2023-as-hottest-month-on-record-ever-since-1880

Odude, V.O., Adesina, A.J., Oyetunde, O.O., Adeyemi, O.O., Ishola, N.B., Etim, A.O. and Betiku, E., 2019. Application of agricultural waste-based catalysts to transesterification of esterified palm kernel oil into biodiesel: A case of banana fruit peel versus cocoa pod husk. *Waste and Biomass Valorization*, 10, pp.877–888.

Ofori-Boateng, C. and Lee, K.T., 2013. The potential of using cocoa pod husks as green solid base catalysts for the transesterification of soybean oil into biodiesel: Effects of biodiesel on engine performance. *Chemical Engineering Journal*, 220, pp.395–401.

Pathak, J., Rajneesh, Maurya, P. K. Singh, S. P., Haeder, D. P., & Sinha, R. P. 2018. Cyanobacterial farming for environment friendly sustainable agriculture practices: innovations and perspectives. *Frontiers in Environmental Science*, *6*, 7.

Prashant, K., Lakhotia, D., Bhattacharyya, T. D., Mahanta, A. K., & Ravoof, A. 2016. A comparative study of two percutaneous pinning techniques (lateral vs medial–lateral) for Gartland type III pediatric supracondylar fracture of the humerus. *Journal of orthopaedics and traumatology*, *17*(3), 223–229.

Qasemi, Z., Jafari, D., Jafari, K., & Esmaeili, H. 2022. Heterogeneous aluminum oxide/calcium oxide catalyzed transesterification of Mespilus germanica triglyceride for biodiesel production. *Environmental Progress & Sustainable Energy*, *41*(2), e13738.

Reddy Yadav, L.S., Raghavendra, M., Udayabhanu, Manjunath, K. and Nagaraju, G., 2018. Photocatalytic, biodiesel, electrochemical sensing properties and formylation reactions of ZnO nanoparticles synthesized via eco-friendly green synthesis method. *Journal of Materials Science: Materials in Electronics*, 29(10), pp.8747–8759.

Saratale, G.D., Saratale, R.G., Banu, J.R. and Chang, J.S., 2019. Biohydrogen production from renewable biomass resources. In *Biohydrogen* (pp. 247–277). Elsevier.

Sarkar, B., Mridha, B., Pareek, S., Sarkar, M. and Thangavelu, L., 2021. A flexible biofuel and bioenergy production system with transportation disruption under a sustainable supply chain network. *Journal of Cleaner Production*, 317, p.128079.

Sathish, S., Supriya, S., Andal, P., Prabu, D., Rajasimman, M., Ansar, S. and Rezania, S., 2022. Effective utilization of azolla filiculoides for biodiesel generation using graphene oxide nano catalyst derived from agro-waste. *Fuel*, 329, p.125412.

Satyanarayana, K.G., Mariano, A.B. and Vargas, J.V.C., 2011. A review on microalgae, a versatile source for sustainable energy and materials. *International Journal of Energy Research*, 35(4), pp.291–311.

Seela, C.R., Raoa, D.V. and Raoa, M.V., 2013. Performance analysis of an air driven engine modified from si engine. *International Journal of Current Engineering and Technology*, 3(4), pp.1440–1446.

Sharma, M., Khan, A.A., Puri, S.K. and Tuli, D.K., 2012. Wood ash as a potential heterogeneous catalyst for biodiesel synthesis. *Biomass and Bioenergy*, 41, pp.94–106.

Shrestha, D.S., Staab, B.D. and Duffield, J.A., 2019. Biofuel impact on food prices index and land use change. *Biomass and Bioenergy*, 124, pp.43–53.

Subramaniam, D., Murugesan, A. and Avinash, A., 2013. Performance and emission evaluation of biodiesel fueled diesel engine abetted with exhaust gas recirculation and Ni coated catalytic converter. *Journal of Renewable and Sustainable Energy*, 5(2).

Takase, M., Essandoh, P.K. and Kipkoech, R., 2021. Modified sodium molybdate as an efficient high yielding heterogeneous catalyst for biodiesel from Ghanaian indigenous Camelina sativa as a non-edible resource. *Discover Sustainability*, 2, pp.1–11.

Vadery, V., Narayanan, B.N., Ramakrishnan, R.M., Cherikkallinmel, S.K., Sugunan, S., Narayanan, D.P. and Sasidharan, S., 2014. Room temperature production of jatropha biodiesel over coconut husk ash. *Energy*, 70, pp.588–594.

Vinukumar, K., Azhagurajan, A., Vettivel, S. C., & Vedaraman, N. 2018. Rice husk as nanoadditive in diesel–biodiesel fuel blends used in diesel engine. *Journal of Thermal Analysis and Calorimetry*, *131*(2), 1333–1343.

Wang, Y., Ou, S., Liu, P., Xue, F. and Tang, S., 2006. Comparison of two different processes to synthesize biodiesel by waste cooking oil. *Journal of Molecular Catalysis A: Chemical*, 252(1–2), pp.107–112.

Wu, S., Song, L., Sommerfeld, M., Hu, Q. and Chen, W., 2017. Optimization of an effective method for the conversion of crude algal lipids into biodiesel. *Fuel*, 197, pp.467–473.

Xu, H., Ou, L., Li, Y., Hawkins, T.R. and Wang, M., 2022. Life cycle greenhouse gas emissions of biodiesel and renewable diesel production in the United States. *Environmental Science & Technology*, 56(12), pp.7512–7521.

Yaashikaa, P.R., Kumar, P.S. and Karishma, S., 2022. Bio-derived catalysts for production of biodiesel: A review on feedstock, oil extraction methodologies, reactors and lifecycle assessment of biodiesel. *Fuel*, 316, p.123379.

Yao, Y., You, Q., Duan, G., Ren, J., Chu, S., Zhao, J., Li, X., Zhou, X. and Jiao, Y., 2020. Quantitative trait loci analysis of seed oil content and composition of wild and cultivated soybean. *BMC Plant Biology*, 20, pp.1–13.

Zhang, Y., Dube, M.A., McLean, D.D.L. and Kates, M., 2003. Biodiesel production from waste cooking oil: 1. Process design and technological assessment. *Bioresource Technology*, 89(1), pp.1–16.

Zheng, S., Kates, M., Dube, M.A. and McLean, D.D., 2006. Acid-catalyzed production of biodiesel from waste frying oil. *Biomass and Bioenergy*, 30(3), pp.267–272.

12 Mycelium Technology
Historical Background, Applications, and Future Prospects

P. Sinu, Nimmi C. Sekhar, and Gazliya Nazimudheen*

12.1 INTRODUCTION

The global population is experiencing an unprecedented and concerning surge in numbers. This demographic explosion necessitates advancements in technology to ensure the well-being of people. However, this imperative for technological progress comes with a significant caveat: if not managed responsibly, it can result in severe environmental consequences, including pollution, excessive waste production, and the depletion of precious natural resources. The consequences of unchecked growth are particularly evident in waste generation. As the world witnesses rapid population expansion and urbanization, it is anticipated that annual waste generation will surge dramatically. In just nine years, from 2016 to 2025, it is projected to increase by a staggering 70%, soaring from 2.01 billion tons to 3.40 billion tons by 2050. This rapid escalation underscores the urgent need for sustainable technological innovations to tackle the impending challenges of waste management and environmental sustainability on a global scale [1, 2].

In contemporary industry, many commonly used materials, including those for construction and packaging, exhibit non-recyclable properties and are often detrimental to the environment. The use of these conventional materials not only depletes valuable natural resources but also exacts a toll on the environment throughout their lifecycle. This includes energy-intensive manufacturing processes, transportation-related emissions, and pollution of air, soil, and water bodies during production and eventual disposal. Alarmingly, a substantial portion of global carbon dioxide emissions, ranging from 8% to 10%, can be attributed to the manufacturing of construction materials alone. These concerns underscore the urgent need for more sustainable and eco-friendly alternatives to mitigate the environmental impact of industrial materials [3–5].

Mycelium-based materials have become more common in the building sector because of their versatility and environmental friendliness. Since they could be manufactured with less energy and could be grown with minimal harm to the environment, mycelium insulation, bricks, and other building materials showed promise as eco-friendly alternatives. The textile and fashion industries have also been impacted by mycelium technology. Designers and manufacturers believe that mycelium has the potential to provide an eco-friendly, degradable alternative to leather and synthetic materials. In addition to appealing to customers who care about the environment, these mycelium textiles helped the fashion industry leave a less ecological footprint [6].

Microplastics, which are defined as plastic particles smaller than 5 mm, have spread across the environment and have been found in a variety of ecosystems, including the soils of terrestrial landscapes, the deep oceans, and even the air we breathe. These tiny plastic fragments mostly result from the aging and breakdown of bigger plastic objects over time. Due to their ability to collect, persist, and interact with a variety of creatures, microplastics pose serious ecological and environmental issues since they may infiltrate the food chain and endanger both animal and human health [7–10].

* All authors having equal contribution.

 DOI: 10.1201/9781003732907-12

By providing sustainable, biodegradable substitutes for conventional plastics, mycelium products represent a possible means of reducing microplastic production and usage. The performance and scalability of mycelium materials must be improved, consumer awareness must be increased, stakeholder cooperation must be fostered, and supportive policies must be put in place to promote the adoption of eco-friendly alternatives across a range of industries [11, 12].

The development of mycelium-based products corresponded to a more pervasive global movement toward environmentally friendly living and sustainable practices. When it became increasingly clear that traditional production methods had a harmful impact on the environment, mycelium technology became a source of hope [13]. Researchers and businesses continue to make investments in creating mycelium culture techniques, expanding scalability, enhancing material performance, and addressing issues in order to make these advances even more accessible and beneficial [14]. In short, the conventional approach to material production falls short of creating a sustainable environment and a circular economy. The advancement of technology in sustainable materials production has become a pivotal focus in biotechnology and civil engineering research. Recent studies highlight the potential of mycelium-based biocomposites as substitutes for traditional materials. Mycelium technology experienced a dramatic resurgence in the 21st century as a result of rising awareness of its ecological and environmental benefits. This resurrection garnered a lot of interest from businesspeople, academics, and inventors due to the enormous potential of mycelium. One of mycelium's most amazing abilities is its capacity to transform into a range of unique and biodegradable materials using a process called "mycotecture." As a result of this increased understanding of mycelium's potential, numerous industries saw revolutionary breakthroughs. For example, mycelium-based goods were employed in packaging as a more environmentally friendly option to traditional plastics. Mycelium packaging demonstrated biodegradability in addition to similar protective characteristics, addressing the pressing issue of plastic waste [2, 13].

Summarizing, mycelium, the vegetative part of fungi, offers a fast-growing, safe, and inert binding matrix. It naturally adheres to various organic substrates, forming a dense network of threads. Mycelium-derived materials offer distinct advantages, including cost-effectiveness, low density, eco-friendliness, and reduced energy consumption compared to synthetic materials [15–17]. This ongoing commitment to innovation suggests that mycelium technology has the potential to revolutionize a wide range of sectors and develop a more sustainable and ecologically conscious society. This chapter aims to explore the potential of fungal mycelium in sustainable materials production and assess its future prospects.

12.1.1 History and Evolution of Mycelium Technology

Mycelium technology has a fascinating past and evolution, from its early use in cultures that valued mushrooms for their therapeutic and gastronomic benefits to the present day of sustainable innovation. The scientific understanding of mycelium and fungi during the 19th century, particularly their role in reproduction, benefited greatly from the work of mycologists like Louis René Tulasne and Anton de Bary [18]. After some time, mycelium began to appear in a variety of industries, first in the production of food and then in medicine, particularly after the discovery of penicillin [2, 19].

The history and development of mycelium technology extend thousands of years, and ancient usage of mycelium for food and medicine stretches back to prehistoric times. Albert Bernhard Frank's discovery of mycorrhizal symbiosis, which revealed the mutualistic relationship between plants and mycorrhizal fungus, was a significant turning point in science in the 1880s [20, 21]. The evolution of fungal cultivation techniques over the 20th century enabled controlled mycelium growth for research and industrial purposes. The ability of mycelium to absorb poisons from the environment is used in mycoremediation, which first gained popularity in the late 20th century [21–23]. Due to its use as a crucial component in the creation of sustainable goods like packaging and building materials, as well as cutting-edge products like meat and leather substitutes, mycelium's potential has substantially expanded in the 21st century. The history and development of mycelium technology are shown in chronological order in Figure 12.1 [24–26].

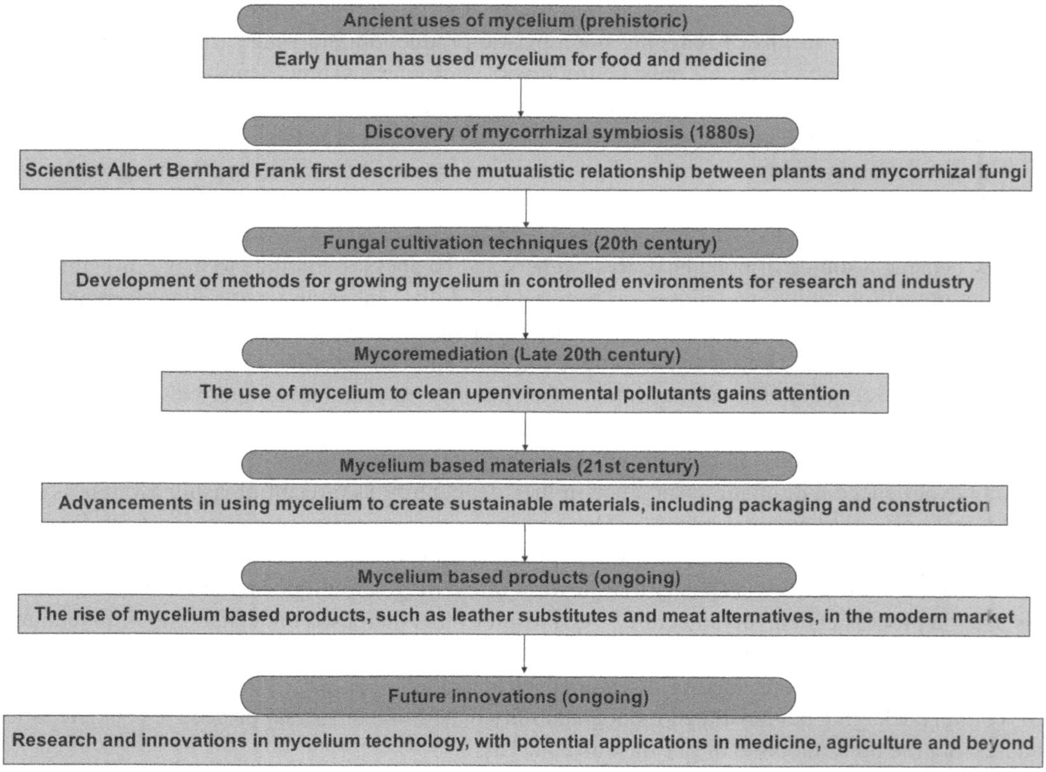

FIGURE 12.1 Chronology of the development and history of mycelium technology.

The tale of mycelium technology is still being written today, too. Our knowledge of and capacity to use mycelium could be improved through research studies that are now in progress and novel applications in a variety of sectors, including agriculture, medicine, and other fields. This technical journey highlights the enormous impact that mushrooms, and more particularly, mycelium, have on the world and the critical role that they play in ecologically friendly and sustainable solutions [27, 28].

12.1.2 Composite Mycelium Materials (CMMs)

Mycelium, an expansive macroscopic network, is eventually produced by the natural growth pattern of filamentous fungi, which is characterized by the continual creation of microscopic tubular cells that are connected to one another. These heterotrophic organisms play important ecological roles in the natural world by efficiently digesting and recycling different types of biomass in both terrestrial and marine habitats. This power is a result of their extraordinary secretory skills, which have been extensively utilized in several biotechnological applications [20, 29]. Harnessing fungal mycelium as a key resource has tremendous promise for providing scalable solutions in the search for sustainable material alternatives, according to clear signs of innovation and a rising trend in patent applications [30]. The first focus was on introducing fungal saprotrophic species to lignocellulosic waste materials like maize husks or sawdust within a mold to create composite mycelium materials (CMM). To stop the growing process, the fungus invaded and bound the substrate particles to create a composite substance [12, 31]. These substances are produced by enabling fungal mycelium to colonize and bind together lignocellulosic waste products, resulting in a natural composite. Composite mycelium materials are an environmentally friendly breakthrough that comes from the fusion of biology and materials science. The hyphal network of fungi, known as mycelium, is incorporated into organic substrates like agricultural waste to create these compounds. As the mycelium matures and binds the substrate together, a robust and biodegradable composite forms [32].

CMMs stand out for their environmental friendliness, versatility, and capacity to replace conventional materials. They are utilized in a variety of industries, including construction, packaging, and fashion. These materials are biodegradable, decreasing their negative environmental consequences, in addition to possessing beneficial properties like fire resistance and insulation. In summary, composite mycelium materials show the potential of harnessing the natural recycling processes to create long-lasting alternatives in our hunt for new and eco-friendly materials [33]. CMMs are useful in a variety of fields, such as eco-friendly packaging, where they serve as protective padding; insulation for buildings, where they can control temperature and save energy; and the creation of light and sustainable furniture, all of which support environmentally friendly design and manufacturing techniques (see Figure 12.2).

Hence, mycelium-based products are vastly preferable to synthetic alternatives, especially in single-use applications, because of their lower carbon footprint and increased sustainability. This advantage is a result of the properties and production process of mycelium materials. In contrast to synthetic materials, which usually need energy-intensive stages and increase carbon emissions during manufacturing, CMMSs are produced from organic substrates in a resource-efficient manner. They utilize agricultural or forestry waste by using them as feedstock and converting them into usable commodities that disintegrate. Mycelium-based single-use products therefore minimize greenhouse gas emissions and garbage disposed in landfills, which is consistent with environmentalists' attempts to lessen the harmful effects of throwaway goods on the environment [2].

The study by Li et al. offers a thorough overview of mycelial pellets, microbial aggregators with outstanding characteristics for metabolite release and pollution absorption. High biological activity, quick sedimentation, and simple solid–liquid separation are a few benefits of these pellets.

FIGURE 12.2 Applications of composite mycelium materials.

Technology using mycelial pellets has potential for using biomass, notably in wastewater treatment and harvesting microalgal biomass. The article discusses their properties, modes of creation, environmental factors, and a variety of applications, providing insightful information on reducing water pollution, collecting microalgae, and industrial fermentation. It also discusses issues with research and provides potential directions for mycelial pellet technology in the future [34]. The investigation of the mechanical properties of bacterial cellulose-reinforced mycelium composite materials by Elise Elsacker and colleagues, which advances the ongoing development of fully biological hybrid materials with efficient mechanical properties, is similar to this. They looked at whether the addition of bacterial cellulose to mycelium composite materials strengthened the internal bonding of the material and provided the material with adjustable mechanical properties. Therefore, our research aids in the continuous development of totally biological hybrid materials with effective mechanical properties [35].

12.2 SYNTHESIS AND STRUCTURAL ORGANIZATION OF MYCELIUM MATRIX

The production of mycelium-based composites is regarded as environmentally friendly because waste materials are recycled and the ecology is preserved by mining natural resources. Fungi employ a wide variety of substrates as feedstock and have the ability to adjust their growth to the surrounding environment. The low-cost bio-fabrication method known as mycelium growth can be used to recycle agricultural waste into sustainable biomaterials. Mycelium-based materials are built upon the three fundamental sets of fungus type and substrate, processing parameters, and applications. These materials are biodegradable, lightweight, and provide a greener alternative to packaging made of petroleum. They are also more successful at insulating against heat and sound than synthetic foam and wood fibers because of their low density, low thermal conductivity, and highly porous structure. They are also sturdy and incredibly fire-resistant, making them suitable alternatives to typical building materials [22, 25, 31, 36, 37]. Fungi can be used to produce biomaterials with potential applications in the pharmaceutical, medical, and cosmetics industries. However, there hasn't been enough research done on the potential contributions of a few fungal species to the study of materials. Interdisciplinary research in collaboration with genomes, metabolomics, and proteomics can disclose the molecular mechanisms of mycelium formation and material biology, increasing the number of fungal species that are exploited in various applications [38].

The novel mycelium-based composites are also naturally biodegradable because they are constructed completely of biocompatible ingredients [39]. Due to their diverse bio-adoptability and habitat, fungi are a promising choice for customizable material production. The various stages of the synthesis of mycelium-based products are schematically depicted in Figure 12.3. Usually, the substrate is inoculated with mycelial fragments or fungal spores to start this process. A web-like network of hyphae (individual fungal threads) is then formed as the mycelium spreads and breaks down the substrate. Mycelium expands its thread-like hyphae across the substrate as it develops, creating a complex network that effectively takes nutrients from the substrate while also supporting the mycelium's structure. The mycelium secretes enzymes and acids that break down the organic substrate material into fungal biomass simultaneously. The mycelium serves as a binding agent throughout this digestive process, which turns the substrate into a thick and linked matrix. As the mycelium develops inside the substrate, it reinforces the matrix and qualifies it for a variety of uses. After harvesting and additional processing, mycelium-based products are praised for their sustainability, biodegradability, and versatility in sectors like construction, insulation, and packaging [40]. The specific structural, physical, chemical, mechanical, and biological properties of the mycelium-based composites are influenced by the strain, feeding substrate, and production procedure. The primary building blocks of the mycelium cell wall are chitin, glucans, proteins, and lipids. Their concentrations depend on the feeding substrate, which in turn defines the final properties of the materials produced [22, 24].

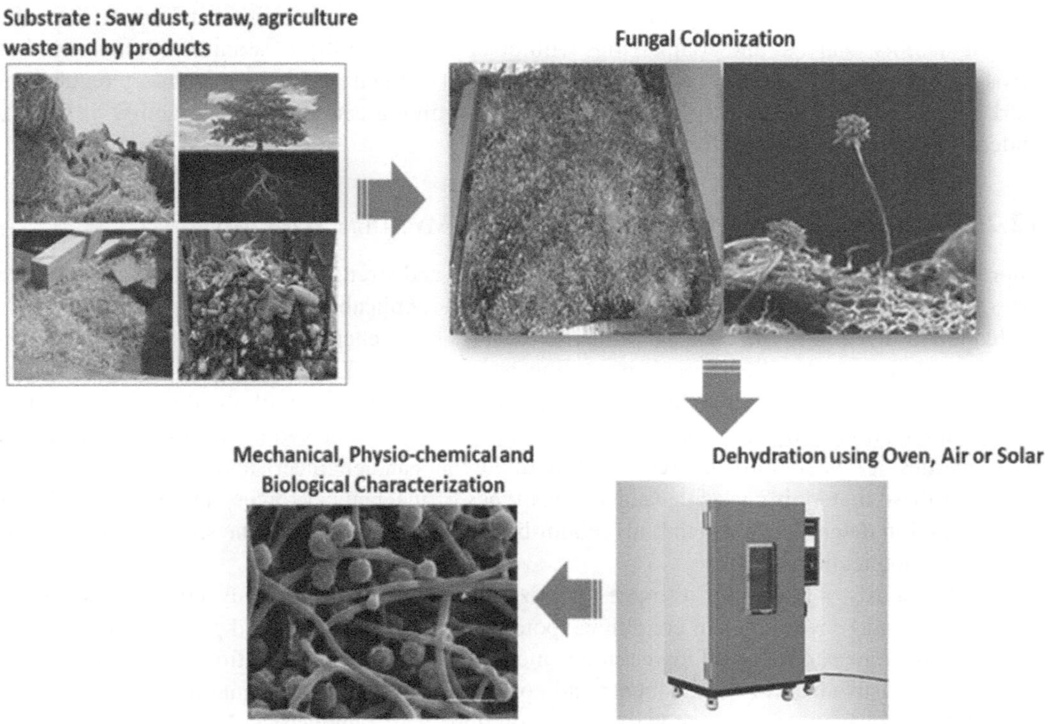

FIGURE 12.3 Schematic diagram of mycelium-based composites' synthesis process.

12.2.1 Growth and Behavior of Fungal Mycelium

A remarkable ability to generate biofilms that attach to solid surfaces and form spherical pellets in liquid cultures is possessed by fungi's delicate thread-like mycelia [41]. Tip extension, a specialized and highly polarized mechanism, is principally responsible for controlling the growth and development of mycelium. Different mathematical models are frequently used by researchers to comprehend the behavior of individual hyphae and the properties of mycelium. Numerous hyphae are produced by mycelia during the first stage of growth, and these hyphae link to form a network on the surface of their substrate [42]. Through a process known as tip extension, these hyphae continuously lengthen at their tips, resulting in tubular structures [43]. Tip extension is a phenomenon that not only affects mycelia but also affects a variety of other species along their cell walls, usually in specialized tissues like root hairs and pollen tubes. Unexpectedly, these developing cells have higher calcium concentration at their tips [44].

The foundation for determining how well mycelium structures may be applied in various fields is a thorough understanding of their mechanical characteristics and strength features. Researchers and engineers can perform quantitative evaluations of mycelium's resistance to a variety of mechanical forces by evaluating its mechanical properties. This evaluation serves as a crucial yardstick for determining if mycelium-based materials are compatible with the specific requirements of different applications [45]. In order to quantitatively measure how well these materials can withstand various types of mechanical forces, performance evaluation can also be accomplished by evaluating the mechanical properties of mycelium structures. This evaluation is crucial in determining whether mycelium-based materials can meet the specific requirements of various applications [46].

In addition, it permits resource optimization by understanding how mycelium reacts to various loads and environmental factors, potentially lowering waste and production costs [47]. Additionally, the mechanical properties of mycelium are evaluated to quantify its environmental impact, assisting in sustainability analyses and comparisons with other materials [48]. Customized design solutions across industries are also made possible by this versatility. When navigating regulatory regulations, market acceptance, and industry standards while determining economic sustainability, it helps to understand how mycelium performs mechanically [49].

12.2.2 Mechanical Properties and Strength of Mycelium Structures

Thorough comprehension of the mechanical properties and strength attributes of mycelium structures is essential, forming the bedrock for assessing their applicability across diverse domains. The evaluation of mycelium's mechanical properties enables researchers and engineers to conduct quantitative analyses of its resilience to a range of mechanical forces. This assessment serves as a critical benchmark for ascertaining whether mycelium-based materials align with the distinct demands of various applications [20]. Also, performance evaluation can be achieved by assessing the mechanical properties of mycelium structures, enabling researchers and engineers to quantitatively measure how well these materials can withstand various types of mechanical forces, and this evaluation is fundamental in determining whether mycelium-based materials can meet the specific requirements of different applications [21].

Simultaneously, it allows for resource optimization by comprehending mycelium's responses to varying loads and environmental conditions, potentially reducing waste and production costs [22]. Furthermore, the evaluation of mycelium's mechanical properties quantifies its environmental impact, aiding sustainability assessments and comparisons with conventional materials [23]. This adaptability also facilitates customized design solutions across industries. Understanding mycelium's mechanical performance helps meet industry standards, gain market acceptance, and navigate regulatory requirements, all while assessing economic viability.

In order to evaluate the mycelium materials' total lifecycle sustainability, their mechanical qualities must be assessed. This evaluation spurs innovation and encourages cross-industry acceptance of mycelium materials as a resource- and environment-wise replacement for conventional materials [50]. It allows for customized compliance with a variety of applications while satisfying particular requirements for strength, flexibility, and durability. This information supports structural integrity and risk reduction, ensuring safety in industries like construction, aerospace, and transportation [51]. By understanding how mycelium responds to various loads and environmental factors, it simultaneously enables resource optimization, potentially lowering waste and output costs. Mycelium's environmental impact is quantified by the measurement of its mechanical properties, which also makes it easier to compare it to other materials and do sustainability analyses. Additionally, this versatility enables industry-specific design solutions. In order to achieve industrial standards, acquire market approval, and comply with regulations while determining economic feasibility, it is important to comprehend the mechanical performance of mycelium. The evaluation of mechanical qualities is essential for determining the mycelium materials' entire lifespan sustainability, spurring innovation and fostering cross-industry adoption as a resource-effective, environmentally benign replacement for conventional materials [13, 52, 53].

12.3 APPLICATIONS OF MYCELIUM-BASED MATERIALS

Recent studies show that around 75% of the global population wants to lead a sustainable life, and ca. 25% (2–3 billion) citizens are already environmentally sustainable. Some renowned global movements, like the World Sustainable Development Summit (WSDS), bring together world leaders, business magnates, the scientific community, and civilians to achieve the common goal of greener and more eco-friendly living. Sustainable living can begin at the individual level and progress to the international level. Material recycling, energy conservation and storage, and waste segregation and management facilitate an eco-friendly lifestyle at the individual scale. Extending these greener approaches to a global level can bring about profound outcomes like cleaner water, reduced carbon footprints, and a magnificent abode for future generations.

The material scientists and research community are thus on a mission to generate greener and more functional materials by incorporating living systems into material science. This is also reflected in the fact that mycelium technology is now being used in almost every aspect of life.

12.3.1 Structural Applications

The building and construction sector holds a major share of global energy consumption, which amounts to around 33%. Cutting down on energy costs can contribute to the economic stability of the world nations. The gravity of this issue can be understood by examining the major policies being implemented by recent international summits like the G20, i.e., sustainability and energy conservation. Sustainability in the area of structural engineering and construction can have numerous benefits, like resource conservation, waste reduction, effective energy utilization, enhanced durability, and a meaningful contribution to corporate social responsibility. Table 12.1 shows an overview of different green materials other than mycelium composites used in engineering structures.

Investigations on the utility of mycelium composites are also progressing in the scientific community. Mechanical strength is one of the important parameters to be tested for building materials. Also, depending on the topography, the deviation in these parameters with a change in relative humidity and moisture absorption must be taken care of. The desirable properties of a thermal insulation material in buildings that mycelium composites possess are low density, low thermal conductivity, high water absorption capacity, and high specific heat capacity. Naturally grown mycelium composites perform well in low and medium humidity conditions, but their flexural strength will decrease in higher humidity [61]. Apart from substrate-grown mycelium, spent mushroom substrate (SMS) can also be effectively used in developing thermal insulators. The study done by M. Fellah et al. in Marrakesh, Morocco, revealed that sandwiching mycelium insulation between brick and mortar in buildings effectively lowered the indoor temperature (by 4°C) in the summer and provided a cozy indoor atmosphere (1°C cooler) during the winter. The remarkable thermal inertia exhibited by the mycelium material is the reason behind this behavior [62]. Comparative studies conducted in eight different climate zones in the United States using mycelium and other conventional insulators also reaffirmed these findings. Thus, besides providing thermal comfort, mycelium composites can assure a greater reduction in energy consumption by cutting heating and cooling loads in colder and hotter climates, respectively. Evidence is available for a drop in overall CO_2 emissions from mycelium composite bricks compared to the existing insulating bricks [63]. An added advantage of mycelium over other conventional thermal insulation materials is its low cost. If a much more durable composite material can be prepared out of mycelium, it will be a breakthrough in sustainable construction materials.

12.3.2 Mycelium Textile or Mycelium Leather

The hydrophobic, heat-resistant, and thermal-resistant properties of mycelium material make it a suitable candidate for wearable applications. It is now being used by renowned fashion brands, as there is a greater urge among consumers to switch to sustainable clothing. By playing with the mycelium strain, the substrate used, and the finishing process adopted, different grades of mycelium leather are patented and manufactured globally.

12.3.3 Electronic Applications

Materials derived from mycelium have great potential in electronic and allied fields. To begin with, it is being used as casings for various devices, thus reducing a share of electronic waste. Moreover, mycelium sheets can be used to fabricate wearable electronics, thus contributing to the era of smart

TABLE 12.1
Sustainable Materials used in Construction Applications

Insulation Material	Key Characteristics	Ref.
PLA and MCF	• A bio-resourced foam showed 12% rise in heating and cooling load • Accelerated biodegradability of foams	[54]
Hemp wool	• 76% reduction in heating and cooling load as compared to non-insulated walls • 71% decrease in greenhouse gas emissions	[55]
Cork fiber	• Low thermal conductivity compared to other thermal insulation materials • Capable of acoustic insulation	[56]
Straw bales	• Good hygrothermal performance • Remarkable thermal and acoustic insulation performance when coupled with coating materials	[57]
Aerogel	• 20% decrease in interior temperature in summer	[58]
Recycled glass	• The life cycle assessment showed that the global warming potential is lower than that of traditional insulator materials	[59]
Plain-woven jute composite	• Superior thermal barrier performance • Good flame-resistance properties • Less water absorption and high durability	[60]

devices. By selecting proper strains of mycelium with biocompatibility, they can be actively used for drug delivery purposes as well. Last but not least, efforts are being taken by the research community to develop batteries with mycelium materials.

12.3.4 Food Industry

Mushrooms are considered a storehouse of flavor. Depending on the mushroom species under consideration, the volatile components responsible for the umami taste may also change. Several kinds of edible mushrooms are developed and consumed presently. Being the vegetative part of the mushroom, this is in fact the major application of mycelium in the food industry. Also, it was reported that a major share of our cuisine will be represented by vegetarian meat by 2040, and bacon strips developed from mycelium is just a start. As discussed earlier, the umami flavor present in mycelium can be extracted and used in various dishes for taste enhancement. Also, the medicinal properties exhibited by several varieties of mushrooms can be exploited to produce nutritional supplements as well.

12.4 LIFE CYCLE ANALYSIS

It is impossible to assess a material's sustainability without considering its life cycle. Optimizing material configurations, calculating payback times, and assessing energy savings are all involved in a comprehensive life cycle assessment (LCA) or cradle-to-grave assessment. The six important phases of a product, service, or technology to be considered while conducting LCA are selection of input resources, processing the raw materials, manufacturing phase, distribution of output, consumption/use of products, and disposal. As per ISO 14040/44, the components of LCA are as shown in Figure 12.4.

In the first stage, the scope, boundaries, and functional units involved in the process are identified. Either a cradle-to-grave or a cradle-to-gate approach is adopted. A cradle-to-grave method comprises all the above phases mentioned in the product life cycle. While in the cradle-to-gate approach, the stages till manufacturing are considered. Usually, the boundary comprises the premises where the activities are carried out. For the cradle-to-gate approach, transportation and

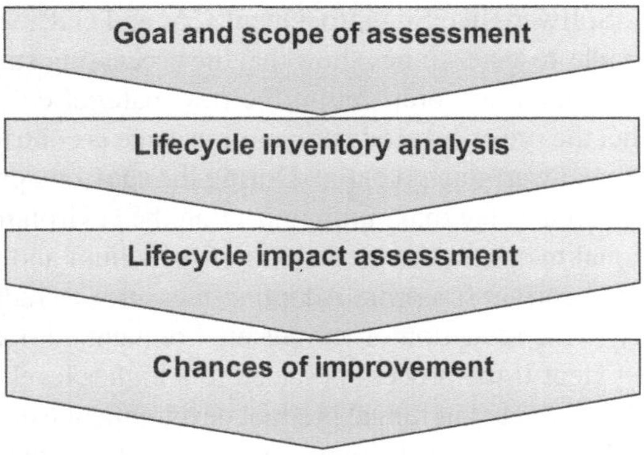

FIGURE 12.4 Components of life cycle assessment.

TABLE 12.2
The Major Environmental Impacts Contributed by the Mycelium Technology and Associated Practices

Impact Criteria Being Assessed	Methods	Measured Value of Impact	Approach Adopted	Ref
Global warming potential	CML	9.20E-01 kg CO_2 equivalent 4.94E-01 kg CO_2 equivalent	Cradle to gate Cradle to gate	[65][66]
Acidification potential	CML	1.40E-02 kg SO_2 equivalent 9.39E-04 kg SO_2 equivalent	Cradle to gate Cradle to gate	[65][66]
Eutrophication potential	CML	6.55E-03 kg PO_4^{3-} equivalent 6.02E-04 kg PO_4^{3-} equivalent	Cradle to gateCradle to gate	[65][66]
Photochemical ozone creation potential	CMLReCiPe	1.15E-03 kg C_2H_4 5.83E-04 kg NO_x equivalent	Cradle to gate Cradle to gate	[65][66]
Soil quality potential	LANCA	1.93E+01	Cradle to gate	[66]
User deprivation potential	AWARE	2.21E-02 m^3 world equivalents	Cradle to gate	[66]
Ozone depletion potential	CML	1.56E-07 kg CFC-11 equivalent	Cradle to gate	[65]
Abiotic resources depletion	CML	1.20E-02 kg Sb equivalent	Cradle to gate	[65]
Human toxicity	CML	6.67E-01 kg 1.4DB equivalent	Cradle to gate	[65]
Freshwater aquatic ecotoxicity	CML	1.08E-01 kg 1.4DB equivalent	Cradle to gate	[65]
Marine aquatic ecotoxicity	CML	7.90E+02 kg 1.4DB equivalent	Cradle to gate	[65]
Terrestrial ecotoxicity	CML	3.58E-03 kg 1.4DB equivalent	Cradle to gate	[65]

Note: The figures included in the impact measurement represent the total impact contributed by substrate preparation, inoculum preparation, and mycelium production.

distribution of products and waste management exist outside the system boundary. The functional unit considered depends upon the end product. It can be a unit of mycelium seeds, a composite brick, or the like.

In the life cycle inventory analysis of mycelium and mycelium composite production, the factors considered are rye or substrate material, the culture medium for mycelium growth, mycelium seed packages, and energy, water, and fuel consumption.

The major environmental impacts contributed by the mycelium technology and associated practices are listed in Table 12.2. In a comprehensive LCA of mycelium and mycelium composites, the impacts were analyzed using methods like CML (Center for Environmental Studies, University of Leiden), ReCiPe, LANCA (LANd use indicator value CAlculation), and AWARE (Available WAter

Remaining per area) [64]. Software like Simapro, OpenLCA, and GaBi serve as aids to conduct the impact assessment. The cradle-to-gate studies show that the processing stage of the mycelium makes a high impact on the environment as compared to the raw material collection and manufacturing phases. To be more specific, the rye or substrate preparation stage is contributing more toward all the impacts assessed except global warming potential. During the substrate preparation stage, the initial mixing and boiling process creates the maximum impact on the environment.

Compared to conventional methods, the production of mycelium and mycelium-based products helps greatly in reducing the carbon footprint. Adopting measures to reduce the fermentation time can contribute beneficially to the reduction of various environmental impacts. Energy consumption can be reduced to a large extent if the process is scaled to a higher level. As a general comment to add, it was found in a study conducted in fungal product development using different substrate materials like molasses, stale bread, and the backslopping method that the environmental impact is rather high when fresh substrate is used. This underlines the need for using alternative materials as a substrate to enhance environmental sustainability and circular economy [66].

12.5 THINGS TO DO

1. Let's have a look at the mycelia network first. You will need a simple microscope, a microscope glass slide (1×3 inch) and cover slip, a dropper, water, and definitely a mushroom. Take a walk in the yard and grab some mushrooms, preferably in 2–3 varieties. Now, cut a thin slice of the portion closer to the soil and place it in the glass slide. Add a few drops of water and place the cover slip above it so that the specimen will remain in position. Now place it under the microscope and watch the thin network-like strands of mycelium.
2. With a recycled bottle, mushroom substrate, and mushroom spore (store-bought), mycelium can be grown at home. Take a bottle or a mason's jar and make sure it is free from contaminants. Different options are available for substrate, such as sawdust, straw, wood chips, coffee grounds, or composted manure, to name a few. The selection of substrate depends on the type of mushroom being grown. Now fill the bottle with substrate and spores layer by layer. Through the cuts made on the periphery of the bottle, the fruiting body will come out, and mycelium will be grown inside.
3. A mycelium block can be prepared by having some grown mycelium and a mold of desired dimensions. Take a nutrient medium, which can be agar solution, malt extract, or yeast extract. Dilute the medium with distilled water and sterilize in an autoclave or pressure cooker. Thereafter, a piece of mushroom is carefully added to the nutrient medium using a scalpel. Later, this mushroom mass is transferred to a substrate for complete growth. After a few days, the fully grown mycelium is taken out and filled into the mold. The block can be thermally cured for better strength.
4. The fully grown mycelium, when placed in a hydraulic press, can result in mycelium sheets. These sheets can be conditioned by applying organic waxes.
   Note: The use of appropriate personal protection measures is highly recommended for the practicing personnel to avoid any kind of sensitive reactions.

12.6 CASE STUDIES

In the words of Ecovative, The Mycelium Technology Company, mycelium is a super material, and they even patented a novel material, which they call "MycocompositeTM." The mycelium fibers are grown here into sheet-like structures in a vertical farm enclosed in an AirMyceliumTM chamber. For the love of nature and sustainable living, the team even created a biological library, which is a storehouse of a wide variety of fungal species. MycocompositeTM, a blend of mycelium and hemp, is widely used for construction purposes, packaging needs, and as form structures. They also marked their presence in the food industry through a partnered venture, MyTMForest Foods, where the product is a non-vegetarian meat, MyTM Bacon. Another vertical from Ecovative is ForagerTM hides, which deals with plastic-free leather production. Likewise, ForagerTM foams are utilized for various end uses like breathable fabric, makeup application, and the like. The company claims that the material degrades 45 days after disposal. The state-of-the-art facilities, innovative concepts, and fool-proof research are key elements in the team's success story. Ecovative also provides an arena for new brains interested in mycelium technology. The Magical Mushroom Company is another group that is associated with Ecovative and has adopted the technology of mycelium composites. With sustainable packaging, they collaborate with many famous brands in the industrial, electronics, and beauty sectors.

Another big name among mycelium technology practitioners is SQIM. Mogu is a venture by SQIM, dealing with interior and architectural products. The main concern of the firm is to transform the low-valued agro-industrial products into value-added products by making them a substrate for the mycelium growth. Their main focus is on acoustic panels with nature-inspired designs, sound-absorbing wall panels, floor flex, and floor tiles. The tradeoff between architectural and design excellence and sustainability is really commendable.

Another prominent company working with mycelium technology is Mycoworks. It is a beautiful blend of art and biotechnology, and one of the co-founders cum artists, Philip Ross, even coined a term Mycotecture – the architectural design with mycelium. The company developed the process Fine MyceliumTM for growing extra-durable, super-strong mycelium fibers. ReishiTM, the premium leather crafted using Fine MyceliumTM technology, is now being used by popular brands in their product line.

12.7 CONCLUSIONS AND FUTURE PROSPECTS

Mycelium technology has a long, illustrious history, since ancient cultures have used fungi for everything from food to medicine to building materials. Mycelium's full potential has recently been unlocked by advances in biotechnology, opening the door to a wide range of creative uses in numerous industries. The uses for mycelium technology are numerous and varied. Mycelium-based building materials offer environmentally friendly substitutes for conventional building materials in construction, lowering carbon footprints, and offering superior insulation qualities. Compounds originating from mycelium are being researched in medicine for their potential in the creation of fresh medications and treatments. Mycelium can be used in the food business to create meat alternatives and other wholesome goods, helping to create a more ethical and sustainable food supply chain.

Future possibilities for mycelium technology are bright and varied. We may anticipate a number of significant advancements in this area as our knowledge of fungi and biotechnology continues to grow:

1. **Environmental and Pollution Remediation**: Mycelium is a useful tool for environmental remediation since it can degrade toxins and pollutants in soil and water. Applications in this field in the future could aid in addressing urgent problems like waste management and pollution.
2. **Continual Innovations in Industry**: Mycelium-based materials will probably be used more frequently in the manufacturing, packaging, and building sectors. These materials are being continuously improved for improved performance and cost-effectiveness by researchers.
3. **Agricultural Yields and Soil Health**: Because of the symbiotic associations that mycelium has with plants, it can be used to enhance soil health and boost agricultural production. This might completely alter the way that sustainable agriculture is done.
4. **Consumer Products**: Furniture, clothing, and other everyday objects are just a few examples of the consumer products that will likely incorporate more mycelium-based components in the future. These eco-friendly options will probably grow increasingly well-liked as people become more concerned about the environment.

5. **Pharmaceutical Industry and Biotechnology**: Mycelium will become more and more important in the production of pharmaceuticals and biotechnology goods in the fields of medicine and biotechnology. Its natural chemicals may be the key to discovering new drugs, particularly in fields like the study of antibiotics.
6. **Packaging Materials**: Mycelium-based packaging materials will continue to replace conventional plastics, helping to significantly reduce plastic waste and its harmful effects on the environment.
7. **Research and Development**: As mycelium technology advances, new opportunities and uses may become available that we had not even thought of at the time.

In summary, mycelium technology has a wide range of applications and has the potential to change sectors while tackling urgent environmental and health concerns at the forefront of sustainable innovation. To fully use mycelium and build a greener, more sustainable future for our world, cooperation between scientists, businesspeople, and legislators will be crucial as we move forward.

REFERENCES

[1] S. Goulette, What a waste, *Proc. Inst. Civ. Eng. Munic. Eng.* 139 (2000) 167–169. https://doi.org/10.1680/muen.2000.139.3.167

[2] D. Alemu, M. Tafesse, A.K. Mondal, Mycelium-based composite: The future sustainable biomaterial, *Int. J. Biomater.* 2022 (2022). https://doi.org/10.1155/2022/8401528

[3] H. Italia, I. Patel, J. Shah, Experimental study of bacterial self-healing effect on concrete: A review, *J. Civ. Eng. Environ. Technol.* 3 (2016) 78–81.

[4] A. Ghazvinian, P. Farrokhsiar, F. Vieira, J. Pecchia, B. Gursoy, Mycelium-based bio-composites for architecture: Assessing the effects of cultivation factors on compressive strength, *Mater. Res. Innov.* 2 (2018) 505–514.

[5] J.M.C. Ongpeng, E. Inciong, V. Sendo, C. Soliman, Using waste in producing bio-composite mycelium bricks, *Appl. Sci.* 10 (2020) 5303. https://doi.org/10.3390/app10155303

[6] R. Rathinamoorthy, T. Sharmila Bharathi, M. Snehaa, C. Swetha, Mycelium as sustainable textile material – review on recent research and future prospective, *Int. J. Cloth. Sci. Technol.* 35 (2023) 454–476. https://doi.org/10.1108/IJCST-01-2022-0003

[7] J.C. Prata, J.P. da Costa, A.C. Duarte, T. Rocha-Santos, Methods for sampling and detection of microplastics in water and sediment: A critical review, *TrAC - Trends Anal. Chem.* 110 (2019) 150–159. https://doi.org/10.1016/j.trac.2018.10.029

[8] H. Du, G. Chen, J. Wang, Highly selective electrochemical impedance spectroscopy-based graphene electrode for rapid detection of microplastics, *Sci. Total Environ.* 862 (2023) 160873. https://doi.org/10.1016/j.scitotenv.2022.160873

[9] J. Sun, X. Dai, Q. Wang, M.C.M. van Loosdrecht, B.J. Ni, Microplastics in wastewater treatment plants: Detection, occurrence and removal, *Water Res.* 152 (2019) 21–37. https://doi.org/10.1016/j.watres.2018.12.050

[10] Z. Chen, W. Wei, X. Liu, B.J. Ni, Emerging electrochemical techniques for identifying and removing micro/nanoplastics in urban waters, *Water Res.* 221 (2022) 118846. https://doi.org/10.1016/j.watres.2022.118846

[11] T. Balaeş, B.M. Radu, C. Tănase, Mycelium-composite materials—A promising alternative to plastics? *J. Fungi.* 9 (2023) 210. https://doi.org/10.3390/jof9020210

[12] S. Manan, M.W. Ullah, M. Ul-Islam, O.M. Atta, G. Yang, Synthesis and applications of fungal mycelium-based advanced functional materials, *J. Bioresour. Bioprod.* 6 (2021) 1–10. https://doi.org/10.1016/j.jobab.2021.01.001

[13] L. Yang, D. Park, Z. Qin, Material function of mycelium-based bio-composite: A review, *Front. Mater.* 8 (2021) 1–17. https://doi.org/10.3389/fmats.2021.737377

[14] A. Ghazvinian, B. Gürsoy, Mycelium-based composite graded materials: Assessing the effects of time and substrate mixture on mechanical properties, *Biomimetics.* 7 (2022). https://doi.org/10.3390/biomimetics7020048

[15] M. Jones, A. Mautner, S. Luenco, A. Bismarck, S. John, Engineered mycelium composite construction materials from fungal biorefineries: A critical review, *Mater. Des.* 187 (2020) 108397. https://doi.org/10.1016/j.matdes.2019.108397
[16] W.A. Elkhateeb, G.M. Daba, The amazing potential of fungi in human life, *ARC J. Pharm. Sci.* 5 (2019) 12–16.
[17] D. Zou, L. Gao, Preparation and properties of fungal mycelium based on garden waste, *Sci. Discov.* 8 (2020) 43–45. https://doi.org/10.11648/j.sd.20200802.14
[18] A. Pepe, M. Giovannetti, C. Sbrana, Lifespan and functionality of mycorrhizal fungal mycelium are uncoupled from host plant lifespan, *Sci. Rep.* 8 (2018) 2–11. https://doi.org/10.1038/s41598-018-28354-5
[19] H.M.C. Azeredo, H. Barud, C.S. Farinas, V.M. Vasconcellos, A.M. Claro, Bacterial cellulose as a raw material for food and food packaging applications, *Front. Sustain. Food Syst.* 3 (2019). https://doi.org/10.3389/fsufs.2019.00007
[20] S. Vandelook, E. Elsacker, A. Van Wylick, L. De Laet, E. Peeters, Current state and future prospects of pure mycelium materials, *Fungal Biol. Biotechnol.* 8 (2021) 1–10. https://doi.org/10.1186/s40694-021-00128-1
[21] J.M. Trappe, AB Frank and mycorrhizae: The challenge to evolutionary and ecologic theory, *Mycorrhiza.* 15 (2005) 277–281. https://doi.org/10.1007/s00572-004-0330-5
[22] V. Meyer, T. Cairns, L. Barthel, R. King, P. Kunz, S. Schmideder, H. Müller, H. Briesen, A. Dinius, R. Krull, Understanding and controlling filamentous growth of fungal cell factories: novel tools and opportunities for targeted morphology engineering, *Fungal Biol. Biotechnol.* 8 (2021) 1–8. https://doi.org/10.1186/s40694-021-00115-6
[23] T. Rämä, C.A. Quandt, Improving fungal cultivability for natural products discovery, *Front. Microbiol.* 12 (2021). https://doi.org/10.3389/fmicb.2021.706044
[24] M.L. Green, L. Espinal, E. Traversa, E.J. Amis, Materials for sustainable development, *MRS Bull.* 37 (2012) 303–308. https://doi.org/10.1557/mrs.2012.51
[25] M. Javaid, A. Haleem, R. Pratap Singh, S. Khan, R. Suman, Sustainability 4.0 and its applications in the field of manufacturing, *Internet Things Cyber-Physical Syst.* 2 (2022) 82–90. https://doi.org/10.1016/j.iotcps.2022.06.001
[26] P. Koltun, Materials and sustainable development, *Prog. Nat. Sci. Mater. Int.* 20 (2010) 16–29. https://doi.org/10.1016/s1002-0071(12)60002-1
[27] N. Attias, A. Livne, T. Abitbol, State of the art, recent advances, and challenges in the field of fungal mycelium materials: a snapshot of the 2021 mini meeting, *Fungal Biol. Biotechnol.* 8 (2021) 3–7. https://doi.org/10.1186/s40694-021-00118-3
[28] G. Kalita, S. Chakravarty, S. Baruah, S. Sen, J. Hazarika, D. Palash, Y. Hajar Arifin, Y. Yusuf, L. Jianga, D. Walczyka, G. Mcintyre, R. Bucinellc, G. Tudrynb, Using mycelium as a building material in India, *ResearchGate.* (2018). https://doi.org/10.13140/RG.2.2.10809.67681
[29] T.C. Cairns, X. Zheng, P. Zheng, J. Sun, Turning inside out: Filamentous fungal secretion and its applications in biotechnology, agriculture, and the clinic, *J. Fungi* 7 (2021) 535.
[30] K. Cerimi, K.C. Akkaya, C. Pohl, B. Schmidt, P. Neubauer, Fungi as source for new bio - based materials: a patent review, *Fungal Biol. Biotechnol.* 6 (2019) 17. https://doi.org/10.1186/s40694-019-0080-y
[31] A. Gandia, J.G. van den Brandhof, F.V.W. Appels, M.P. Jones, Flexible fungal materials: Shaping the future, *Trends Biotechnol.* 39 (2021) 1321–1331. https://doi.org/10.1016/j.tibtech.2021.03.002
[32] K. Rafiee, G. Kaur, S.K. Brar, Fungal biocomposites: How process engineering affects composition and properties? *Bioresour. Technol. Reports.* 14 (2021) 100692. https://doi.org/10.1016/j.biteb.2021.100692
[33] S. Nanda, B.R. Patra, R. Patel, J. Bakos, A.K. Dalai, Innovations in applications and prospects of bioplastics and biopolymers: a review, *Environ. Chem. Lett.* 20 (2022) 379–395. https://doi.org/10.1007/s10311-021-01334-4
[34] L. Li, T. Liang, W. Liu, Y. Liu, F. Ma, A comprehensive review of the mycelial pellet: Research status, applications, and future prospects, *Ind. Eng. Chem. Res.* 59 (2020) 16911–16922. https://doi.org/10.1021/acs.iecr.0c01325
[35] E. Elsacker, S. Vandelook, B. Damsin, A. Van Wylick, E. Peeters, L. De Laet, Mechanical characteristics of bacterial cellulose-reinforced mycelium composite materials, *Fungal Biol. Biotechnol.* 8 (2021) 18. https://doi.org/10.1186/s40694-021-00125-4

[36] M. Duque-Acevedo, I. Lancellotti, F. Andreola, L. Barbieri, L.J. Belmonte-Ureña, F. Camacho-Ferre, Management of agricultural waste biomass as raw material for the construction sector: an analysis of sustainable and circular alternatives, *Environ. Sci. Eur.* 34 (2022). https://doi.org/10.1186/s12302-022-00655-7

[37] J. Jose, K.N. Uvais, T.S. Sreenadh, A. V. Deepak, C.R. Rejeesh, Investigations into the development of a mycelium biocomposite to substitute polystyrene in packaging applications, *Arab. J. Sci. Eng.* 46 (2021) 2975–2984. https://doi.org/10.1007/s13369-020-05247-2

[38] M.S. Birajdar, H. Joo, W.G. Koh, H. Park, Natural bio-based monomers for biomedical applications: a review, *Biomater. Res.* 25 (2021) 8. https://doi.org/10.1186/s40824-021-00208-8

[39] F.S. Chambergo, E.Y. Valencia, Fungal biodiversity to biotechnology, *Appl. Microbiol. Biotechnol.* 100 (2016) 2567–2577. https://doi.org/10.1007/s00253-016-7305-2

[40] M.E. Antinori, M. Contardi, G. Suarato, A. Armirotti, R. Bertorelli, G. Mancini, D. Debellis, A. Athanassiou, Advanced mycelium materials as potential self-growing biomedical scaffolds, *Sci. Rep.* 11 (2021) 1–14. https://doi.org/10.1038/s41598-021-91572-x

[41] G.K. Villena, T. Fujikawa, S. Tsuyumu, M. Gutiérrez-Correa, Structural analysis of biofilms and pellets of Aspergillus niger by confocal laser scanning microscopy and cryo scanning electron microscopy, *Bioresour. Technol.* 101 (2010) 1920–1926. https://doi.org/10.1016/j.biortech.2009.10.036

[42] M. Nopharatana, D.A. Mitchell, T. Howes, Use of confocal scanning laser microscopy to measure the concentrations of aerial and penetrative hyphae during growth of Rhizopus oligosporus on a solid surface, *Biotechnol. Bioeng.* 84 (2003) 71–77. https://doi.org/10.1002/bit.10752

[43] N.N. Levina, R.R. Lew, The role of tip-localized mitochondria in hyphal growth, *Fungal Genet. Biol.* 43 (2006) 65–74. https://doi.org/10.1016/j.fgb.2005.06.008

[44] L. Steinhorst, J. Kudla, Calcium - A central regulator of pollen germination and tube growth, *Biochim. Biophys. Acta - Mol. Cell Res.* 1833 (2013) 1573–1581. https://doi.org/10.1016/j.bbamcr.2012.10.009

[45] L. Gou, S. Li, J. Yin, T. Li, X. Liu, Morphological and physico-mechanical properties of mycelium biocomposites with natural reinforcement particles, *Constr. Build. Mater.* 304 (2021). https://doi.org/10.1016/j.conbuildmat.2021.124656

[46] X.Y. Chan, N. Saeidi, A. Javadian, D.E. Hebel, M. Gupta, Mechanical properties of dense mycelium - bound composites under accelerated tropical weathering conditions, *Sci. Rep.* 11 (2021) 22112. https://doi.org/10.1038/s41598-021-01598-4

[47] E. Elsacker, S. Vandelook, A. Van Wylick, L. De Laet, E. Peeters, A comprehensive framework for the production of mycelium-based lignocellulosic composites, *Sci. Total Environ.* 725 (2020) 138431. https://doi.org/10.1016/j.scitotenv.2020.138431

[48] M.R. Islam, G. Tudryn, R. Bucinell, L. Schadler, R.C. Picu, Morphology and mechanics of fungal mycelium, *Sci. Rep.* 7 (2017) 13070. https://doi.org/10.1038/s41598-017-13295-2

[49] M. Varl, J. Duhovnik, J. Tavčar, Customized product development supported by integrated information, *J. Ind. Inf. Integr.* 25 (2022). https://doi.org/10.1016/j.jii.2021.100248

[50] C. Girometta, A.M. Picco, R.M. Baiguera, D. Dondi, S. Babbini, M. Cartabia, M. Pellegrini, E. Savino, Physico-mechanical and thermodynamic properties of mycelium-based biocomposites: A review, *Sustainability* 11 (2019) 281. https://doi.org/10.3390/su11010281

[51] M. Jones, A. Mautner, S. Luenco, S. John, Engineered mycelium composite construction materials from fungal biorefineries: A critical review, *Mater. Des.* 187 (2019) 108397. https://doi.org/10.1016/j.matdes.2019.108397

[52] N. Idayu, N. Kai, S. Chua, A. Fadziyana, M. Roshanida, A.R. Jau, C. Lai, Effect of growth factors on the production of mycelium - based biofoam, *Clean Technol. Environ. Policy.* 24 (2022) 351–361. https://doi.org/10.1007/s10098-021-02146-4

[53] Y.J. Grobman, Mycelium bio-composites in industrial design and architecture: Comparative review and experimental analysis, *J. Clean. Prod.* 246 (2019) 119037. https://doi.org/10.1016/j.jclepro.2019.119037

[54] K. Oluwabunmi, N.A. D'Souza, W. Zhao, T.Y. Choi, T. Theyson, Compostable, fully biobased foams using PLA and micro cellulose for zero energy buildings, *Sci. Rep.* 10 (2020) 17771. https://doi.org/10.1038/s41598-020-74478-y

[55] M. Dlimi, O. Iken, R. Agounoun, A. Zoubir, I. Kadiri, K. Sbai, Energy performance and thickness optimization of hemp wool insulation and air cavity layers integrated in Moroccan building walls', *Sustain. Prod. Consum.* 20 (2019) 273–288. https://doi.org/10.1016/j.spc.2019.07.008

[56] S. Sair, B. Mandili, M. Taqi, A. El Bouari, Development of a new eco-friendly composite material based on gypsum reinforced with a mixture of cork fibre and cardboard waste for building thermal insulation, *Compos. Commun.* 16 (2019) 20–24. https://doi.org/10.1016/j.coco.2019.08.010

[57] B. Marques, A. Tadeu, J. Almeida, J. António, J. de Brito, Characterisation of sustainable building walls made from rice straw bales, *J. Build. Eng.* 28 (2020) 101041. https://doi.org/10.1016/j.jobe.2019.101041

[58] J. Yang, H. Wu, X. Xu, G. Huang, T. Xu, S. Guo, Y. Liang, Numerical and experimental study on the thermal performance of aerogel insulating panels for building energy efficiency, *Renew. Energy.* 138 (2019) 445–457. https://doi.org/10.1016/j.renene.2019.01.120

[59] L. Cozzarini, L. Marsich, A. Ferluga, C. Schmid, Life cycle analysis of a novel thermal insulator obtained from recycled glass waste, *Dev. Built Environ.* 3 (2020) 100014. https://doi.org/10.1016/j.dibe.2020.100014

[60] M.A.R. Bhuiyan, A. Ali, H. Akter, M.A.R. Dayan, M.J. Hossen, M.J. Abden, A.N. Khan, Flame resistance and heat barrier performance of sustainable plain-woven jute composite panels for thermal insulation in buildings, *Appl. Energy.* 345 (2023) 121317. https://doi.org/10.1016/j.apenergy.2023.121317

[61] X. Zhang, J. Hu, X. Fan, X. Yu, Naturally grown mycelium-composite as sustainable building insulation materials, *J. Clean. Prod.* 342 (2022) 130784. https://doi.org/10.1016/j.jclepro.2022.130784

[62] M. Fellah, S. Ouhaibi, N. Belougadia, K. Mansouri, H. Naji, Towards thermal insulation and energy performance's assessment of a mycelium-based composite wall for eventual sustainable use in buildings, 2023. https://doi.org/10.2139/ssrn.4569784

[63] S. Al-Qahtani, M. Koç, R.J. Isaifan, Mycelium-based thermal insulation for domestic cooling footprint reduction: A review, *Sustain.* 15 (2023). https://doi.org/10.3390/su151713217

[64] L. Stelzer, F. Hoberg, V. Bach, B. Schmidt, S. Pfeiffer, V. Meyer, M. Finkbeiner, Life cycle assessment of fungal-based composite bricks, *Sustain.* 13 (2021) 1–17. https://doi.org/10.3390/su132111573

[65] F.J. Leiva, J.C. Saenz-Díez, E. Martínez, E. Jiménez, J. Blanco, Environmental impact of *Agaricus bisporus* mycelium production, *Agric. Syst.* 138 (2015) 38–45. https://doi.org/10.1016/j.agsy.2015.05.003

[66] P. Brancoli, R. Gmoser, M.J. Taherzadeh, K. Bolton, The use of life cycle assessment in the support of the development of fungal food products from surplus bread, *Fermentation* 7 (2021) 173. https://doi.org/10.3390/fermentation7030173

13 Chitosan Nanoparticles as a Promising Tool in Nanomedicine

A. Annam Renita, S. Lalithakumari, J. Aravind Kumar, and V. Sivasubramanian

13.1 INTRODUCTION

Nanotechnology is the stream of science that deals with nanometre-scale materials (at least one of the dimensions of the material is less than 100 nm) and their applications. The key advantage of nanomaterials is that they possess a large surface area, and hence they are preferred over conventional bulk materials (Auffan et al., 2009; Ealia & Saravanakumar, 2017; Khan et al., 2019). Nanotechnology is developing at a fast pace in the global market. The market value of nanotechnology was 1.8 billion USD in 2020 and is expected to surpass 33 million USD in 2030 (Roco, 2017). Every nanomaterial differs in its optical, electrical or magnetic properties and biological, chemical and other physicochemical properties. Hence, according to their properties, the nanomaterials differ in their end applications (Lue, 2001; Yokoyama et al., 2008). For the past 20 years, nanomaterials have been very much utilised in imaging diagnostics, targeted delivery and therapeutics in the field of medicine (Lane et al., 2015). Nanomaterials are crucial in the field of medicine as they can deliver drugs inside the human body for diagnostics or therapeutics without any trouble (Mout et al., 2012; Yohan & Chithrani, 2014). Chitosan is one among the nanomaterials and is significant because it is bio-degradable, non-toxic, biocompatible and easily available from chitin, which is present in the cell walls of fungi, shellfish, shrimps, crabs, and some sorts of algae, microbes and invertebrates. Also, chitosan is cost-effective, hydrophilic and carries a positive charge, thus favouring easy fabrication with other nanomaterials (da Silva et al., 2020, Ye et al., 2015). Also, chitosan is highly flexible to any natural tissue (Jee et al., 2012; Peniche & Peniche, 2011; Sharifi-Rad et al., 2021). Chitosan nanoparticles possess antiviral, antimicrobial and immune adjuvant properties; hence, they have tremendous clinical and pharmaceutical applications (Dacrory et al., 2022; Nair et al., 2022; Yan et al., 2017). The broad-spectrum applications of chitosan and its derivatives are illustrated in Figure 13.1.

Despite the advantageous features of chitosan nanoparticles, they are not widely used in medical applications. This is because of their low solubility. Besides, chitosan nanoparticles are weak in nature (Cheung et al., 2015; Liu et al., 2022). Hence, efforts are made to stabilise and strengthen chitosan nanoparticles. Hydroxyl and amino groups are used to formulate various chitosan derivatives with good solubility. Also, other metallic and non-metallic nanoparticles are conjugated with chitosan to improve their mechanical properties (Zhang et al., 2010). This makes them be utilised in a wide range of pharmaceutical and medical applications. Thus, this chapter presents an overview of chitosan, its derivatives, and applications of chitosan in tissue engineering, cancer therapy, drug delivery, vaccine adjuvant and wound healing. In addition, a detailed investigation on the usage of chitosan and its derivatives in drug delivery pertaining to ocular, brain, pulmonary and targeted delivery of bio-markers is also mentioned. Modification in structure is mainly carried out to enhance the solubility properties of chitosan. One of the prominent derivatives of CS is quaternised alkyl chitosan

DOI: 10.1201/9781003732907-13

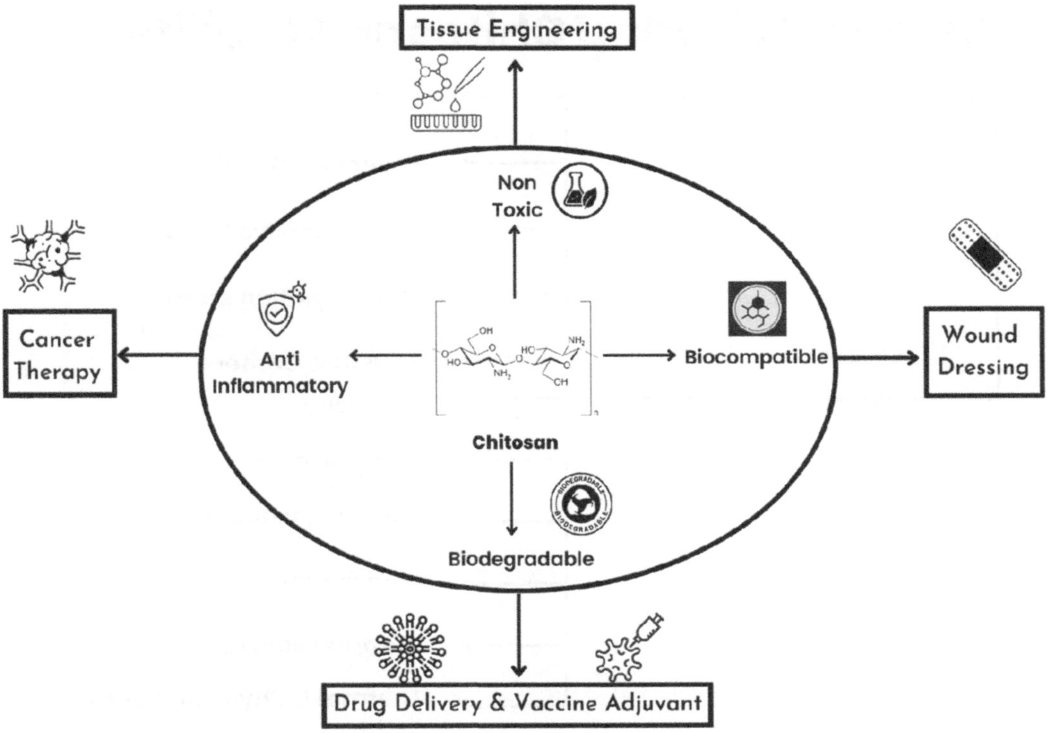

FIGURE 13.1 Attributes of chitosan and its applications.

(Boles et al. 2019), followed by N-alkyl, N-benzyl, N-acyl, N-carboxyalkyl, aryl, O carboxyl, N-carboxyl, phosphorylated and thiolated chitosan. Also, this chapter states the future scope of assessing the cytotoxicity of upcoming novel chitosan-based nanoparticles utilising chitosan stability and its derivatives in nanomedicine. Finally, this chapter concludes that in spite of the challenges faced in using chitosan, such as poor solubility, stability and storage issues, there is a worldwide need for chitosan-based nanomaterials owing to their broad range of applications for the benefit of humans.

13.2 CHITOSAN AND ITS CHEMICAL MODIFICATIONS

In general, polymers have multifaceted applications and thus occupy a prominent part of everyday life. Polymers that are available naturally are beneficial when compared with synthetic ones as they are biodegradable, biocompatible and flexible in their end application. Chitosan is a natural biodegradable polymer obtained from chitin that is easily available after cellulose. Chitosan is linear in nature and is composed of beta-(1-4)-linked D-glucosamine, and N-acetyl-D-glucosamine distributed randomly. Concerning its solubility, it is insoluble in organic solvents, while it is rapidly soluble in acidic solutions, which are dilute in nature. This can be reasoned from the fact that chitosan is believed to be a strong base, as it possesses primary amino groups with an acid dissolution value of 6.3. Also, chitosan's amino groups indicate that the pH greatly affects the charged state and other properties. When the pH is lower, amines gain a negative charge, thus enabling chitosan as a cationic polyelectrolyte soluble in water. Hence, chitosan has low solubility at pH >6, and the applicability of chitosan concerning drug release is noticeably reduced. Another major disadvantage is its usage in the sustained release of drugs. This is because chitosan is adsorbed in water quickly, and hence the swelling rate of chitosan is very high, resulting in the rapid release of drugs. Thus, modifications are performed in chitosan, discussed in the sections below. Factors affecting the stability of chitosan are illustrated in Figure 13.2.

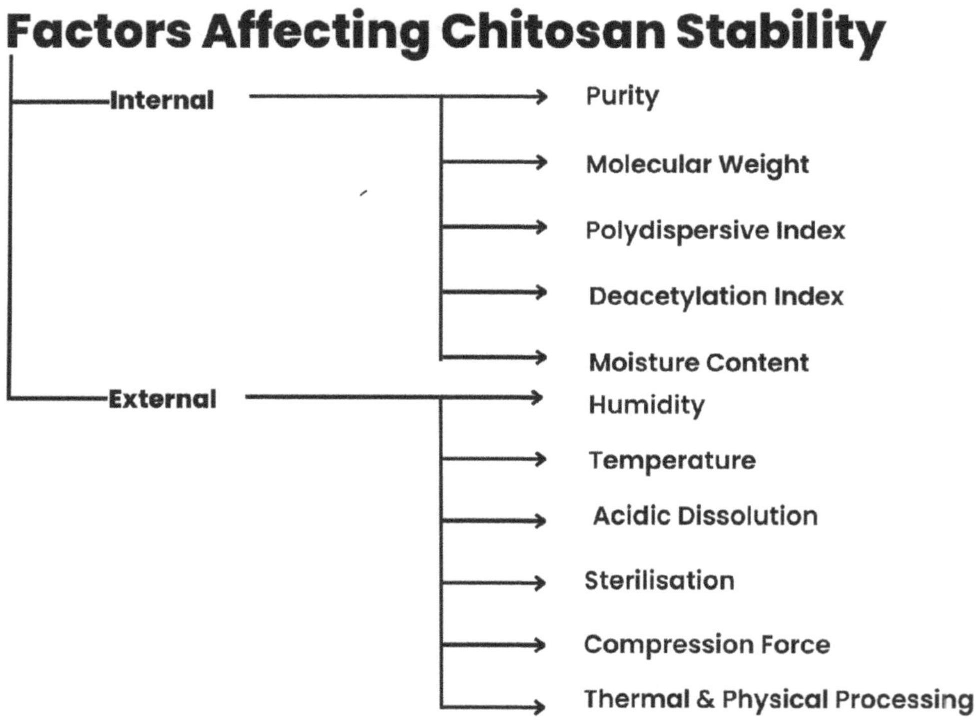

FIGURE 13.2 Factors affecting the stability of chitosan.

13.3 CHEMICAL MODIFICATION

Chitosan has hydroxyl, acetamido and amine functional groups; hence, chemical modification is possible in CS structure (Zhu et al., 2022). Besides, addition or conjugation in the chitosan structure helps in overcoming its limitations, thus enabling effectual transfection. chitosan, after the modification, has variations in its drug-carrying features to the desired locations in the human body. As described above, its structure has three functional groups: reactive. An amino group is present in the C-2 position, and primary and secondary hydroxyl groups are present in C-6 and C-3 positions. Modification in structure is mainly carried out to enhance the solubility properties of chitosan. One of the prominent derivatives of CS is quaternised alkyl chitosan, followed by N-alkyl, N-benzyl, N-acyl, N-carboxyalkyl, aryl, O carboxyl, N-carboxyl, phosphorylated and thiolated chitosan. Trimethyl chitosan is an example of quaternised alkyl chitosan. Other examples of its derivatives are propionyl, palmitoyl, N-carboxymethyl and N-carboxyethyl chitosan.

Some of the derivatives of chitosan are semi-synthetic in nature, possessing significant properties and functionalities with a vast range of applications in nanomedicine, mainly drug delivery and cancer therapy (Kumar et al., 2022). Chemical modifications to chitosan structure subsequently break down the crystal structure, thus increasing the solubility of the altered polymer. It is noteworthy to observe that the derived nanoparticles of chitosan with enhanced physicochemical and biological properties have major applications with respect to carrying the drugs to a specific location with sustained release (Min et al. 2008). Besides, they have numerous lone pairs of electrons possessing great binding power with other materials (Mourya & Inamdar, 2008; Rajasree & Rahate, 2013; Zhang et al., 2010). Apart from substitution in the functional group, crosslinking is also performed to modify the structure, thus enhancing solubility and stability. Crosslinking of chitosan results in the formation of chemical bonds, thus resulting in a strong three-dimensional network. The agent used in crosslinking plays a crucial role in stabilising the crosslinking and forming ionic bonds and physical crosslinking to help form hydrogen bonds. The agent used in crosslinking also depends on the chemical structure and molecular weight of chitosan and the active groups present in chitosan. Modifying its structure using crosslinking helps improve the stability and mechanical properties. Different crosslinking approaches include emulsion, ionic, solvent evaporation, spray drying, precipitation, flocculation and solution coating (Wang et al., 2005; Desai & Park, 2005; Negm et al., 2020). Modifications of chitosan and its derivatives enable them to be applicable in different arenas, as presented in the upcoming sections of this chapter. Some of the applications of chitosan nanoparticles in the field of nanomedicine are listed in Table 13.1.

TABLE 13.1
Chitosan Nanoparticles and Their Derivative, Along With Their Applications in Nanomedicine

Chitosan Nanoparticles and Their Modifications	Applications	Results After Utilisation of chitosan	References
Trivalent Fe (III) chitosan nanoparticle n-benzaldehyde Schiff base	Antimicrobial, antibacterial and antifungal	The materials were tested for their antimicrobial properties against *Bacillus subtilis, Bacillus cereus* and *Staphylococcus aureus*, *Pseudomonas aeruginosa*. With respect to fungi, they were tested against *Candida albicans, Aspergillus niger*. They were tested against drugs such as Meropenem, gentamicin, vancomycin, amoxicillin/clavulanic acid, ciprofloxacin, sulfamethoxazole/trimethoprim, ceftazidime, erythromycin and fluconazole	Okunzuwa et al. (2024)
Chitosan nanoparticles	Delivery of siRNA in cancer therapy	Effects of proliferation, metastasis and chemoresistance were reduced using siRNA-loaded chitosan nanoparticles—the nanoparticles aid in delivering siRNA and antitumour agents with significant stability. The outcome of the preclinical studies stated that the chitosan-based siRNA delivery is biocompatible and safe.	Ashrafizadeh et al. (2021), Bharathiraja et al. (2018), Saravanakumar et al. (2018)
Chitosan nanoparticles with Hyaluronic acid and nAg	Wound dressing in diabetic foot ulcers	Antimicrobial sponges are produced and used to treat diabetic foot ulcers. Antibacterial properties of the sponge were analysed by testing against *E. coli*, *S. aureus*, methicillin-resistant *S. aureus* (MRSA), *P. aeruginosa* and *Klebsiella pneumonia*. Sponges possessing higher concentrations of the nanoparticles exhibited excellent antibacterial activity.	Mohandas et al. (2018)
Chitosan nanoparticles CS-NPs	Delivery of doxorubicin (DOX) in cancer treatment	Delivery of doxorubicin with curcumin and cisplatin for effective cancer treatment. Further, doxorubicin was loaded along with siRNA, shRNA and miRNA, and this can reduce the progression of cancer development. Also, they provide chemotherapeutic effects.	Ashrafizadeh et al. (2023), Bharathiraja et al. (2018), Saravanakumar et al. (2018)
Carboxymethyl cellulose chitosan hydrogel	Wound healing, bioimaging, drug delivery and tissue engineering	The carboxymethyl cellulose and chitosan gel combination was tested in 3T3-L1 fibroblast cells. Also, *in vivo*, toxicity was assessed using subcutaneous injection. The study's outcome showed that the hydrogel has good biocompatibility and can be used for tissue regeneration.	Shah et al. (2023)

(*Continued*)

TABLE 13.1 (Continued)

Chitosan Nanoparticles and Their Modifications	Applications	Results After Utilisation of chitosan	References
Cinnamomum cassia oil (CEO) combined with chitosan nanoparticles	Treatment of cancer	When the nanoparticles are viewed under a Transmission Electron Microscope (TEM), the nanoparticles are regular and uniform and form either a spherical or a quasi-spherical structure. The synthesised nanoparticles were stable when stored at 4°C. Further, they showed an increase in the antitumour activity of 52 μg/mL compared with that of Cinnamomum cassia oil alone.	Xu et al. (2023)
Athyrium sinense combined with chitosan nanoparticles	Antibacterial activity	The antibacterial properties were assessed using in vitro tests (within 48 h), time-kill curve (within 60 h) and *in vivo* studies (within 72 h). The minimum bactericidal concentrations & minimum inhibitory concentration of the nanoparticles were 0.63 mg/mL & 0.16 mg/mL, respectively.	Cai et al. (2023)
Platelet-rich fibrin impregnated into chitosan nanoparticles loaded into the gelatin-chitosan hydrogel	Wound healing	The hydrogel's antimicrobial properties were assessed using gram-positive and gram-negative bacteria. The hydrogel showed the lowest scavenging capacity of 5.82 μgmL^{-1} and maximum radical scavenging activity of about 83%. This is reasoned because of the presence of chitosan nanoparticles in the hydrogel. Further, the resultant hydrogel showed good cell viability and proliferation.	Mirjalili and Mahmoodi (2023)
Fumaria officinalis impregnated with chitosan into calcium alginate hydrogel	Diabetes wound healing	A rat model infected with diabetes was used to test the wound-healing properties of the synthesised hydrogel. The results were compared with that of the commercially viable wound care product GranuGEL®. The synthesised nanoparticles showed a comparable wound-healing property. *In vitro* studies showed that the hydrogel aided skin cells' viability and migration.	Yang et al. (2023)
Chitosan/bacterial cellulose nanofibres with a novel silver-based metal-organic framework	Wound dressing	Antibacterial properties were tested against samples containing *Escherichia coli* and *Staphylococcus aureus*. In vivo studies were conducted, which showed an outstanding result of about 74 % wound closure when treated with the synthesised nanoparticles.	Barjasteh et al. (2023)

13.4 CHITOSAN-BASED NANOCARRIERS IN DRUG DELIVERY

Since the 1980s, numerous efforts have been attempted to employ nanomaterials in delivering drugs within the human body, and challenges include the incorporation of drugs and their release; drug formulation; stability; biocompatible; specific targeting; and shelf life. Chitosan-based nanoparticles are widely used to deliver drugs to the human body (Couvreur, 2013). Among the natural polymers, chitosan-based nanoparticles stood out, and they are widely used to deliver drugs to specific targets in the human body. Though the insolubility of chitosan in water is the main disadvantage of being used in clinical applications, it is a linear polymer with three functional groups. Thus, modifications in its structure are possible, as mentioned in the previous section, resulting in better solubility (Bernkop-Schnürch & Dünnhaupt, 2012; Sonia & Sharma, 2011). Administration of drugs orally neither causes infection nor causes pain. Yet, a suitable drug delivery system is highly needed for better encapsulation and the protection of drugs, as the gastrointestinal tract is highly acidic. Proteins and other polysaccharides form polyanions and polycations, which combine to form polyelectrolyte complexes. Besides, the reactions among the charged proteins or polysaccharides are affected by the concentration of the formed polyelectrolyte complexes. One example of polyelectrolyte formation is the nanocarrier, which combines chitosan and caseinate, which can form anionic micelles in aqueous solutions. This anionic micellar nano-colloid combines chitosan and polycationic chitosan to form a polyelectrolyte complex. The size of the polyelectrolyte complex is 200 nm, which has good stability features.

The efficacy of the formed chitosan-caseinate nanoparticles in insulin encapsulation is 75%. Also, the carrier system is non-toxic, safe and biodegradable for drug-sustained release in specific targets (Hamman, 2010; Jana & Jana, 2020). In another instance, chitosan is used along with levocetirizine by the spray drying technique. This system delivers drugs orally in a sustained manner. When levocetirizine alone is used, a burst effect is created, causing irritation and pain in the gastrointestinal tract. Also, the bioavailability of the drug is affected. On the other hand, when levocetirizine is used along with chitosan nanoparticles, it not only enhances the sustained release but also improves the bioavailability of the drug (Kirimlioğlu & Öztürk, 2020). Interestingly, in one of the earlier studies, it was noticed that chitosan with soy lecithin was used for the sustained release of raloxifene. The combination, as mentioned above, with raloxifene at low dosage is not effective towards reducing cell viability. The nanocarrier with raloxifene is evaluated in female Wistar rats, and the loaded drug's bioavailability is noticeably increased at higher dosages.

Besides, the endocytic process is utilised for ingesting the NPs and the drug. Hence, they are attached to the mucus cover in the intestine (Murthy et al., 2020). Another research work demonstrated using the chitosan-based nanocarrier system for transferring the drug, 'rotigotine', from the nose to the brain. The ionic gelation technique prepares the nanocarrier system composed of chitosan nanoparticles and the drug rotigotine. The above delivery system is completely non-toxic without damaging the mucous membrane present in the nose (Samprasit et al., 2022; Tzeyung et al., 2019). Chitosan is also frequently used for targeted drug delivery in colon cancer. For instance, alpha-Mangostin is used with chitosan and thiolated chitosan for specific drug delivery for colorectal cancer (Samprasit et al., 2022).

13.5 GENE DELIVERY

Almost 4000 diseases are caused due to the malfunctioning of genes present in a single type of cell. Gene therapy is a method used to cure, prevent or treat any genetic disorders. Gene therapy is a treatment method that involves introducing genes into cells to correct genetic diseases. Though it seems simple, there are challenges, as numerous physiological and intracellular barriers exist to reaching the target cell. Also, very less vectors having low toxicity, good stability and transfection efficacy are available for gene therapy. Hence, intracellular gene delivery is hampered (Cao et al., 2019). Chitosan, as discussed earlier, is a versatile natural polymer and a non-viral vector that is highly suitable for gene delivery (Abdelhamid, 2022). A few studies that employed chitosan-based nanoparticles in gene delivery are presented below. Chitosan is used to deliver protein p10 of avian reovirus DNA. This helps in the in vivo and in vitro treatment of murine B16 melanoma. Mild cytotoxicity is observed in the above case. Also, it was witnessed that, after the administration of the system composed of avian reovirus–p10 with chitosan nanoparticles in mice with B16 melanoma, the progression of the tumour was delayed. Besides, lymphoid populations in both the tumour tissue and spleen did not change. Thus, chitosan forms a promising bone-viral vector to transfer DNA. Furthermore, it was stated that chitosan is used in the treatment of melanoma as it helps deliver salmon anaemia virus fusion protein cDNA (an infectious virus) into cells (Robles-Planells, Barrera-Avalos, et al., 2020b; Robles-Planells, Sánchez-Guerrero, et al., 2020a). Chitosan nanoformulations are also evaluated as potent anti-HIV therapeutic systems (Ramana et al. (2014).

13.6 WOUND HEALING

Wounds are omnipresent conditions produced due to different factors such as surgery, grafting, ulcers and other conditions. Healing of skin wounds involves several factors, and the physiological process in wound healing involves five steps: coagulation, haemostasis, inflammation, proliferation and remodelling. The right treatment of the wound requires special wound care material, which should be stable, flexible and biodegradable. Besides, the materials should be able to keep the wounds moist, stop the bleeding from wounds and adsorb the fluid that leaks from the blood vessels from the wounds (Guo & DiPietro, 2010). Nanomaterials enable scientists to achieve an easy feat in the wound healing process because of their excellent adsorption ability and antimicrobial and drug incorporation properties. Chitosan has exceptional features pertaining to anti-inflammatory, antimicrobial, biodegradability and biocompatible (Bai et al., 2022). Sami et al. (2020) studied the wound healing process using nanofibres prepared using 5% turmeric, 1% oregano and 1% chitosan in ulcer models of diabetic rats. It was noteworthy to observe that the healing process was completed on the 15th day when the above nanofibre was used. At the same time, the wound was partially healed even during the 21st day using a commercial dressing. Thus, it was inferred that chitosan used in the nanofibre plays a more vital role in wound healing than the commercial dressing using silver alginate (Sami et al., 2020). Another research work developed a dressing made up of sponge-like chitosan materials incorporated into in-turn-silver nanoparticles. The prepared dressing helps by improving collagen formation, and a new epithelium is formed on the wound surface. Thus, the wound healing process occurs in mice, thus increasing the antibacterial attributes (Pourali & Yahyaei, 2016). The vaccine's efficacy after incorporating chitosan nanoparticles in wound healing is investigated for both in vitro and in vivo studies. The above-developed complex, composed of chitosan-vaccarin, has excellent biocompatibility and antibacterial properties, which help in rapid wound healing. In a nutshell, wound dressing formulated using chitosan nanoparticles has significant efficacy with respect to antimicrobial, anti-inflammatory and antioxidant properties (Correa et al. 2020; Hou et al., 2020).

13.7 TISSUE ENGINEERING

Loss of organs or tissues is one of the common issues faced by humans. This problem of organ or tissue loss can be overcome by a technique called tissue engineering. Tissue engineering is considered trouble-free as well as an effective way when compared with conventional treatment or organ transplantation. Though tissue engineering seems easier for scientists and doctors over traditional methods, selecting appropriate materials for tissue engineering is crucial (Rouchi & Mahdavi-Mazdeh, 2015). Chitosan is a good candidate as it possesses a hydrophilic structure that enables cell adhesion, proliferation and differentiation. Chitosan is used in various forms, such as hydrogels, sponges, fibres, sheets, films and other structures (Croisier & Jérôme, 2013). A scaffold that is biodegradable in nature has been developed for corneal endothelial tissue engineering. The transparent scaffold is prepared by loading chitosan nanoparticles into the membrane composed of chitosan-polycaprolactone. The resultant scaffold was tested for cytotoxicity and found that no cytotoxicity was associated with the developed scaffold. Also, the scaffold composed of chitosan and polycaprolactone was appropriate for corneal endothelial tissue regeneration (Prasad et al., 2015). Alendronate is incorporated into chitosan nanoparticles to regenerate human adipose stem cells. It was observed that the incorporated nanoparticles have a prominent effect on the osteogenic differentiation of adipose stem cells. Hence, this is suitable for bone tissue engineering scaffolds (Moradikhah et al., 2020).

13.8 VACCINE ADJUVANT

Conventional vaccination depends on attenuated and inactivated vaccine preparations. Recent techniques help in the development of a vaccine with higher efficacy. Also, safety features are taken into consideration. Though vaccines based on nucleic acids are being introduced, they have low immune response and effectiveness. Hence, there is a current requirement for the development of novel vaccines as well as delivery systems to specific sites in the human body. This opens new avenues in the field of vaccination using nanomaterials, thus improving the overall immune response (Pulendran et al., 2021). Chitosan nanoparticles have great potential as both vaccine adjuvants and delivery systems. Porcine epidemic diarrhoea virus in attenuated live vaccine form is encapsulated into chitosan nanoparticles. The above formulation can be stored for a long period, thus making it easy for transportation. Also, the live attenuated vaccine exhibited a tremendous performance with respect to immunity towards the infection. Inactivation of the virus is carried out using different concentrations of acid or alkaline solution, peroxyacetic acid and formaldehyde (at a given temperature range). The mentioned solution allows the reaction with chitosan and the virus, which in turn inactivates the virus (Hosseini et al., 2018). Another study used chitosan to inactivate the virus in antiviral vaccines (Lopes et al., 2018). It is noteworthy to observe a significant enhancement in the loading ability of antigens in the vaccine when the live Newcastle disease virus was loaded along with chitosan nanoparticles. Besides enhancing the antigen-loading capacity, chitosan helps improve the mucosa's absorption ability (Yang et al., 2020).

In summary, chitosan nanoparticles play an important role in increasing the immune response of the formulated vaccine. Also, chitosan enables desired activity irrespective of low antigen concentrations with minimal side effects. Furthermore, it can be concluded that with the newly evolving disease, such as coronavirus disease (COVID), chitosan proves to be a promising vaccine adjuvant (Mao et al., 2021).

13.9 CURRENT CHALLENGES AND PROSPECTIVE PROSPECTS

Though chitosan is extensively used in drug delivery, tissue engineering and other fields, it needs better stability and is a great disadvantage in clinical applications. Besides, during long-term storage, chitosan experiences degradation of its chain, after which the functional groups are broken down. As a result, there is an irreversible loss of its properties related to medical and other biological applications. Though efforts are made to enhance the stability of chitosan nanoparticles, there are no ubiquitous procedures to store products based on chitosan; hence, it is vital to work on optimising the storage conditions of chitosan (No et al., 2006; Szymańska & Winnicka, 2015). Another major challenge is its toxicity. Since chitosan derivatives find applications in drug delivery and cancer therapeutics, it is essential to study the toxicity properties of chitosan nanoparticle derivatives (Shanmuganathan et al., 2019).

Furthermore, with respect to complexes involving chitosan and proteins, the interaction, phase behaviour and colloidal properties between polysaccharide (chitosan) and protein have to be studied (Bekale et al., 2015). Many new chitosan derivatives are produced; hence, there is a pressing need to select suitable nanomaterials depending on the end application. Also, elaborate research that encompasses the sustainable production of chitosan-based nanomaterials is necessary for the betterment of humans and the environment (Lowry et al., 2019; Sharifi-Rad et al., 2021).

13.10 CONCLUSION

Nanotechnology offers a great advantage to the field of medicine. Nanomedicine can overcome clinical and biological barriers and is highly effective in delivering drugs to target sites. Though only a few nanomaterials were approved to be used for clinical and pharmaceutical purposes, chitosan is one such biopolymer extensively used in the field of nanomedicine (Bashir et al., 2022).

This chapter briefly demonstrates the properties of chitosan and its modifications to form chitosan derivatives. In addition, a detailed study was conducted on the applications of chitosan nanoparticles, such as in drug delivery, vaccine delivery, tissue engineering, gene delivery, antibacterial agents and wound healing. The advantages of chitosan derivatives in the medical field are that it is biocompatible, highly porous, soft, flexible and nontoxic. The disadvantages are that it is laborious, has low mechanical strength and shrivels, affecting crosslinking (Croisier and Jérôme, 2013). Additionally, challenges faced while using chitosan and its derivatives in nanomedicine, such as stability and solubility issues, are discussed. The toxicity associated with chitosan nanoparticles is also reviewed. Numerous novel chitosan-based nanomaterials have been developed recently; hence, their application prospects in the field of medicine must be thoroughly studied. Overall, chitosan is a vital biopolymer that can form its derivatives that have a tremendous role in nanomedicine. Also, in the future, green production of chitosan nanoparticles has to be focused on for the benefit of humans.

REFERENCES

Abdelhamid, H. N. (2022). Chitosan-based nanocarriers for gene delivery. In S. Jana & S. Jana (Eds.) *Nanoengineering of Biomaterials*, 91–105, Wiley-VCH.

Ashrafizadeh, M., Delfi, M., Hashemi, F., Zabolian, A., Saleki, H., Bagherian, M., ... Varma, R. S. (2021). Biomedical application of chitosan-based nanoscale delivery systems: Potential usefulness in siRNA delivery for cancer therapy. *Carbohydrate Polymers*, *260*, 117809.

Ashrafizadeh, M., Hushmandi, K., Mirzaei, S., Bokaie, S., Bigham, A., Makvandi, P., ... Orive, G. (2023). Chitosan-based nanoscale systems for doxorubicin delivery: Exploring biomedical application in cancer therapy. *Bioengineering & Translational Medicine*, *8*(1), e10325.

Auffan, M., Rose, J., Bottero, J.-Y., Lowry, G. V., Jolivet, J.-P., & Wiesner, M. R. (2009). Towards a definition of inorganic nanoparticles from an environmental, health and safety perspective. *Nature Nanotechnology*, *4*(10), 634–641.

Bai, Q., Zheng, C., Chen, W., Sun, N., Gao, Q., Liu, J., ... Zhang, Y. (2022). Current challenges and future applications of antibacterial nanomaterials and chitosan hydrogel in burn wound healing. *Materials Advances*, *3*(17), 6707–6727.

Barjasteh, M., Dehnavi, S. M., Seyedkhani, S. A., Rahnamaee, S. Y., & Golizadeh, M. (2023). Improved biological activities of dual nanofibrous chitosan/bacterial cellulose wound dressing by a novel silver-based metal-organic framework. *Surfaces and Interfaces*, *36*, 102631.

Bashir, S. M., Ahmed Rather, G., Patrício, A., Haq, Z., Sheikh, A. A., Shah, M. Z. u. H., ... Ahmad, S. B. (2022). Chitosan nanoparticles: A versatile platform for biomedical applications. *Materials*, *15*(19), 6521.

Bekale, L., Agudelo, D., & Tajmir-Riahi, H. (2015). Effect of polymer molecular weight on Chitosan–protein interaction. *Colloids and Surfaces B: Biointerfaces*, *125*, 309–317.

Bernkop-Schnürch, A., & Dünnhaupt, S. (2012). Chitosan-based drug delivery systems. *European Journal of Pharmaceutics and Biopharmaceutics*, *81*(3), 463–469.

Bharathiraja, S., Bui, N. Q., Manivasagan, P., Moorthy, M. S., Mondal, S., Seo, H., ... Lee, K. D. (2018). Multimodal tumor-homing chitosan oligosaccharide-coated biocompatible palladium nanoparticles for photo-based imaging and therapy. *Scientific Reports*, *8*(1), 1–16.

Boles, L. R., Bumgardner, J. D., Fujiwara, T., Haggard, W. O., Guerra, F. D., & Jennings, J. A. (2019). Characterisation of trimethyl chitosan/polyethylene glycol derivatised Chitosan blend as an injectable and degradable antimicrobial delivery system. *International Journal of Biological Macromolecules*, *133*, 372–381.

Cai, J., Yang, D., & Wang, Q. (2023). Preparation and characterisation of chitosan nanoparticles loaded with Athyrium sinense essential oil with antibacterial properties against Pectobacteriumcarotovorum subsp. carotovorum. *Industrial Crops and Products*, *195*, 116382.

Cao, Y., Tan, Y. F., Wong, Y. S., Liew, M. W. J., & Venkatraman, S. (2019). Recent advances in chitosan-based carriers for gene delivery. *Marine Drugs*, *17*(6), 381.

Cheung, R. C. F., Ng, T. B., Wong, J. H., & Chan, W. Y. (2015). Chitosan: an update on potential biomedical and pharmaceutical applications. *Marine Drugs*, *13*(8), 5156–5186.

Correa, V. L. R., Martins, J. A., de Souza, T. R., Rincon, G. d. C. N., Miguel, M. P., de Menezes, L. B., & Amaral, A. C. (2020). Melatonin loaded lecithin-chitosan nanoparticles improved the wound healing in diabetic rats. *International Journal of Biological Macromolecules*, *162*, 1465–1475.

Couvreur, P. (2013). Nanoparticles in drug delivery: past, present and future. *Advanced Drug Delivery Reviews*, *65*(1), 21–23.

Croisier, F., & Jérôme, C. (2013). Chitosan-based biomaterials for tissue engineering. *European Polymer Journal*, *49*(4), 780–792.

da Silva, A. B., Rufato, K. B., de Oliveira, A. C., Souza, P. R., da Silva, E. P., Muniz, E. C., … Martins, A. F. (2020). Composite materials based on chitosan/gold nanoparticles: From synthesis to biomedical applications. *International Journal of Biological Macromolecules*, *161*, 977–998.

Dacrory, S., Hashem, A. H., & Kamel, S. (2022). Antimicrobial and antiviral activities with molecular docking study of chitosan/carrageenan@ clove oil beads. *Biotechnology Journal*, *17*(2), 2100298.

Desai, K., & Park, H. (2005). Preparation of crosslinked chitosan microspheres by spray drying: Effect of crosslinking agent on the properties of spray dried microspheres. *Journal of Microencapsulation*, *22*(4), 377–395.

Ealia, S. A. M., & Saravanakumar, M. (2017). A review on the classification, characterisation, synthesis of nanoparticles and their application. Paper presented at the *IOP conference series: materials science and engineering*.

Guo, S. A., & DiPietro, L. A. (2010). Factors affecting wound healing. *Journal of Dental Research*, *89*(3), 219–229.

Hamman, J. H. (2010). Chitosan based polyelectrolyte complexes as potential carrier materials in drug delivery systems. *Marine Drugs*, *8*(4), 1305–1322.

Hosseini, Z. S., Amani, J., Baghbani Arani, F., Nazarian, S., Motamedi, M. J., Shafighian, F., … Cho, I.-S. (2018). Efficacy of inactivated variant porcine epidemic diarrhea virus vaccines in growing pigs. *Clinical and Experimental Vaccine Research*, *7*(1), 61–69.

Hou, B., Qi, M., Sun, J., Ai, M., Ma, X., Cai, W., … Xu, F. (2020). Preparation, characterisation and wound healing effect of vaccarin-chitosan nanoparticles. *International Journal of Biological Macromolecules*, *165*, 3169–3179.

Jana, S., & Jana, S. (2020). *Functional Chitosan: Drug delivery and biomedical applications*: Springer.

Jee, J.-P., Na, J. H., Lee, S., Kim, S. H., Choi, K., Yeo, Y., & Kwon, I. C. (2012). Cancer targeting strategies in nanomedicine: Design and application of chitosan nanoparticles. *Current Opinion in Solid State and Materials Science*, *16*(6), 333–342.

Khan, I., Saeed, K., & Khan, I. (2019). Nanoparticles: Properties, applications and toxicities. *Arabian Journal of Chemistry*, *12*(7), 908–931.

Kirimlioğlu, G. Y., & Öztürk, A. A. (2020). Levocetirizine dihydrochloride-loaded chitosan nanoparticles: formulation and in vitro evaluation. *Turkish Journal of Pharmaceutical Sciences*, *17*(1), 27.

Kumar, S., Gogoi, A. S., Shukla, S., Trivedi, M., & Gulati, S. (2022). Conclusion and future prospects of chitosan-based nanocomposites. In S. Gulati (Ed.) *Chitosan-based nanocomposite materials: Fabrication, characterisation and biomedical applications* (pp. 305–341). Springer.

Lane, L. A., Qian, X., Smith, A. M., & Nie, S. (2015). Physical chemistry of nanomedicine: understanding the complex behaviors of nanoparticles in vivo. *Annual Review of Physical Chemistry*, *66*, 521–547.

Liu, X.-Q., Zhao, X.-X., Liu, Y., & Zhang, T.-A. (2022). Review on preparation and adsorption properties of chitosan and chitosan composites. *Polymer Bulletin*, *79*(4), 2633–2665.

Lopes, P. D., Okino, C. H., Fernando, F. S., Pavani, C., Casagrande, V. M., Lopez, R. F., … Montassier, H. J. (2018). Inactivated infectious bronchitis virus vaccine encapsulated in chitosan nanoparticles induces mucosal immune responses and effective protection against challenge. *Vaccine*, *36*(19), 2630–2636.

Lowry, G. V., Avellan, A., & Gilbertson, L. M. (2019). Opportunities and challenges for nanotechnology in the agri-tech revolution. *Nature Nanotechnology*, *14*(6), 517–522.

Lue, J.-T. (2001). A review of characterisation and physical property studies of metallic nanoparticles. *Journal of Physics and Chemistry of Solids*, *62*(9–10), 1599–1612.

Mao, L., Chen, Z., Wang, Y., & Chen, C. (2021). Design and application of nanoparticles as vaccine adjuvants against human corona virus infection. *Journal of Inorganic Biochemistry*, *219*, 111454.

Min, K. H., Park, K., Kim, Y.-S., Bae, S. M., Lee, S., Jo, H. G., … Kim, K. (2008). Hydrophobically modified glycol chitosan nanoparticles-encapsulated camptothecin enhance the drug stability and tumor targeting in cancer therapy. *Journal of Controlled Release*, *127*(3), 208–218.

Mirjalili, F., & Mahmoodi, M. (2023). Controlled release of protein from gelatin/chitosan hydrogel containing platelet-rich fibrin encapsulated in chitosan nanoparticles for accelerated wound healing in an animal model. *International Journal of Biological Macromolecules*, *225*, 588–604.

Mohandas, A., Deepthi, S., Biswas, R., & Jayakumar, R. (2018). Chitosan based metallic nanocomposite scaffolds as antimicrobial wound dressings. *Bioactive Materials*, *3*(3), 267–277.

Moradikhah, F., Doosti-Telgerd, M., Shabani, I., Soheili, S., Dolatyar, B., & Seyedjafari, E. (2020). Microfluidic fabrication of alendronate-loaded chitosan nanoparticles for enhanced osteogenic differentiation of stem cells. *Life Sciences*, *254*, 117768.

Mourya, V., & Inamdar, N. N. (2008). Chitosan-modifications and applications: Opportunities galore. *Reactive and Functional Polymers*, *68*(6), 1013–1051.

Mout, R., Moyano, D. F., Rana, S., & Rotello, V. M. (2012). Surface functionalisation of nanoparticles for nanomedicine. *Chemical Society Reviews*, *41*(7), 2539–2544.

Murthy, A., Ravi, P. R., Kathuria, H., & Vats, R. (2020). Self-assembled lecithin-chitosan nanoparticles improve the oral bioavailability and alter the pharmacokinetics of raloxifene. *International Journal of Pharmaceutics*, *588*, 119731.

Nair, A., Mallya, R., Suvarna, V., Khan, T. A., Momin, M., & Omri, A. (2022). Nanoparticles—Attractive carriers of antimicrobial essential oils. *Antibiotics*, *11*(1), 108.

Negm, N. A., Hefni, H. H., Abd-Elaal, A. A., Badr, E. A., & Abou Kana, M. T. (2020). Advancement on modification of chitosan biopolymer and its potential applications. *International Journal of Biological Macromolecules*, *152*, 681–702.

No, H. K., Kim, S. H., Lee, S. H., Park, N. Y., & Prinyawiwatkul, W. (2006). Stability and antibacterial activity of chitosan solutions affected by storage temperature and time. *Carbohydrate Polymers*, *65*(2), 174–178.

Okunzuwa, G., Enaroseha, O., & Okunzuwa, S. (June 2024). Synthesis and characterization of Fe (III) chitosan nanoparticles n-benzaldehyde Schiff base for biomedical application. *Chemical Papers*, *78*(5), 3253–3260.

Peniche, H., & Peniche, C. (2011). Chitosan nanoparticles: a contribution to nanomedicine. *Polymer International*, *60*(6), 883–889.

Pourali, P., & Yahyaei, B. (2016). Biological production of silver nanoparticles by soil isolated bacteria and preliminary study of their cytotoxicity and cutaneous wound healing efficiency in rat. *Journal of Trace Elements in Medicine and Biology*, *34*, 22–31.

Prasad, T., Shabeena, E., Vinod, D., Kumary, T., & Anil Kumar, P. (2015). Characterisation and in vitro evaluation of electrospun chitosan/polycaprolactone blend fibrous mat for skin tissue engineering. *Journal of Materials Science: Materials in Medicine*, *26*, 1–13.

Pulendran, B., Arunachalam, P. S., & O'Hagan, D. T. (2021). Emerging concepts in the science of vaccine adjuvants. *Nature Reviews Drug Discovery*, *20*(6), 454–475.

Rajasree, R., & Rahate, K. (2013). An overview on various modifications of Chitosan and it's applications. *International Journal of Pharmaceutical Sciences and Research*, *4*(11), 4175–4193.

Ramana, L. N., Sharma, S., Sethuraman, S., Ranga, U., & Krishnan, U. M. (2014). Evaluation of chitosan nanoformulations as potent anti-HIV therapeutic systems. *Biochimica et Biophysica Acta (BBA)-General Subjects*, *1840*(1), 476–484.

Robles-Planells, C., Barrera-Avalos, C., Rojo, L. E., Spencer, E., Martin, C.-S., Matiacevich, S., … Martínez, B. A. (2020b). Chitosan-based delivery of avian reovirus Fusogenic protein P10 gene: in vitro and in vivo studies towards a new vaccine against melanoma. *BioMed Research International*, *2020*, 4045760.

Robles-Planells, C., Sánchez-Guerrero, G., Barrera-Avalos, C., Matiacevich, S., Rojo, L. E., Pavez, J., … Milla, L. A. (2020a). Chitosan-based nanoparticles for intracellular delivery of ISAV fusion protein cDNA into melanoma cells: A path to develop oncolytic anticancer therapies. *Mediators of Inflammation*, 2020, 8680692.

Roco, M. C. (2017). Overview: Affirmation of nanotechnology between 2000 and 2030. In T. O. Mensah, B. Wang, G. Bothun, J. Winter, & V. Davis (Eds.), *Nanotechnology commercialisation: Manufacturing processes and products* (pp. 1–23).

Rouchi, A. H., & Mahdavi-Mazdeh, M. (2015). Regenerative medicine in organ and tissue transplantation: shortly and practically achievable? *International Journal of Organ Transplantation Medicine*, *6*(3), 93.

Sami, D. G., Abdellatif, A., & Azzazy, H. M. (2020). Turmeric/oregano formulations for treatment of diabetic ulcer wounds. *Drug Development and Industrial Pharmacy*, *46*(10), 1613–1621.

Samprasit, W., Opanasopit, P., & Chamsai, B. (2022). Alphamangostin and resveratrol, dual drugs loaded mucoadhesive thiolated chitosanbased nanoparticles for synergistic activity against colon cancer cells. *Journal of Biomedical Materials Research Part B: Applied Biomaterials*, *110*(6), 1221–1233.

Saravanakumar, K., Jeevithan, E., Chelliah, R., Kathiresan, K., Wen-Hui, W., Oh, D.-H., & Wang, M.-H. (2018). Zinc-chitosan nanoparticles induced apoptosis in human acute T-lymphocyte leukemia through activation of tumor necrosis factor receptor CD95 and apoptosis-related genes. *International Journal of Biological Macromolecules*, *119*, 1144–1153.

Shah, S. A., Sohail, M., Karperien, M., Johnbosco, C., Mahmood, A., & Kousar, M. (2023). Chitosan and carboxymethyl cellulose-based 3D multifunctional bioactive hydrogels loaded with nano-curcumin for synergistic diabetic wound repair. *International Journal of Biological Macromolecules*, *227*, 1203–1220.

Shanmuganathan, R., Edison, T. N. J. I., LewisOscar, F., Kumar, P., Shanmugam, S., & Pugazhendhi, A. (2019). Chitosan nanopolymers: an overview of drug delivery against cancer. *International Journal of Biological Macromolecules*, *130*, 727–736.

Sharifi-Rad, J., Quispe, C., Butnariu, M., Rotariu, L. S., Sytar, O., Sestito, S., & Krishna, A. (2021). Chitosan nanoparticles as a promising tool in nanomedicine with particular emphasis on oncological treatment. *Cancer Cell International*, *21*(1), 1–21.

Sonia, T., & Sharma, C. P. (2011). Chitosan and its derivatives for drug delivery perspective. In R. Jayakumar, M. Prabaharan, & R. A. A. Muzzarelli (Eds.), *Chitosan for biomaterials I* (Vol. 243, pp. 23–53. Springer.

Szymańska, E., & Winnicka, K. (2015). Stability of Chitosan—a challenge for pharmaceutical and biomedical applications. *Marine Drugs*, *13*(4), 1819–1846.

Tzeyung, A. S., Md, S., Bhattamisra, S. K., Madheswaran, T., Alhakamy, N. A., Aldawsari, H. M., & Radhakrishnan, A. K. (2019). Fabrication, optimisation, and evaluation of rotigotine-loaded chitosan nanoparticles for nose-to-brain delivery. *Pharmaceutics*, *11*(1), 26.

Wang, S., Shen, L., Tong, Y., Chen, L., Phang, I., Lim, P., & Liu, T. (2005). Biopolymer chitosan/montmorillonite nanocomposites: preparation and characterisation. *Polymer Degradation and Stability*, *90*(1), 123–131.

Xu, X., Li, Q., Dong, W., Zhao, G., Lu, Y., Huang, X., & Liang, X. (2023). Cinnamon cassia oil chitosan nanoparticles: Physicochemical properties and anti-breast cancer activity. *International Journal of Biological Macromolecules*, *224*, 1065–1078.

Yan, T., Zhang, H., Huang, D., Feng, S., Fujita, M., & Gao, X.-D. (2017). Chitosan-functionalised graphene oxide as a potential immunoadjuvant. *Nanomaterials*, *7*(3), 59.

Yang, X., Mo, W., Shi, Y., Fang, X., Xu, Y., He, X., & Xu, Y. (2023). Fumaria officinalis-loaded chitosan nanoparticles dispersed in an alginate hydrogel promote diabetic wounds healing by upregulating VEGF, TGF-β, and b-FGF genes: A preclinical investigation. *Heliyon*, *9*(7), e17704.

Yang, Y., Xing, R., Liu, S., Qin, Y., Li, K., Yu, H., & Li, P. (2020). Chitosan, hydroxypropyltrimethyl ammonium chloride chitosan and sulfated chitosan nanoparticles as adjuvants for inactivated Newcastle disease vaccine. *Carbohydrate Polymers*, *229*, 115423.

Ye, Y., Ding, S., Ye, Y., Xu, H., Cao, X., Liu, S., & Sun, H. (2015). Enzyme-based sensing of glucose using a glassy carbon electrode modified with a one-pot synthesised nanocomposite consisting of Chitosan, reduced graphene oxide and gold nanoparticles. *Microchimica Acta*, *182*, 1783–1789.

Yohan, D., & Chithrani, B. D. (2014). Applications of nanoparticles in nanomedicine. *Journal of Biomedical Nanotechnology*, *10*(9), 2371–2392.

Yokoyama, T., Masuda, H., Suzuki, M., Ehara, K., Nogi, K., Fuji, M., & Hayashi, K. (2008). Basic properties and measuring methods of nanoparticles. In M. Naito, T. Yokoyama, K. Hosokawa, & K. Nogi (Eds.), *Nanoparticle technology handbook* (pp. 3–48). Elsevier.

Zhang, J., Xia, W., Liu, P., Cheng, Q., Tahirou, T., Gu, W., & Li, B. (2010). Chitosan modification and pharmaceutical/biomedical applications. *Marine Drugs*, *8*(7), 1962–1987.

Zhu, C.-Y., Li, F.-L., Zhang, Y.-W., Gupta, R. K., Patel, S. K., & Lee, J.-K. (2022). Recent strategies for the immobilisation of therapeutic enzymes. *Polymers*, *14*(7), 1409.

14 Antimicrobial Nanocomposite Food Packaging Films for Sustainable Applications

Habtamu Shebabaw Kassa, S. Anuradha Jabasingh, Shegaw Ahmed Mohammed, Soo-Young Park, and Seung-Yeop Baek

14.1 INTRODUCTION

The materials used for packaging are petroleum-based synthetic polymers, and they are non-biodegradable and harmful to the environment. The globe, as well as Ethiopia, is trying to achieve a green energy economy, so replacing non-biodegradable petroleum-based food packaging materials with biodegradable biopolymers is a future necessity and evolving trend. The mechanical properties exhibited by the bioplastics are fabulous and hence could be further strengthened by the incorporation of fillers. Thus, the composite materials made of polymers and natural fibers are critical to future applications (Dey and Ray, 2018). Cellulose nanocrystals (CNCs) are capable of improving the mechanical, thermal, barrier, and electrical conductivity of starch films for packaging applications (Azeredo et al., 2016). The lower weight, surface area to volume ratio, presence of hydroxyl groups, excellent mechanical strength, good aspect ratio, lower density, and acceptable thermal and electrical properties allow the CNCs to be a good reinforcement filler for the composite materials (Kassa et al., 2022; Mishra et al., 2018).

CNCs found in a stabilized form prevent the movement of the bioactive components in the prepared biopolymers, and this limits the release of CNCs. CNCs further increase the tortuosity of starch-based films, decrease the water vapor permeability, improve the food quality, increase the food shelf life, and resist infections. The high compatibility that exists between the starch matrix and nanocellulose fillers has been found to improve the water sensitivity of the film for food packaging applications (Bagde and Nadanathangam, 2019).

Antimicrobial packaging encompasses the integration of antimicrobials into the packaging to prevent microbial development to increase the shelf life (Mlalila et al., 2018). Rosemary (*Rosmarinus officinalis L.*) is a spicy, medicinal herb with the highest presence of antioxidant activity, and it is also found to be an excellent antibacterial and antifungal agent. The antibacterial effect of rosemary has been widely demonstrated in beef meatballs, cooked beef, and pork sausages (Gómez-Estaca et al., 2010).

Coccinia abyssinica, or anchote, is a native tuber crop found in Ethiopia, and it belongs to the Cucurbitales and Cucurbitaceae order and family, respectively (Fekadu, 2014). The main chemical composition of anchote is starch (Beruk et al., 2015). The overall starch content of anchote is 78% (Abera et al., 2019). The physicochemical characteristics of thermoplastic anchote starch films showed promising results for the edible films by several researchers (Tessema and Admassu, 2021).

Research on the preparation of biopolymers from anchote starch requires intensive study regarding the improvisation of its several properties. By adding reinforcing elements and antimicrobial compounds to anchote starch polymer matrix, multifunctional biocomposites might be created. Intensive study regarding the analysis of the chemical components of anchote bagasse (AnchB) is essential. Therefore, the overall aim of the study is to develop and characterize CNCs reinforced anchote

DOI: 10.1201/9781003732907-14

starch-based antimicrobial food packaging film with rosemary oil. Research on the creation of biopolymers from anchote starch to produce biopolymers with various properties is still the need of the hour.

The thermoplastic starch (TPS) films have properties that are somehow limitations to their application for the production of biopolymers. This includes their hydrophilicity, brittleness, low mechanical strength, and decrease in the barrier properties (Címtia et al., 2009). Fillers and reinforcing agents could be added to the TPS to overcome these limitations.

The decay of food is usually caused by exposure to contaminated unclean environments during cooking, processing, and storage. The foodborne pathogens are found to mainly grow on the surfaces of the containers and transfer vessels (Arfat et al., 2014). The development of antimicrobial packaging films is critical in that the antimicrobial agents carried by the polymers will interact with the microorganism and possibly act negatively on them, thereby leading to a control in the diffusion rate of antimicrobial agents.

Microorganisms are present everywhere, and they contaminate food at various stages during manufacturing, handling, and storage. Even though there are different preservation methods that increase the shelf life of food products, some microorganisms may escape and contaminate the food again. Microorganisms present in the canned and pasteurized foods result in food spoilage, food loss, economic effects, and environmental effects. Microorganisms cause food spoilage, which results in the undesirable or unacceptable nature, abrupt variations in the sensory characteristics, and consequent effect on the economic situation of the country.

Incorporation of antimicrobial components on the food surface may lead to limited efficacy of antimicrobial migration, loss of freshness, and inactivation of certain important nutrients in the food components. The findings of this study will have a significant effect on supplementing the literature related to the idea of replacing non-biodegradable food packaging materials with biodegradable food packaging materials and thus a reduction in plastic waste.

The result of this research will provide additional information on the biopolymer production from anchote starch by combining it with CNCs and rosemary essential oils. The study will offer packaging materials based on anchote starch, focusing on the enhanced mechanical strength and antibacterial applications for the food products. Processing of anchote, to isolate starch, generates a large amount of solid waste like peels, unusable roots, and bagasse. The result of the study will help to assess the potential of anchote bagasse for cellulose and cellulose nanocrystal applications and as a reinforcing material in biopolymer films. The outcome of this research will create job opportunities for different stakeholders. Anchote producers, investors, professionals, laborers, and researchers are expected to benefit from this study.

14.2 BIOPOLYMER FILMS

Biopolymers are made up of monomers covalently bound to one another to produce chains. Biopolymers are indicated by the prefix "bio" as being biodegradable and thus could be broken down by microbes, leading to the byproducts that are likely organic in nature. Biopolymers, also referred to as biodegradable polymers, require precise conditions such as optimum values of pressure, temperature, relative humidity, and aerobic atmospheres for the enhancement of their biodegradable abilities (Sorrentino et al., 2007). Due to their biodegradability, abundance in nature, sustainability, and ability to regenerate, biopolymers have been viewed as a possible alternative to materials derived from petroleum (Chandra and Rustgi, 1998).

Biopolymers could be either thermosetting or thermoplastic and could be created by polymerizing synthetic monomers derived from green materials (Shogren et al., 2004). Polylactic acid, polyhydroxybutyrate, plant-originated polysaccharides, cellulose, alginate, animal-originated polysaccharides, and chitin, thermoplastic starches and hyaluronates are examples of bio-thermoplastics that are frequently utilized in green composites. Polymers based on carbohydrates (starch) have so far been employed most frequently for coatings or films used in food packaging. In addition, numerous researches have demonstrated the mechanical, optical, and thermal properties of the biopolymers. And this includes color, glass transition, melting, durability, conductivity, photosensitivity, gloss, flexibility, strength, and optical qualities.

14.3 STARCH-BASED BIOPOLYMERS

Starches are one of the types of biodegradable polymers and are a low-cost polysaccharide. Another name for it is hydrocolloid biopolymer. It is made up of branching and amorphous polymer-amylopectin and linear and crystalline polymer-amylose. Depending on the source, starch contains between 10–20% and 80–90%, respectively, of amylose and amylopectin, respectively (Weber et al., 2002). Amylose has a helical structure and is soluble in water (Leja and Lewandowicz, 2010). Biopolymers are made using various starch resources such as cassava, tapioca, potato, corn, and rice (Balakrishnan et al., 2017). Typically, starch is utilized as a thermoplastic. It is broken down into plastic through de-structuration by the application of optimum heat, water, or plasticizers and is subsequently extruded. Because of mechanical limitations and low resistance to moisture, starches have limited applications. Starches are combined with various biopolymers and additives to enhance these qualities (Mangaraj et al., 2019).

14.4 CELLULOSE NANOCRYSTAL

Cellulose nanocrystals (CNCs) are rod-shaped nanoparticles of cellulose that are produced during the hydrolysis of lignocellulosic biomass. Many biomasses have been chosen for the extraction of CNCs utilizing the mineral acid hydrolysis technique, which creates nano-sized cellulose crystals of various sizes and crystallinities. In recent years, both the scientific and industrial communities have become interested in the manufacture of cellulose nanocrystals from various lignocellulosic fibers. This is because of their cost-effectiveness, adaptability, and mechanical and thermal characteristics. High solidity and crystallinity are typical characteristics of nanocrystals, which enable sustainability. Adding cellulose nanoparticles (CNPs) to polymers broadens the range of potential applications. The mechanical qualities of CNC have improved as a result of its aspect ratio. It is investigated whether the characteristics of polymers may be improved by using bio-reinforcing agents like nanocellulose. In order to produce fully green materials, a lot of research has to be carried out on using bio-reinforcing materials, instead of inorganic reinforcing materials (Moustafa et al., 2019).

14.5 CELLULOSE NANOCRYSTALS REINFORCED STARCH FILMS

The plant-based fillers in starch-based thermoplastics improve the mechanical performance, water resistance, and the sensitivity of the starch biopolymers (Halimatul et al., 2019). The application of these fillers resulted in the "green composites" or "bio composites." A composite material is one that has many constituent materials mixed to obtain significant chemical and physical behaviors. Composites have been used from the beginning of time. The cellulose-reinforced fibrils in the wood are bound by a hemicellulose and lignin matrix. Composite materials known as "natural fiber composites" are those in which at least some of the reinforcing fibers are derived from renewable resources. Additionally, because bio-composites are renewable resources for new, environmentally friendly goods, they are advantageous from an economic aspect as well, given that the price of petroleum is always rising owing to depletion. When compared to macrocomposite materials, nanocomposites have improved optical transparency in addition to their physical properties.

The native starch and its derivative thermoplastic starch could be employed as a foundation material due to their abundant natural supply, renewability, affordability, biodegradability, and processability for the environmentally friendly packaging films (Cao et al., 2008). The starch-based composites have drawn interest for a long time and have been successfully manufactured and marketed due to their preferable interface. The tensile strength and Young's modulus of the starch-based materials have both been enhanced by the inclusion of cellulose nanocrystals (Soykeabkaew et al.,

2012). The creep resistance and creep recovery rate of starch film may both be improved by the use of cellulose nano-fibrils, according to reports (Li et al., 2015). According to Panaitescu et al., starch films reinforced with cellulose nanocrystals were even comparable to polyolefin (Panaitescu et al., 2015). Cellulose crystals are more thermally stable than starch films; they can give more processability and greater mechanical qualities to starch films. As all of the ingredients utilized in these films come from food sources, they are biodegradable, safe, and create edible films. Depending on the kind of polymeric matrix, a low or moderate nanocrystal concentration from green coconut fiber was shown to be more efficient, while larger concentrations caused unfavorable nanocrystal agglomeration.

The tensile strength of the bioplastics increases for pure starch, and the elongation at break decreases in contrast to tensile strength. The hydrogen bonding in the hydroxyl groups at the interface of both fillers and the matrix is critical as well. As MCC content increased, bioplastics' density dropped. Moreover, adding MCC content reduced the water absorption of bioplastics. The flax cellulose nanocrystal (FCN) was used to strengthen pea starch (PS)-based film, and the properties, including the mechanical and water resistance capabilities, were tested (Cao et al., 2008). The authors observed that the tensile and Young's modulus increased when the FCN content was increased, indicating the formation of a robust and less deformed plastic. This improvement of mechanical properties was due to the homogeneous dispersion of nanocrystals into the matrix, and the strong hydrogen bonding as well.

Chang et al. developed nanocomposite films containing glycerol thermoplastic wheat starch and cellulose nanoparticles in order to investigate the mechanical characteristics, thermal properties, and permeability to water vapor (Chang et al., 2010). By adding nanoparticles to the wheat starch solution, films were produced. The increase in the filler content increased the tensile strength of the cast film, and the elongation at break decreased. As a result of cellulose and starch having similar fundamental structures, the results showed a positive matrix–filler relationship. The creation of a tortuous channel, which could be created by properly dispersing nanoparticles into the matrix, was the cause of the declining water vapor permeability (WVP). Contrarily, excessive CNP addition caused agglomeration, which decreased the films' WVP.

In order to reinforce composite films made of potato starch, Chen and his team (Chen et al., 2012) produced cellulose nanoparticles made of cotton and potato peel waste. Thermal characteristics were shown to be unaffected by CNC source information. With CNC, there was little variation in water vapor permeability. On the other hand, adding PCN resulted in a greater Young's modulus for thermoplastic starch. However, in both situations, the increase in Young's modulus accelerated as the quantity of CNCs increased. In a different study, thermoplastic films made of potato starch were developed using cellulose and CNCs as fillers (Aila-Suárez et al., 2013). The thermal abilities, mechanical properties, permeability, and water solubility are some of the critical factors to be evaluated. Due to the incorporation of cellulose and CNCs, there was a decrease in the composite films' solubility, while the melting temperature, melting enthalpy, tensile strength, and elastic modulus were found to increase. The appropriate interactions between the starches and fillers, and the incorporation of filler into films were responsible for the increase in melting temperature. Also, they claimed that adding cellulose and CNCs improved the mechanical qualities, although the effect was greater for the CNCs than the cellulose. It was obvious that the CNCs were evenly distributed throughout the matrix and formed tortuous paths, which decreased the rate of water transmission (Chang et al., 2010).

14.6 ANTIMICROBIAL BIOPOLYMER PACKAGING FILMS INCORPORATED WITH ESSENTIAL OILS

Several researchers have investigated how various plant essential oils affect the antibacterial properties of biopolymer packaging sheets. It was found that adding cinnamon essential oil (CEO) to pectin or whey protein edible film produced an antimicrobial agent. This material can be used in the packaging industry to coat foods like bakery goods, dairy goods, meat, and vegetables without compromising their quality (Sharma et al., 2017).

Strong antibacterial effects were shown on the addition of clove essential oil to starch-based edible films. A 16–27 mm-diameter zone of inhibition was visible in clove oils at 3%. An increase in oil content decreased the tensile break strength of antimicrobial starch films containing *Mentha pulegium* essential oil and *Zataria multiflora Boiss oil*, but concurrently increased elongation at break. The moisture barrier qualities of the film also had a more notable reduction in WVP. The outcomes also showed that ZEO or MEO-containing films effectively slowed the growth of microbes.

The antimicrobial activity of cinnamon and clove essential oils is widely used in cassava starch films to produce an active packaging based on the results of their minimum inhibitory concentration. The results were compared to those of control films, and this showed that the essential oil content influenced the film properties (Souza et al., 2013).

Šuput et al. (2016) studied features of edible films with oregano oil addition. The addition of essential oils to films had a positive impact on their mechanical, water vapor barrier, and swelling properties. The formation of the inhibition zone against *S. typhimurium*, *E. coli*, and *L. monocytogenes* varied for the samples containing oregano essential oil, according to an analysis of the microbial activity of the edible films. It was found that the application of oregano essential oil to the matrix had a higher antibacterial impact.

Chitosan and rosemary essential oil's effects on the characteristics of starch films were investigated, and peanuts were wrapped in them. Both chitosan and rosemary improved the film's tensile strength. Both of these elements decreased the film's solubility, moisture absorption, and water vapor permeability, with the interaction impact of the rosemary and chitosan being particularly important in reducing solubility. Although chitosan in this situation lacked antibacterial properties, rosemary essential oil decreased *A. flavus* growth and aflatoxin B1 development. Moreover, chitosan concentrations and other physical testing conditions for the film were developed. It appears that rosemary plays a significant role in enhancing the films' physical and microbial qualities, whereas chitosan has a significant impact on the physical qualities (Sayyahi et al., 2017).

14.7 NANOCOMPOSITE FILMS PROCESSING METHODS

Chemical and mechanical processes can be used to create a polymer matrix. The uniform and homogenous dispersion of nanoparticles in the polymer matrix is a huge challenge to be addressed, though. The nature of nanofillers' propensity to form micron-sized filler clusters restricts the dispersion of nanoparticles and thus influences the characteristics of nanocomposites and leads to further deterioration as well. Many attempts have been made to uniformly disperse nanofillers in the matrix through various chemical reactions, polymerization, and surface modification of filler materials. The four techniques used to create polymer nanocomposites are the intercalation method, in situ polymerization, sol–gel method, and the direct mixing of polymer and other nanofillers.

14.8 INTERCALATION METHOD

The intercalation method disperses nanomaterials into the polymer matrix and thus improves the qualities, including the stiffness, shrinkage, and flammability, incorporated into polymer matrices. For uniform dispersion of nanofillers in the matrix, intercalation, a top-down strategy, using the following two methods, can be used to evenly distribute the nanoplatelets (Alexandre and Dubois, 2000):

Chemical technique: This approach uses in-situ polymerization, in which monomer is mixed with nanoparticles to initiate a polymerization reaction. This technique involves dispersing nanoplatelets into a polymer and then performing further polymerization. In a monomer solution, the nanoplatelets swell, and a polymerization process creates a polymer between the intercalated sheets.

Mechanical technique: This is practiced by the fluid mixing process and by the direct intercalation of polymer with nanoplatelets. When the two solutions are combined, the polymer chains in the solution intercalate into the nanoplatelet layers and displace the solvent. The polymer is dissolved in a co-solvent, and nanoplatelet sheets are allowed to swell in the solvent.

14.9 MELT INTERCALATION METHOD

Melt intercalation is a technique that is frequently employed in business. Using this technique, molten polymer matrix and nanofillers (clays) are combined. This technique involves either static or shear annealing of a polymer and nanofiber combination. This technology uses the polymers that are not suited for in-situ polymerization or solution intercalation. A comparable procedure is melt blending. This method incorporated the nanofillers into the polymer solution via shear rate and temperature diffusion by melting the polymer powder/pellets to create a viscous solution. Compression and injection molding, or fiber production techniques, are the further steps that could be employed.

14.10 IN-SITU POLYMERIZATION METHOD

This involves allowing the swelling of the nanofillers in the monomer solution. The polymerization of the resultant solution is initiated mostly by diffusion, radiation, heat, or an organic substance. Following this, the monomer is thus polymerized between interlayers to create either exfoliated or intercalated nanocomposites. The process is comparable to in-situ template synthesis. In this procedure, polymer chains are used to create the clay layers. Aqueous solutions are used to dissolve the polymer matrix at a higher temperature. Clay layer nucleation and growth occur at high temperatures, trapping the polymer chains inside the clay layers. High temperature synthesis results in polymer decomposition, which is the only shortcoming of this technique.

14.11 SOL–GEL METHOD

The sol–gel method is a bottom-up strategy that operates on the polar opposite of all the prior techniques. Sol and gel are the first two steps in a sol–gel relationship. Gel is a three-dimensional interconnected network created between phases, and the term "sol" refers to a colloidal suspension of nanoparticles in monomer solution. This technique involves dispersing solid nanoparticles in a monomer solution to create a colloidal suspension of solid nanoparticles (sol), which then undergoes polymerization and hydrolysis to create a gel that connects the phases. The liquid is filled with a 3D network of polymer nanoparticles. In addition to acting as a nucleating agent, the polymer also encourages the development of multilayer crystals. The polymer enters the spaces between the layers as the crystals expand, creating a nanocomposite.

14.12 DIRECT MIXING OF POLYMER AND NANOFILLERS

The top-down method of creating nanocomposite materials, which relies on the breakdown of aggregated nanofillers during mixing, involves directly combining a polymer matrix with nanofillers. The two common techniques to mix the polymer and nanofillers are "melt compounding," involving the mixing of a polymer with nanofillers above the glass transition temperature of the polymer, without the use of any solvents. And the other method, known as the solvent method or solution mixing method, employs the mixing of polymers and nanofillers in a solution.

14.13 MELT COMPOUNDING

Melt compounding involves the addition of nanofibers to the polymer above the glass transition temperature. The viscous drag induces shear stress in the polymer melt, and this shear stress is employed to break down the aggregates, promoting homogenous and nanofiller dispersion in the polymer matrix.

14.14 SOLVENT METHOD

In this process, the polymer is dissolved in a cosolvent, and the nanoparticles are dispersed in the solvent. Using solvent evaporation or a coagulation process, the created nanocomposites are removed from the solvent. The melt compounding approach decreases the shear stresses created in the polymer matrix. Since nanocellulose has a polar surface and is challenging to distribute in nonpolar media, a number of research have sought to understand the obstacles in using nanocellulose as filler in polymer matrix. Blending or mixing is one of the crucial steps in the creation of nanocomposite materials. Hence, for nanocellulose's performance to be properly optimized, it needs to be evenly distributed throughout the polymer matrix. Since nanocellulose has a high surface area and a low nanofiller content, it tends to collect rather than disseminate the matrix, making dispersion levels of cellulose reinforcement in the resin crucial (Jumaidin et al., 2020). Films made of polysaccharide nanocomposite can be produced using either the solvent casting process, which uses either organic solvent or water evaporation, or the extrusion method, which uses freeze–dried nanocellulose particles.

The frequently used approach, in the creation of starch-based nanocomposites, depends on the matrix polymer nature. Examples include, polymer emulsions, non-hydro-soluble and water-soluble polymers. A polymer starch-based nanocomposite could be created by combining the polymer solution with nanocellulose suspension, which further results in the creation of a uniform dispersion of nanocellulose in the nanocomposite film.

The second approach uses extrusion or melt compounding to combine nanocellulose with a composite (Siqueira et al., 2010). However, the drawback rests in the poor dispersion and distribution of nanocellulose. Nanocomposites prepared from a solution casting approach have greater mechanical properties compared to those prepared by the freeze–drying and extrusion processes. This is the result of the solution casting technique's strong hydrogen bonding effect, which is used to create rigid nanocellulose networks. The solution casting approach has the benefit of requiring less processing time than other methods. The extrusion approach, on the other hand, is constrained in terms of nanoscale reinforcement since solid hydrogen bonds can only be created after the polysaccharide nanocellulose has dried.

14.15 ANTIMICROBIAL AND ANTIOXIDANT ACTIVITY OF ROSEMARY ESSENTIAL OIL

Rosmarinus officinalis L. is a common spicy, medicinal herb used all over the world. It is one of the spices with a high antioxidant activity among the naturally occurring antioxidants (Peng et al., 2005). The oil has antifungal and antibacterial properties (Fernández-López et al., 2005). Assay techniques based on MIC or MBC, namely the minimum inhibitory concentration or the minimum bactericidal concentration, are used by several researchers to determine the antibacterial activity of rosemary. Sienkiewicz et al. (2013) showed the antibacterial properties of rosemary and basil. The essential oils were found to inhibit microbial growth. Disc diffusion was used to test antibiotic susceptibility. According to the findings, the essential oils are effective against *Escherichia coli*. Additional research has demonstrated the antibacterial effectiveness of rosemary oil on *Salmonella choleraesuis*, *Clostridium perfringens*, *Bacillus cereus*, *Staphylococcus aureus*, and *E. coli* (Burt, 2004).

The acids in the Rosemary essential oil, including the Rosmarinic acid and the carnosic acid, and the other compounds, including the isorosmanol, rosmaridiphenol, epirosmanol, carnosol, and rosmanol, are responsible for the inhibitory effect of the oil. These components have various roles, including the generation of fatty acids, production of genetic material, and synthesis and assimilation of nutrients, transfer of electrons, prevention of cellular component leakage, and fatty acid transport. Moreover, the protein interaction with the membrane results in the loss of membrane structure and functionality. According to Vegara et al. (2011), carnosic acid is more effective than any other significant extract component, including rosmarinic acid, at inhibiting harmful germs. On the other hand, different scientific studies have reached different conclusions regarding any connections that might exist between the polyphenolic extract's chemical makeup and its antimicrobial action. It was shown by researchers that the efficiency of rosemary may be related to a potential synergy between the rosmarinic phenolic acid and the carnosic acid diterpene (Ivanovic et al., 2012). Yet, the antibacterial activity of these extracts is closely related to the concentrations of carnosic acid and carnosol diterpenes. According to Zaouali et al. (2010), the presence of dipinene as a significant component enhances antibacterial activity against *S. aureus*. This action might be explained by the fact that terpenes disrupt the cell membrane and lead to lysis (Bajpai et al., 2012).

The effectiveness of rosemary oil against microbes is due to the combination of various small components contained in its volatile fraction (Tornuk et al., 2011). The various chemical composition and proportions of the oil components has a serious effect on the effectiveness of *E. coli*, *L. monocytogenes*, and *S. aureus*. Several food tests, including those on meatballs (Fernández-López et al., 2005), beef (Ahn et al., 2007), and sausage, have adequately proven the antibacterial action of rosemary oil (Pandit and Shelef, 1994). According to Gomez-Estaca et al. (Gómez-Estaca et al., 2010), essential oil from rosemary will inhibit the growth of several microbes that cause harmful effects on foodstuffs.

According to Camo et al. (2008), rosemary extract applied to lamb meat had an inhibitory effect on the growth of psychrotrophic bacteria. In veal meatballs, the antibacterial activity of rosemary extracts was examined by Fernández-López et al. (2005). In comparison to the other extracts tested, the findings indicated that rosemary extracts had a stronger antibacterial activity against 11 types of bacteria.

14.16 GELATINIZATION TEMPERATURES AND FILM FORMATION USING ANCHOTE STARCH

Gelatinization temperatures and endothermic gelatinization enthalpy (ΔH) are used to describe significant properties of starch granules related to the effect of temperature. These temperatures can also be divided into the onset (T_o), peak (T_p), and conclusion (T_c) gelatinization temperatures. Granules of starch undergo a process called gelatinization, the process that leads to the transformation of the semi-crystalline structure of starch to an amorphous one. Heating starch in the presence of water or various plasticizers could lead to a severe transition (Charles et al., 2016). For the thermal study of starch, differential scanning calorimetry (DSC) is frequently utilized. The temperature at which samples start to gelatinize is known as the "onset temperature," while the "peak temperature" is the temperature at which gelatinization reaches its highest intensity, and the "conclusion temperature" is the temperature at which it stops. Findings suggest that anchote starch has higher crystallinity than potato and wheat starch due to the higher gelatinization temperature values. The overall anchote starch content is 78%, and the majority of its granules had a dome form and a geometry that was polygonal or irregular. The size of the granules ranged from 1.3 to 18 µm. X-ray powder diffraction analysis revealed a B-type pattern. Because of its greater gelatinization temperature, anchote starch granules must have a higher level of crystallinity. Anchote paste exhibited good heat and resistance to shear. Anchote starch is a possible contender for the thermoplastic products due to its thermal stability (Abera et al., 2019).

A solution casting technique was used to create anchote starch films with different types of plasticizers, and the morphological, thermal, and mechanical characteristics of these films were examined. In comparison to films made using polyols, those plasticized with 1-ethyl-3-methylimidazolium acetate had greater flexibility. Films made with a 40% (w/w) plasticizer concentration showed a more uniform microstructure than those made with a 30% (w/w) plasticizer concentration for all plasticizer types. Similar anchote starch absorption bands were visible in all films. Due to the high hydroxyl group and molecular structure, the 40% (w/w) sorbitol film had good tensile strength and modulus of elasticity. Increasing plasticizer concentration decreased the tensile strength and elastic modulus regardless of the type of plasticizer used. Due to the high hydroxyl group content and molecular structure, the 40% (w/w) sorbitol film had the maximum tensile strength and modulus of elasticity. Increasing plasticizer concentration decreased tensile strength and elastic modulus regardless of the type of plasticizer used. Tessema et al. also developed edible films from anchote (*Coccinia abyssinica*) starch. The result showed that edible films with desirable physical properties can be effectively developed from starch and further could be used in the food packaging industry (Tessema and Admassu, 2021).

14.17 ANCHOTE FILM CASE STUDY

Anchote CNCs (AnchCNC) were extracted and characterized from the anchote bagasse (AnchB). Antimicrobial nanocomposite anchote starch films incorporated with anchote cellulose nanocrystals (AnchCNC) and rosemary essential oil were prepared by the solvent casting method. Anchote cellulose (AnchC) was isolated by alkali treatment with NaOH and bleaching with NaOCI. After acid hydrolysis and ultrasonic disintegration, anchote cellulose nanocrystals were produced. The chemical treatments removed hemicelluloses and lignin. The thermal stability was reduced after hydrolysis. XRD measurements showed an increase in crystallinity. SEM indicated the removal of amorphous fractions by the chemical treatments. Zetasizer analysis revealed the average diameter of AnchCNC as 92 nm and the zeta potential value as −21.0 mV. Solution casting method was employed in this study with water and glycerol as a solvent and plasticizer. The tensile strength and water vapor permeability of films were improved by increasing the amount of AnchCNC. The films showed satisfactory antimicrobial efficiency against microorganisms, too. The antimicrobial nanocomposite anchote starch films could be used for active packaging, as revealed by the research findings (Kassa et al., 2022, 2023).

14.18 EXTRACTION AND CHARACTERIZATION OF CELLULOSE NANOCRYSTALS FROM ANCHOTE (*COCCINIA ABYSSINICA*) BAGASSE

AnchB is an environmentally friendly and low-cost cellulose raw material. A fairly good amount of cellulose in AnchB and the alkaline and bleaching methods used to decrease the lignin and hemicellulose enhanced the production strategies employed in the film making (Andrade-Mahecha et al., 2015). The alkali treatment enhanced the stretching ability of the fibers, the surface area of the same, and the hydrolyzability of the polymeric components. The yield of AnchCNC from AnchB was only 18% in that several other authors, including Bondeson et al. (Bondeson et al., 2006), reported a yield of about 30 wt% from microcrystalline cellulose and a 50–58 wt% yield from cassava.

The FTIR spectroscopy revealed that AnchB is a lignocellulosic substance (Kassa et al., 2022, 2023). The primary FTIR bands detected in the FTIR spectra were associated with cellulose –OH stretching (Leticia et al., 2017), and this was more predominant for AnchC and AnchCNC. There was an increase in cellulose content after chemical pre-treatments and acid hydrolysis. Peak intensities at 2900 cm^{-1} indicated the C–H stretching. Peak at 1430 cm^{-1} was found to be associated with symmetric bending of CH_2. The intense peak at 1166 cm^{-1} indicated the C–O–C asymmetric stretching vibrations. The peaks at 760 cm^{-1} indicated the C–H stretching deformations, and the peak at 860 cm^{-1} indicated the aromatic ring and C–H bonds stretching in lignin after the chemical treatments.

TGA was used to study the thermal stability, and this revealed water evaporation, thermal degradation, and carbonaceous residue degradation (Kassa et al., 2022, 2023). The crystal structures examination with XRD revealed that the primary diffraction peaks of AnchB occur at 16.8° and 22.4°, for AnchC at 16.2°, 22.6°, and 35.8°. AnchCNC had diffraction peaks at 16.2° (Widiarto et al., 2019). SEM revealed the presence of starch leftover from starch and bagasse separation. Alkalization and bleaching treatments removed the starch granules. Chemical treatments removed the lignin and hemicellulose attached to the celluloses (Sai Prasanna and Mitra, 2020). Acid hydrolysis and sonication reduced the fiber size and hence the fiber diameter. The acid treatment decreased the fiber aggregation, dissolved the fibers, removed the amorphous sections, and created the ordered crystalline sections. The average diameter of 92 nm observed by this study was comparatively thicker due to aggregations. The acid hydrolysis and ultrasonic treatments resulted in AnchCNCs with needle-like morphology. Acid hydrolysis clearly resulted in the breakdown of the hydrogen bonds in the polymers and thus reduced fiber diameter. Sonication resulted in depolymerization and shorter fiber length due to the breaking down of the bonds between glucose units (Asrofi et al., 2018). The average zeta potential of the AnchCNC suspension as observed in this study was found to be −21.0 mV (Widiarto et al., 2019).

14.19 MECHANICAL, BARRIER, AND ANTIMICROBIAL PROPERTIES OF THERMOPLASTIC ANCHOTE (*COCCINIA ABYSSINICA*) STARCH

Nanocomposite antimicrobial films created by solution casting using AnchCNC and REO had good film-forming ability. Figure 14.1 shows the anchote nanocomposite films produced with different concentrations of AnchCNC and REO. The nanocomposite films were found to be

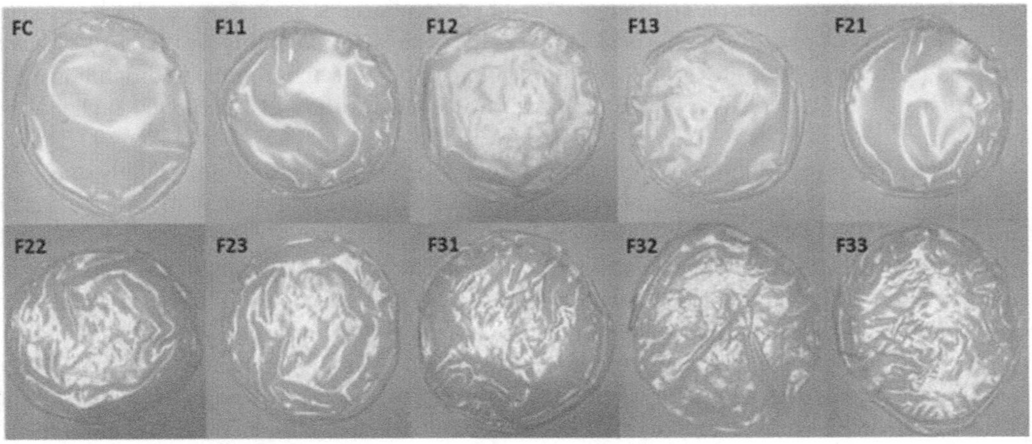

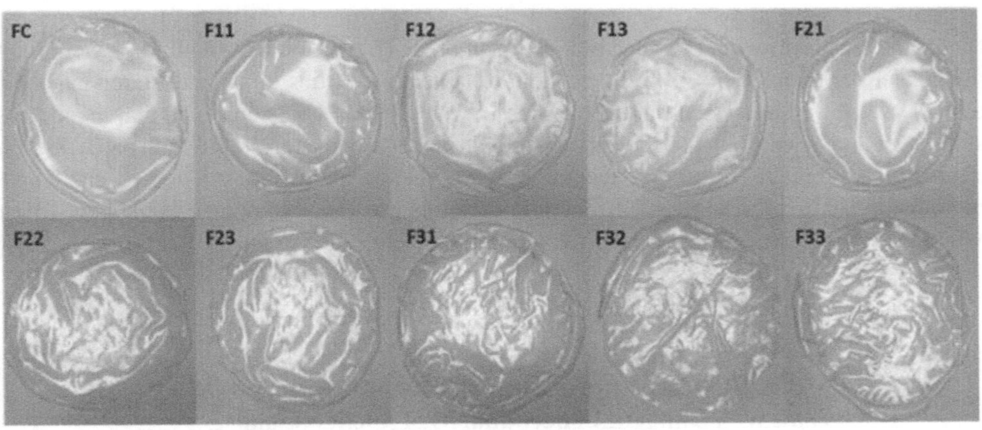

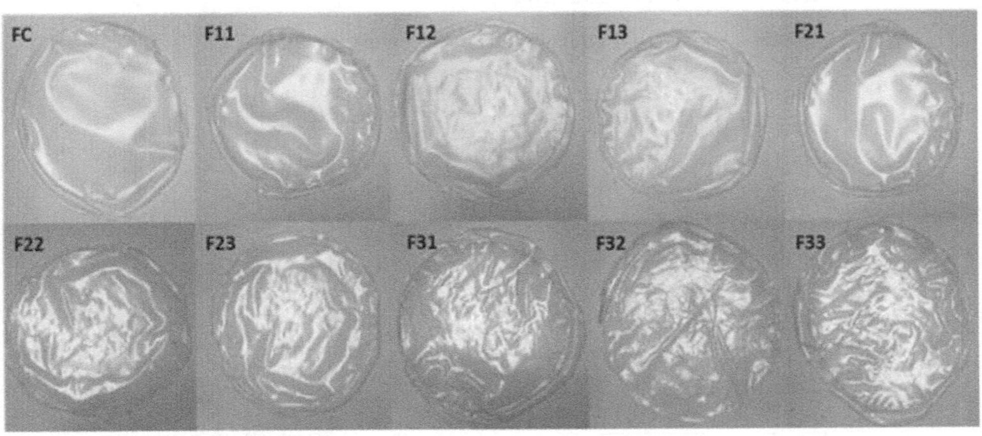

FIGURE 14.1 Anchote nanocomposite films with different AnchCNC and REO concentrations.

translucent, homogenous, and flexible with a thickness between 0.12 and 0.22 mm (Kassa et al., 2022, 2023).

Additionally, it was observed that the essential oil concentration influenced the thickness of active nanocomposite films (Hosseini et al., 2016). The CNC solution dispersion method also influences the nanocomposite film thickness, moisture content, and water solubility of the films. Melanin nanoparticle inclusion in the agar films significantly increased the thickness (Roy and Rhim, 2019). The solubility of films is chiefly affected by the type of ingredients, their ratios, and the hydrophilicity and hydrophobicity indices. Hydrophobic and hydrophilic substances make a film less soluble and more soluble, respectively (Kavoosi et al., 2013). The water solubility of the anchote starch films and their nanocomposite films was observed to exhibit a decrease in the hydrophilicity. The interaction of the nanocrystals with starch chains resulted in films that exhibited higher water solubility.

The water vapor permeability (WVP) ranged from 3 to 9×10^{-11} gH_2O/Pa m s in this study. The film with the highest concentration of AnchCNC and REO had the lowest WVP. An increase in AnchCNC and REO showed a film with lower WVP values. CNCs from cotton linters significantly reduced the WVP (Bagde and Nadanathangam, 2019). The WVP for thermoplastic starch is high due to the hydrophilic nature of starch, and in addition, the presence of CNCs in the matrix reduces WVP. Water vapor permeability is important in food packaging as it indicates the moisture transfer reduction and the moisture transfer rate in the food from its surrounding environment. Conventionally, the films with high water vapor permeability necessarily reduce the applicability for their intended usage. The ratio of hydrophilic components to hydrophobic components is very important in that this factor actually determines the water vapor permeability of films (Rojas-Graü et al., 2006).

Data for the tensile strength (TS) and elongation at break (E) obtained from all films were found to show greater variation. Table 14.1 shows the mechanical properties of anchote starch nanocomposite films with different concentrations of AnchCNC and REO. It was observed that the increase in AnchCNC contents was found to show a tremendous increase in the tensile strength and a decrease in the elongation at break. The control anchote starch film showed the lowest tensile strength and a higher elongation at break. The tensile strength of the films was found to reduce from about 15% to 11% when the REO was increased from 2 to 6%.

TABLE 14.1
Mechanical Properties of Anchote Starch Nanocomposite Films with Different Concentrations of AnchCNC and REO (Kassa et al., 2022, 2023)

Samples	Tensile Strength (MPa)	Elongation at break (%)
Sample FC	$9.42 \pm 0.54_g$	$73.91 \pm 0.72_a$
Sample F11	$14.93 \pm 0.47_d$	$61.54 \pm 0.60_b$
Sample F12	$12.55 \pm 0.60_e$	$62.52 \pm 0.703_b$
Sample F13	$10.81 \pm 0.44_f$	$63.27 \pm 0.45_b$
Sample F21	$23.41 \pm 0.27_a$	$55.32 \pm 0.76_d$
Sample F22	$22.52 \pm 0.37_{ab}$	$57.25 \pm 0.67_{cd}$
Sample F23	$21.43 \pm 0.35_{bc}$	$58.49 \pm 0.47_c$
Sample F31	$23.44 \pm 0.80_a$	$27.71 \pm 0.650_e$
Sample F32	$22.32 \pm 0.56_{ab}$	$28.23 \pm 0.590_e$
Sample F33	$20.12 \pm 0.75_c$	$29.23 \pm 1.002_e$

Note: No significant difference between values with the same subscript letter in a column ($p > 0.05$), according to Tukey's test with 95% confidence.

14.20 POLYPHENOL MIGRATION AND THE ANTIMICROBIAL APPLICATION OF AS/ANCHCNC/REO NANOCOMPOSITE FILMS

The polyphenol concentration, DPPH*-scavenging activity, and migration results showed the polyphenol content of the films in the range of 2–6 mg GAE/g dried films. The DPPH inhibition percentage was found to be in the range of 15–50%. Increasing the oil content created films with higher polyphenol content and enhanced DPPH*-scavenging activity. DPPH-radical scavenging activities were found to be directly proportional to the polyphenol content of the samples. This is very significant in that various applications could be tailored from an industrial standpoint with different rates of antioxidant activity. The phenolic compounds in the plant extracts are found to be responsible for high effective antioxidant activity and as well as the free radical scavenging activity. This enhancement in the total phenols of this study was found to be consistent with the results reported in fabricating the chitosan films with tea polyphenols (Wang et al., 2013).

The polyphenols migrated from the antimicrobial films to the food simulants based on the days of film exposition. The migration assay is responsible for calculating the transfer of phenolic compounds from the aqueous medium. However, a low amount of phenolic compounds was used with ethanol as a simulant (Buonocore et al., 2003). The quick water penetration is the cause for the release of rosemary polyphenols into the aqueous medium, thereby leading to the diffusion of the active compounds into the anchote starch/glycerol/CNC matrix.

Several studies have investigated the release of polyphenols from the film into food simulants. Table 14.2 shows the polyphenol content and migration test results obtained for the antimicrobial films (Kassa et al., 2022, 2023). However, comparatively fewer authors have investigated the polyphenol release into the atmosphere as a function of storage conditions, time, and temperature. The release of polyphenols from the films was determined by testing the polyphenol content of films at 25°C for 30 days. The loss of polyphenol from the AS/AnchCNC/REO nanocomposite films was found to

TABLE 14.2
Polyphenol Content and Migration Test Results Obtained for the Antimicrobial Films (Kassa et al., 2022, 2023)

			Migration Test (25°C)	
Film Sample	mg GAE/g Dried Film	DPPH Inhibition (%)	Food Simulant	mg GAE/kg Simulant
F11	1.8	15	Water	17
			Ethanol 95%	<7
F12	3.7	29	Water	35
			Ethanol 95%	<7
F13	6.2	51	Water	60
			Ethanol 95%	<7
F21	1.9	16	Water	18
			Ethanol 95%	<7
F22	3.9	32	Water	37
			Ethanol 95%	<7
F23	6.3	52	Water	60
			Ethanol 95%	<7
F31	1.9	16	Water	18
			Ethanol 95%	<7
F32	3.9	32	Water	37
			Ethanol 95%	<7
F33	6.4	53	Water	60
			Ethanol 95%	<7

occur during storage, and an equilibrium polyphenol concentration was observed after the 20th day. Moreover, the release of polyphenol is attributed to its high volatility and the presence of voids.

14.21 ANTIMICROBIAL EFFECT OF FILMS ON MEAT PRODUCTS

The antibacterial activities of control film and nanocomposite films were tested against *E. coli, L. monocytogenes*, and *S. aureus*. The application of antimicrobial films was performed on raw meat (beef) preservation. Using meat products to evaluate the tested films showed that film FC (control) has no antibacterial activity. In addition, untreated meat products show a high colony number of *E. Coli, L. Monocytogenes*, and *S.* Aurous (Gill & Holley, 2006). Figure 14.2 shows the antimicrobial films for meat preservation. On the antimicrobial activity for all concentrations tested, on the other hand, anchote starch films incorporated with AnchCNC and REO have good antimicrobial activity.

The microbial count was significantly reduced in the meat products by increasing REO (Kassa et al., 2023). Tables 14.3–14.5 show the antibacterial activity of anchote starch films incorporated

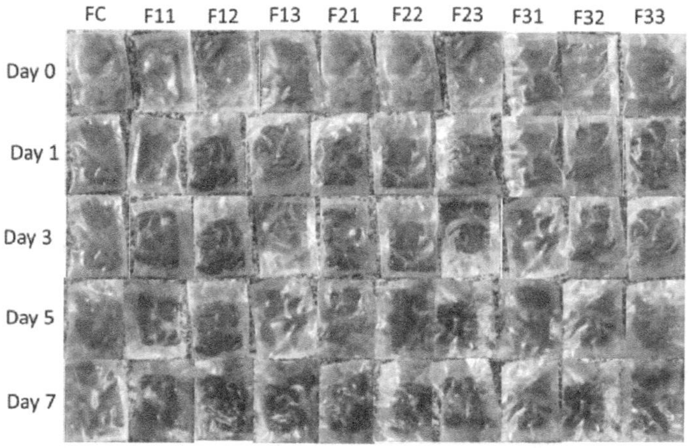

FIGURE 14.2 Antimicrobial films for meat preservation.

TABLE 14.3
Antibacterial Activity of Anchote Starch Films Incorporated with AnchCNC and REO Against *E. coli* on Meat Products (Kassa et al., 2022, 2023)

	E. coli (Number of Colonies)				
Samples	**Day 0**	**Day 1**	**Day 3**	**Day 5**	**Day 7**
Control	415	423	416	448	464
FC	451	459	443	452	448
F11	314	301	299	302	294
F12	114	103	107	102	108
F13	31	25	22	19	17
F21	304	293	287	293	290
F22	101	98	89	94	95
F23	25	20	16	11	17
F31	282	273	275	272	278
F32	88	83	74	98	93
F33	9	5	0	0	0

Note: Control treatment refers to meat slices uncovered with films.

TABLE 14.4
Antibacterial Activity of Anchote Starch Films Incorporated with AnchCNC and REO Against *L. Monocytogenes* on Meat Products (Kassa et al., 2022, 2023)

	L. Monocytogenes (Colony Count)				
Samples	**Day 0**	**Day 1**	**Day 3**	**Day 5**	**Day 7**
Control	315	334	342	361	372
FC	327	322	324	321	319
F11	285	276	271	276	278
F12	111	116	108	112	107
F13	51	46	42	47	43
F21	240	239	242	238	233
F22	81	89	95	91	87
F23	30	25	21	29	27
F31	219	212	205	209	211
F32	65	58	61	54	64
F33	0	0	0	0	0

Note: Control treatment refers to meat slices uncovered with films.

TABLE 14.5
Antibacterial Activity of Anchote Starch Films Incorporated with AnchCNC and REO Against *L. Monocytogenes* on Meat Products (Kassa et al., 2022, 2023)

	S. Aurous (Colony Count)				
Samples	**Day 0**	**Day 1**	**Day 3**	**Day 5**	**Day 7**
Control	326	345	353	371	380
FC	338	332	335	331	330
F11	296	285	280	286	288
F12	120	125	128	122	118
F13	60	56	53	55	54
F21	251	249	253	248	244
F22	87	95	99	96	92
F23	35	31	27	32	29
F31	224	215	209	213	214
F32	67	61	64	57	60
F33	0	0	0	0	0

Note: Control treatment refers to meat slices uncovered with films.

with AnchCNC and REO against *E. coli, L. Monocytogenes*, and *L. Monocytogenes*, respectively. The noteworthy decrease in the improvement of the bacterial and the long-term control of antibacterial inhibition was characteristic of the antimicrobial activity of ingredients in the essential oil.

14.22 CONCLUSIONS

Cellulose nanocrystals (CNCs) are an emerging class of nanomaterials having unique physical and chemical properties. Anchote bagasse, a byproduct of the processing of anchote starch, has been identified as a potential source of CNCs. However, the literature suggests that there is a lack of comprehensive studies on the extraction and characterization of CNCs from anchote bagasse and investigations into their potential applications. Addressing these gaps could lead to the development of cost-effective and sustainable methods for producing CNCs from anchote bagasse and facilitate their use in a range of industrial applications. The use of CNCs has shown significant improvement in the properties of polymer matrices due to their high strength, stiffness, and large specific surface area. Similarly, rosemary essential oil has been found to contain natural antimicrobial compounds that could protect food packaging against microbial growth. Current literature indicates that there is a significant gap in research regarding the reinforcing effect of CNCs and the antimicrobial effect of rosemary essential oil on thermoplastic anchote starch films. CNCs are widely recognized as strong nanofillers, capable of reinforcing the mechanical strength and improving the barrier properties of polymer-based materials. However, there is no research on their use in thermoplastic anchote starch films and their combined use with rosemary essential oil for improved antimicrobial efficacy. In this case study, CNCs reinforced anchote starch-based antimicrobial food packaging film containing rosemary essential oil was developed and characterized. Anchote CNCs (AnchCNC) were extracted and characterized from anchote bagasse (AnchB). Antimicrobial nanocomposite anchote starch films incorporated with anchote cellulose nanocrystals and rosemary essential oil were prepared by the solvent casting method. The polyphenol content and antioxidant activity (2,2-Diphenyl-1-Picrylhydrazyl, DPPH inhibition) of films were determined. The migration of polyphenol from the films to water and ethanol food simulant was studied at 30°C after 7 days. The release rate and diffusion coefficient of polyphenol migration from film sample F33 (5% AnchCNC, 6% REO) into water food simulant were determined by overall kinetics and diffusion modeling at 30, 40, 50, and 60°C. The stability of polyphenol at ambient storage of the films, as well as the antimicrobial application of the films for meat preservation, was evaluated. In summary, anchote CNCs reinforced anchote starch-based antimicrobial food packaging film containing rosemary essential oil was developed, characterized, and found to be promising for meat preservation.

REFERENCES

Abera, G., Woldeyes, B., Dessalegn, H., and Miyake, G. M. (2019). Comparison of physicochemical properties of indigenous Ethiopian tuber crop (*Coccinia Abyssinica*) starch with commercially available potato and wheat starches. *International Journal of Biological Macromolecules*, *140*, 43–48.

Ahn, D. U., Kim, S. M., & Lee, E. J. (2007). Natural Antioxidants for Meat and Meat Products. In C. T. Ho, C. Y. Lee, & M. T. Huang (Eds.), *Natural Antioxidants: Chemistry, Health Effects, and Food Applications*. American Chemical Society.

Aila-Suárez, S., Palma-Rodríguez, H. M., Rodríguez-Hernández, A. I., Hernández-Uribe, J. P., Bello-Pérez, L. A., and Vargas-Torres, A. (2013). Characterization of films made with chayote tuber and potato starches blending with cellulose nanoparticles. *Carbohydrate Polymers*, *98*(1), 102–107.

Alexandre, M., and Dubois, P. (2000). Polymer-layered silicate nanocomposites: Preparation, properties and uses of a new class of materials. *Materials Science and Engineering R: Reports*, *28*(1), 1–63.

Andrade-Mahecha, M. M., Pelissari, F. M., Tapia-Blácido, D. R., and Menegalli, F. C. (2015). Achira as a source of biodegradable materials: Isolation and characterization of nanofibers. *Carbohydrate Polymers*, *123*, 406–415.

Arfat, Y. A., Benjakul, S., Prodpran, T., Sumpavapol, P., and Songtipya, P. (2014). Properties and antimicrobial activity of fish protein isolate/fish skin gelatin film containing basil leaf essential oil and zinc oxide nanoparticles. *Food Hydrocolloids*, *41*, 265–273.

Asrofi, M., Abral, H., Kasim, A., Pratoto, A., Mahardika, M., Park, J. W., and Kim, H. J. (2018). Isolation of nanocellulose from water hyacinth fiber (WHF) produced via digester-sonication and its characterization. *Fibers and Polymers*, *19*(8), 1618–1625.

Azeredo, H. M. C., Rosa, R. M. F., & Rangel, E. M. S. C. (2016). Nanocellulose in thin films, coatings, and plies for packaging applications: A review. *BioResources*, *11*(4), 10565–10580.

Bagde, P., and Nadanathangam, V. (2019). Mechanical, antibacterial and biodegradable properties of starch film containing bacteriocin immobilized crystalline nanocellulose. *Carbohydrate Polymers*, *222*(June), 115021.

Bajpai, V. K., Baek, K.-H., and Kang, S. C. (2012). Control of Salmonella in foods by using essential oils: A review. *Food Research International*, *45*(2), 722–734.

Balakrishnan, P., Sreekala, M. S., Kunaver, M., Huskić, M., and Thomas, S. (2017). Morphology, transport characteristics and viscoelastic polymer chain confinement in nanocomposites based on thermoplastic potato starch and cellulose nanofibers from pineapple leaf. *Carbohydrate Polymers*, *169*, 176–188.

Beruk, B. D., Tadesse, F. T., and Dereje, H. (2015). Physical and proximate characterization of Anchote (Coccinia Abyssinica) accessions grown under Hawassa and Wondo genet conditions, Southern Ethiopia. *Food Science and Quality Management*, *42*, 62–75.

Bondeson, D., Mathew, A., and Oksman, K. (2006). Optimization of the isolation of nanocrystals from microcrystalline cellulose by acid hydrolysis. *Cellulose*, *13*(2), 171–180.

Buonocore, G. G., Del Nobile, M. A., Panizza, A., Corbo, M. R., and Nicolais, L. (2003). A general approach to describe the antimicrobial agent release from highly Swellable films intended for food packaging applications. *Journal of Controlled Release*, *90*(1), 97–107.

Burt, S. (2004). Essential oils: Their antibacterial properties and potential applications in foods - A review. *International Journal of Food Microbiology*, *94*(3), 223–253.

Camo, J., Beltrán, J. A., and Roncalés, P. (2008). Extension of the display life of lamb with an antioxidant active packaging. *Meat Science*, *80*(4), 1086–1091.

Cao, X., Chen, Y., Chang, P. R., Muir, A. D., and Falk, G. (2008). Starch-based nanocomposites reinforced with flax cellulose nanocrystals. *Express Polymer Letters*, *2*(7), 502–510.

Chandra, R., and Rustgi, R. (1998). Biodegradable polymers. *Progress in Polymer Science (Oxford)*, *23*(7), 1273–1335.

Chang, P. R., Jian, R., Zheng, P., Yu, J., and Ma, X. (2010). Preparation and properties of glycerol plasticized-starch (GPS)/cellulose nanoparticle (CN) composites. *Carbohydrate Polymers*, *79*(2), 301–305.

Charles, A. L., Cato, K., Huang, T. C., Chang, Y. H., Ciou, J. Y., Chang, J. S., and Lin, H. H. (2016). Functional properties of arrowroot starch in cassava and sweet potato composite starches. *Food Hydrocolloids*, *53*, 187–191.

Chen, D., Lawton, D., Thompson, M. R., and Liu, Q. (2012). Biocomposites reinforced with cellulose nanocrystals derived from potato peel waste. *Carbohydrate Polymers*, *90*(1), 709–716.

Cíntia, M., Victória, M., and Grossmann, E. (2009). Effect of plasticizers on properties of oat starch films effects of plasticizers on the properties of oat starch films. *Materials Science and Engineering C*, *29*(2), 532–538.

Dey, P., and Ray, S. (2018). ScienceDirect an overview of the recent trends in manufacturing of green composites – Considerations and challenges. *Materials Today: Proceedings*, *5*(9), 19783–19789.

Fekadu, H. (2014). Nutritional composition, antinutritional factors and effect of boiling on nutritional composition of Anchote (Coccinia Abyssinica) tubers. *Food Science and Quality Management*, *26*, 25–39.

Fernández-López, J., Zhi, N., Aleson-Carbonell, L., Pérez-Álvarez, J. A., & Kuri, V. (2005). Antioxidant and antibacterial activities of natural extracts: Application in beef meatballs. *Meat Science*, *69*(3), 371–380.

Gill, A. O., and Holley, R. A. (2006). Inhibition of membrane bound ATPases of Escherichia coli and Listeria monocytogenes by plant oil aromatics. *International Journal of Food Microbiology*, *111*(2), 170–174.

Gómez-Estaca, J., López de Lacey, A., López-Caballero, M. E., Gómez-Guillén, M. C., and Montero, P. (2010). Biodegradable gelatin-chitosan films incorporated with essential oils as antimicrobial agents for fish preservation. *Food Microbiology*, *27*(7), 889–896.

Halimatul, M. J., Sapuan, S. M., Jawaid, M., Ishak, M. R., and Ilyas, R. A. (2019). Effect of sago starch and plasticizer content on the properties of thermoplastic films: Mechanical testing and cyclic soaking-drying. *Polimery/Polymers*, *64*(6), 422–431.

Hosseini, S. F., Rezaei, M., Zandi, M., and Farahmandghavi, F. (2016). Development of bioactive fish gelatin/chitosan nanoparticles composite films with antimicrobial properties. *Food Chemistry*, *194*, 1266–1274.

Ivanovic, J., Misic, D., Zizovic, I., and Ristic, M. (2012). In vitro control of multiplication of some food-associated bacteria by thyme, rosemary and sage isolates. *Food Control*, *25*(1), 110–116.

Jumaidin, R., Khiruddin, M. A. A., Asyul Sutan Saidi, Z., Salit, M. S., and Ilyas, R. A. (2020). Effect of cogon grass fibre on the thermal, mechanical and biodegradation properties of thermoplastic cassava starch biocomposite. *International Journal of Biological Macromolecules*, *146*, 746–755.

Kassa, H. S., Jabasingh, S. A., Mohammed, S. A., Park, S-Y., and Baek, S. Y. (2022). Extraction and characterization of cellulose nanocrystals from Anchote (Coccinia Abyssinica) bagasse. *Macromolecular Research*, *30*(11), 776–782.

Kassa, H. S., Jabasingh, S. A., Mohammed, S. A., Park, S-Y., and Baek, S. Y. (2023). Mechanical, barrier, and antimicrobial properties of Anchote (*Coccinia abyssinica*) starch films containing cellulose nanocrystals and rosemary essential oil. *Biomass Conversion and Biorefinery*, *13*, 7333–7347.

Kavoosi, G., Dadfar, S. M. M., and Purfard, A. M. (2013). Mechanical, physical, antioxidant, and antimicrobial properties of gelatin films incorporated with thymol for potential use as nano wound dressing. *Journal of Food Science*, *78*(2), E244–E250.

Leja, K., and Lewandowicz, G. (2010). Polymer biodegradation and biodegradable polymers - A review. *Polish Journal of Environmental Studies*, *19*(2), 255–266.

Leticia, A., Pereira, M., Zanon, C. D., and Menegalli, F. C. (2017). Isolation and characterization of cellulose nanofibers from cassava root bagasse and peelings. *Carbohydrate Polymers*, *157*, 962–970.

Li, M., Li, D., Wang, L. J., and Adhikari, B. (2015). Creep behavior of starch-based nanocomposite films with cellulose nanofibrils. *Carbohydrate Polymers*, *117*, 957–963.

Mangaraj, S., Yadav, A., Bal, L. M., Dash, S. K., and Mahanti, N. K. (2019). Application of biodegradable polymers in food packaging industry: A comprehensive review. *Journal of Packaging Technology and Research*, *3*(1), 77–96.

Mishra, R. K., Ha, S. K., Verma, K., and Tiwari, S. K. (2018). Recent progress in selected bio-nanomaterials and their engineering applications: An overview. *Journal of Science: Advanced Materials and Devices*, *3*(3), 263–288.

Mlalila, N., Hilonga, A., Swai, H., Devlieghere, F., and Ragaert, P. (2018). Antimicrobial packaging based on starch, poly (3-hydroxybutyrate) and poly (lactic-co-glycolide) materials and application challenges. *Trends in Food Science & Technology*, 74(February), 1–11.

Moustafa, H., Youssef, A. M., Darwish, N. A., and Abou-Kandil, A. I. (2019). Eco-friendly polymer composites for green packaging: Future vision and challenges. *Composites Part B*, *172*, 16–25.

Panaitescu, D. M., Frone, A. N., Ghiurea, M., and Chiulan, I. (2015). Influence of storage conditions on starch/PVA films containing cellulose nanofibers. *Industrial Crops and Products*, *70*, 170–177.

Pandit, V. A., & Shelef, L. A. (1994). Antilisterial Activity of Spices and Food Components in Minced Beef. *Journal of Food Protection*, 57(8), 677–680.

Peng, Y., Yuan, J., Liu, F., & Ye, J. (2005). Determination of active components in rosemary by capillary electrophoresis with electrochemical detection. *Journal of pharmaceutical and biomedical analysis*, *39*(3-4), 431–437.

Rojas-Graü, M. A., Avena-Bustillos, R. J., Friedman, M., Henika, P. R., Martín-Belloso, O., and McHugh, T. H. (2006). Mechanical, barrier, and antimicrobial properties of apple puree edible films containing plant essential oils. *Journal of Agricultural and Food Chemistry*, *54*(24), 9262–9267.

Roy, S., and Rhim, J.-W. (2019). Agar-based antioxidant composite films incorporated with melanin nanoparticles. *Food Hydrocolloids*, *94*, 391–398.

Sai Prasanna, N., and Mitra, J. (2020). Isolation and characterization of cellulose nanocrystals from Cucumis sativus peels. *Carbohydrate Polymers*, *247*, 3–6.

Sayyahi, Z., Beigmohammadi, F., and Shoaiee, S. (2017). Optimization of starch biopolymer enriched with chitosan containing rosemary essential oil and its application in packaging of peanuts. *Nutrition and Food Sciences Research*, *4*(3), 19–28.

Sharma, D., Dhanjal, D. S., and Mittal, B. (2017). Development of edible biofilm containing cinnamon to control food-borne pathogen. *Journal of Applied Pharmaceutical Science*, *7*(1), 160–164.

Shogren, R. L., Petrovic, Z., Liu, Z., and Erhan, S. Z. (2004). Biodegradation behavior of some vegetable oil-based polymers. *Journal of Polymers and the Environment*, *12*(3), 173–178.

Sienkiewicz, M., Łysakowska, M., Pastuszka, M., Bienias, W., and Kowalczyk, E. (2013). The potential of use basil and rosemary essential oils as effective antibacterial agents. *Molecules*, *18*(8), 9334–9351.

Siqueira, G., Bras, J., and Dufresne, A. (2010). Cellulosic bionanocomposites: A review of preparation, properties and applications. *Polymers*, *2*(4), 728–765.

Sorrentino, A., Gorrasi, G., and Vittoria, V. (2007). Potential perspectives of bio-nanocomposites for food packaging applications. *Trends in Food Science and Technology*, *18*(2), 84–95.

Souza, A. C., Goto, G. E. O., Mainardi, J. A., Coelho, A. C. V., and Tadini, C. C. (2013). Cassava starch composite films incorporated with cinnamon essential oil: Antimicrobial activity, microstructure, mechanical and barrier properties. *LWT - Food Science and Technology*, *54*(2), 346–352.

Soykeabkaew, N., Laosat, N., Ngaokla, A., Yodsuwan, N., and Tunkasiri, T. (2012). Reinforcing potential of micro- and nano-sized fibers in the starch-based biocomposites. *Composites Science and Technology*, *72*(7), 845–852.

Šuput, D., Lazić, V. L., Pezo, L., Markov, S., Vastag, Z., Popović, L., … & Popović, S. (2016). Characterization of starch edible films with different essential oils addition. *Polish Journal of Food and Nutrition Sciences*, *66*(4), 277–285.

Tessema, A., and Admassu, H. (2021). Extraction and characterization of starch from Anchote (Coccinia abyssinica): physico-chemical, functional, morphological and crystalline properties. *Journal of Food Measurement and Characterization*, *15*(4), 3096–3110.

Tornuk, F., Cankurt, H., Ozturk, I., Sagdic, O., Bayram, O., and Yetim, H. (2011). Efficacy of various plant hydrosols as natural food sanitizers in reducing Escherichia coli O157:H7 and Salmonella Typhimurium on fresh cut carrots and apples. *International Journal of Food Microbiology*, *148*(1), 30–35.

Vegara, S., Funes, L., Martí, N., Saura, D., Micol, V., and Valero, M. (2011). Bactericidal activities against pathogenic bacteria by selected constituents of plant extracts in carrot broth. *Food Chemistry*, *128*(4), 872–877.

Wang, L., Dong, Y., Men, H., Tong, J., and Zhou, J. (2013). Food hydrocolloids preparation and characterization of active fi lms based on chitosan incorporated tea polyphenols. *Food Hydrocolloids*, *32*(1), 35–41.

Weber, C. J., Haugaard, V., Festersen, R., and Bertelsen, G. (2002). Production and applications of biobased packaging materials for the food industry. *Food Additives and Contaminants*, *19*, 172–177.

Widiarto, S., Pramono, E., Suharso, Rochliadi, A., and Arcana, I. M. (2019). Cellulose nanofibers preparation from cassava peels via mechanical disruption. *Fibers*, *7*(5), 2–11.

Zaouali, Y., Bouzaine, T., and Boussaid, M. (2010). Essential oils composition in two Rosmarinus officinalis L. varieties and incidence for antimicrobial and antioxidant activities. *Food and Chemical Toxicology*, *48*(11), 3144–3152.

15 Greener Production of Biosurfactants and Its Applications in Healthcare and Food Sector

A Review

C. Suganya, C. Vigneshwaran, V. Dhineshkumar, and V. Sivasubramanian

15.1 INTRODUCTION

In the pursuit of sustainable advancements in science and technology, the focus has unmistakably shifted toward safeguarding our environment, protecting human populations, and preserving delicate ecosystems. The traditional landscape of "brown chemistry," characterized by the indiscriminate use of synthetic surfactants, requires a fundamental transformation toward "green chemistry" principles [1]. To catalyze this paradigm shift, the scientific community has increasingly turned to natural resources, recognizing them as potent and dependable alternatives for a more sustainable future.

Biosurfactants are a fascinating class of surface-active compounds produced by microorganisms, characterized by their unique molecular structure. These compounds consist of a hydrophobic "tail," typically composed of fatty acids or β-hydroxyl fatty acids with varying carbon chain lengths, and a hydrophilic "head" [2]. These bio-alternative substances boast numerous attractive qualities, including reduced toxicity, biodegradability, chemical stability, environmental safety, and ease of recycling [3–5]. It has garnered substantial attention worldwide due to its potential as an environmentally friendly alternative to synthetic surfactants. Beyond their surface-active properties, biosurfactants exhibit a diverse range of functions, including acting as bioemulsifiers, antibiotics, antifungals, heavy metal binders, and even antitumor agents. As a result, biosurfactants find applications across various sectors, including industry, environmental protection, medicine, and agriculture, making them highly versatile and valuable bioproducts [6].

Their potential in drug design and delivery has recently gained recognition, further solidifying their role in industrial and pharmaceutical applications [3]. Biosurfactants stand at the forefront of this development, serving as emulsifiers, de-emulsifiers, wetting and foaming agents, edible food additives, and integral components of cosmetics and detergents [7]. They find diverse applications in various sectors, including petrochemical, wastewater treatment, soil remediation, agriculture, and more. Despite their multifaceted utility and eco-friendly attributes, biosurfactants still face challenges related to production costs. Researchers are actively exploring economically and environmentally friendly methods for biosurfactant production, with a particular focus on utilizing waste materials like agricultural and food processing waste, dairy waste, olive mill wastewater, brewery and distillery waste, sewage sludge, lignocellulosic biomass, oil and fat waste, industrial effluents, algae biomass, and agro-industrial waste to promote a circular bio-economy [8]. The choice of waste material for biosurfactant production hinges on its composition, determining the type of biosurfactants

 DOI: 10.1201/9781003732907-15

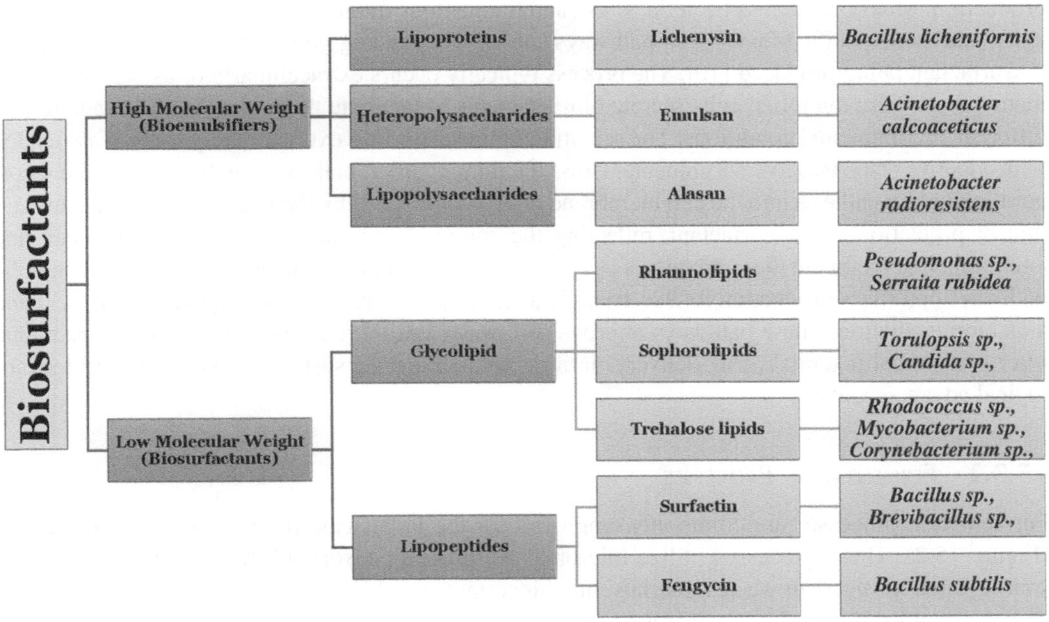

FIGURE 15.1 Types of biosurfactants and microorganisms involved.

produced. Availability of abundant waste streams is crucial for cost-effective, sustainable production, and compatibility with specific microorganisms ensures optimal synthesis, emphasizing the importance of aligning waste characteristics with microbial preferences. This strategic approach transforms waste into a valuable resource, addressing environmental concerns while promoting efficient biosurfactant production.

With applications spanning pharmaceuticals, cosmetics, detergents, agriculture, heavy metal removal, and oil recovery, biosurfactants hold immense promise in addressing both environmental and industrial challenges.

15.2 CLASSIFICATION OF BIOSURFACTANTS

Biosurfactants, produced by a variety of microorganisms like bacteria, fungi, and yeasts, are categorized based on their chemical composition and microbial origin. They can be broadly classified into high molecular weight (HMW) and low molecular weight (LMW) groups. Figure 15.1 represents the different types of biosurfactants and the microorganisms involved in the production process. LMW biosurfactants include glycolipids, lipopeptides, phospholipids, fatty acids, and neutral lipids, while HMW biosurfactants encompass polymeric substances like polysaccharides, lipopolysaccharides, lipoproteins, and particulate surfactants [9–11]. Notable examples include glycolipids like rhamnolipids and sophorolipids, as well as lipopeptides such as surfactin and fengycin [12]. These diverse biosurfactants play crucial roles in various industrial applications, thanks to their unique surface-active properties, including emulsification, foam formation, detergency, and oil dispersion.

15.3 BIOSURFACTANT PRODUCTION

15.3.1 Microbial Biosynthesis

Biosurfactants are primarily synthesized by microorganisms through various metabolic pathways. Microbial biosynthesis of biosurfactants involves the utilization of different carbon and nitrogen sources, which are transformed into the hydrophilic and hydrophobic moieties required for surfactant production. Microbes such as bacteria, yeast, and filamentous fungi are known to produce biosurfactants, and their specific biosynthetic pathways can vary depending on the organism and the type of biosurfactant being produced [10]. The process typically occurs extracellularly or as part of the cell membrane, where microbial cells secrete biosurfactants to facilitate their survival and adaptation to different environmental conditions. The secretion of biosurfactants extracellularly is often associated with adaptive responses to environmental stress, facilitating microbial survival. In contrast, the incorporation of biosynthesis into the cell membrane may be influenced by the need for immediate utilization or protection of the surfactant, reflecting the complex interplay of biological, chemical, and environmental factors in these processes [13]. Understanding and enhancing biosurfactant metabolic pathways optimizes production for eco-friendly alternatives in industries (detergents, cosmetics) and aids bioremediation. These pathways improve soil, water health, reduce environmental impact, and offer biomedical potential (drug delivery, antimicrobials), driving sustainable practices and technological advancements.

15.3.2 Fermentation Processes

Fermentation processes are commonly employed for the large-scale production of biosurfactants (Figure 15.2). These processes utilize microbial cultures to convert substrates, often carbon-rich sources such as sugars or waste materials, into biosurfactants [14]. Fermentation tanks provide controlled conditions, including temperature, pH, and oxygen levels, to support microbial growth and biosurfactant production. The choice of microorganism, fermentation medium, and process parameters can significantly impact the efficiency and yield of biosurfactant production [15]. Fermentation processes are versatile and can be tailored to different biosurfactant types, making them a key component of biosurfactant manufacturing. Various raw materials, such as agricultural byproducts, waste oils, sugars, and renewable resources like vegetable oils, have been utilized as sources for microbial species to produce biosurfactants and their fermentation process parameters as listed in Table 15.1 [28].

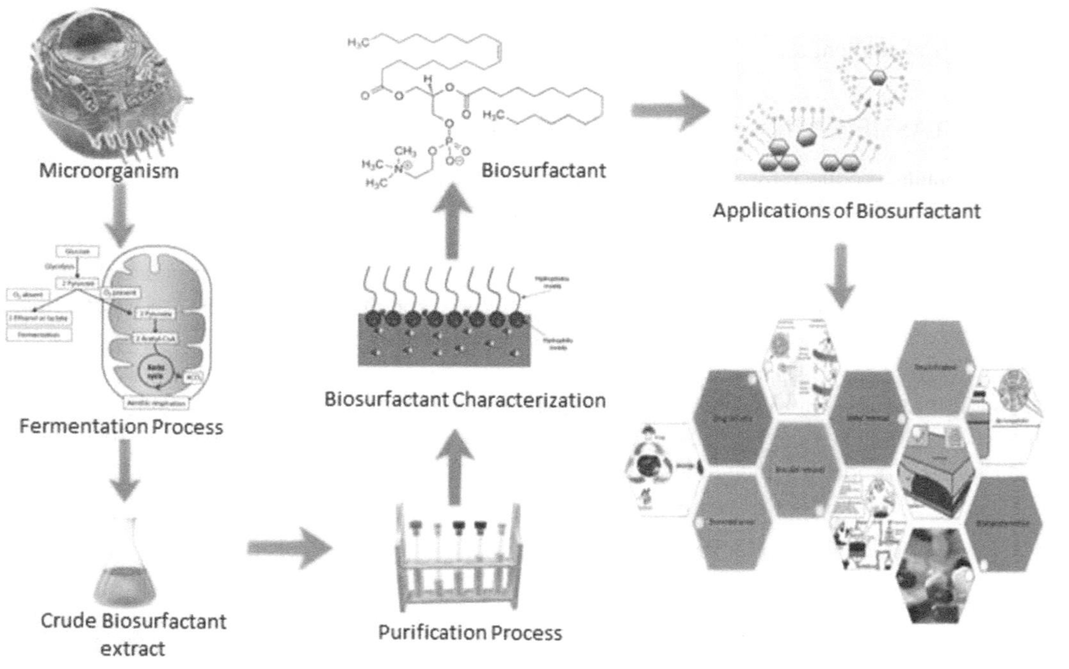

FIGURE 15.2 Green synthesis and application of biosurfactant.

TABLE 15.1
Raw Materials used for Biosurfactant Production and Their Fermentation Mode and Process Parameters

Biosurfactant Type	Raw Material	Microbe	Nitrogen Source	Fermentation Mode and Conditions	Fermentation Scale	Biosurfactant Concentration (g/L)	Reference
Glycolipids	Sugarcane molasses (5%) and waste soybean oil (5.0%)	*Candida bombicola* URM 3718	Corn steep liquor (3.0%) and	Batch mode, 28°C, 180 rpm, 120 h at	Shake flasks	25.00	[16]
	Olive oil (4.0%)	*Pseudomonas aeruginosa* M408	$NaNO_3$ (0.50%)	Batch mode, 28°C, 200 rpm, 96 h	Shake flasks	12.00	[17]
	Waste canola oil (2%)	*Pseudomonas cepacia* CCT6659	Corn steep liquor (3%) and $NaNO_3$ (0.20%)	Batch mode, 28°C, 250 rpm, 60 h	Shake flasks	8.00	[18]
	Sunflower oil soapstock waste (3.0%)	*Pseudomonas aeruginosa* LB1	$NaNO_3$ (0.40%)	Batch mode, 30°C, 150 rpm, 72 h	Shake flasks	7.3	[19]
Glycolipopeptide	Whey (15.0%), vinasse (1.0%), and sucrose (2.0%)	*Lactococcus lactis* CECT-4434	Yeast extract (0.75%)	Batch mode, 37°C, 100 rpm, 24 h	Shake flasks	0.11	[20]
Lipopeptides	Waste soybean oil (2.0%)	*Bacillus cereus* UCP 1615	Peptone (0.12%)	Batch mode, 28°C, 1.0 vvm, 48 h, 250 rpm	50-L bioreactor	4.70	[21]
	Sugarcane molasses (5%)	*Bacillus subtilis* RSL-2	No nitrogen sources	Batch mode, 41°C, 180 rpm, 168 h	Shake flasks	12.34	[22]
	Waste sunflower oil (4.0%)	*Bacillus amyloliquefaciens* RHNK 22	$NaNO_3$ (0.05%)	Batch mode, 37°C, 180 rpm, 48 h	Shake flasks	0.80	[23]
	Clarified cashew apple juice (4–9%)	*Bacillus subtilis* LAMI005	$(NH_4)_2SO_4$ (0.10%)	Batch mode, 30°C, 180 rpm, 72 h	Shake flasks	0.32	[24]
	Used Engine oil	*Brevibacillus sp. AVN 13*	1% v/v and 0.5% w/v	Batch mode, 40°C, pH 7, 72 h	Shake flasks	1.26	[25]
	Trub (residue from brewing industry) (2.0%)	*Bacillus subtilis*	Yeast extract (0.7%) and peptone (0.09%)	Batch mode, 1.0 vvm, 30°C, 225 rpm, 24 h	5-L Bioreactor	1.11	[26]
	Vinasse (5.0%)	*Bacillus pumilus* CCT 2487	$NaNO_3$ (0.7%)	Batch mode, 30°C, 200 rpm, 96 h	Shake flasks	27.70	[27]

Diverse biosurfactants such as glycolipids, rhamnolipids, sophorolipids, and lipopeptides can be produced through fermentation using different raw materials, showcasing multifaceted applications. From environmental roles like oil spill cleanup and bioremediation to diverse industrial sectors including agriculture, cosmetics, and medicine, these compounds demonstrate versatility in emulsification, detergency, and microbial control. They find application across various industries, from cleaning and textiles to nanotechnology, offering functionalities such as emulsification, wetting, and dispersion across multiple applications [28].

15.3.3 Genetic Engineering Approaches

Advances in genetic engineering have opened up new avenues for improving biosurfactant production. Researchers can modify the genetic makeup of microorganisms to enhance their biosurfactant-producing capabilities by introducing or overexpressing specific genes responsible for biosurfactant synthesis, optimizing metabolic pathways, or enhancing the microbial strains' overall productivity.

Engineered hyper-producing microbial strains, combined with optimized cultivation parameters, drive "green" biosurfactant innovation, impacting the global economy [4]. Emphasizing global biosurfactant synthesis, genetic engineering's role in biosurfactant gene expression, recent advancements, and future prospects highlight molecular dynamics as pivotal biotechnological tools. These tools foster development, enhance synthesis, and bolster biosurfactant applications, exemplified by efficiently designed mutant recombinants elevating biosurfactant nature, quantity, and quality [29].

Genetic engineering allows for the development of high-yield biosurfactant-producing strains that can efficiently convert substrates into biosurfactants [30]. Ambaye et al. [31] concluded that leveraging genetically engineered microbial strains, cost-effective substrates, optimized media, enhanced fermentation processes, and refined downstream processing with well-developed static models can offer commercially viable biological and engineering solutions.

Biosynthesis and genetic engineering strategies are extensively employed to boost surfactin production and create new surfactin variants, aiming to overcome its limited commercial viability due to low productivity [32]. Recently, Wu et al. [33] achieved a milestone in modularizing metabolic pathways, employing a systematic engineering approach to enhance surfactin biosynthesis. Their work led to a remarkable increase in surfactin production, reaching 12.8g/L with a yield of 65.0 mmol/mol sucrose (42% of the theoretical yield) in a metabolically engineered strain. These advancements aim to achieve cost-effective large-scale industrial biosurfactant production, contributing significantly to environmental sustainability.

15.3.4 Waste-Based Biosurfactant Production

An emerging trend in biosurfactant production is the utilization of waste materials as substrates. Agro-industrial waste streams, vegetable oil waste, fruit waste, and other organic residues can serve as cost-effective carbon sources for biosurfactant production [34]. This approach aligns with the principles of a circular bio-economy by repurposing waste materials and reducing environmental impact. Waste-based biosurfactant production not only offers economic advantages but also contributes to sustainable development goals by minimizing waste generation and promoting resource efficiency [8]. Microorganisms involved in biosurfactant production using different waste-based substrates are listed in Table 15.2.

An emerging trend in biosurfactant production involves utilizing waste materials as substrates, such as agro-industrial waste streams and organic residues, aligning with circular bio-economy principles. This approach not only reduces environmental impact but also yields economic advantages by significantly cutting production costs and enhancing profitability through sustainable resource utilization.

TABLE 15.2
Microorganisms in Biosurfactant Production from Low-Cost Waste Substrates

Different Low-Cost Waste Substrates	Microorganism Involved in Biosurfactant Production	References
Animal Waste (slaughterhouse waste, animal fats, fish processing waste)	*Pseudomonas gessardii, Nocardia higoensis, Aneurinibacillus migulanus*	[35], [36], [37]
Food and Agro-industrial residue (date molasses, sugarcane bagasse, orange peel, sesame peel flour, tuna fish residue, banana peel, potato peel, corn steep liquor, peanut oil cake, cassava waste, moringa residue)	*Halobacteriaceae archaeon, Bacillus pumilis, Bacillus licheniformis, Cunninghamella phaeospora, Candida tropicalis, Pseudomonas aeruginosa*	[38], [39], [40], [41], [42], [43]
Waste cooking oil (waste frying coconut oil, waste cooking oil)	*Pseudomonas aeruginosa, Candida lipolytica*	[44], [45], [46]
Agro-industrial and mill waste (including refinery waste) (olive mill waste, tannery pretreated effluent, palm oil industry waste, soybean oil industry waste)	*Brachybacterium paraconglomeratum, Pseudomonas aeruginosa, Bacillus pseudomycoides, Pseudomonas aeruginosa, Bacillus subtilis*	[47], [48], [49]

15.4 GREENER PRODUCTION METHODS

15.4.1 Utilization of Renewable Resources

Renewable resources for eco-friendly biosurfactant production include abundant agricultural waste like fruit peels, vegetable scraps, and starchy residues, along with microbial feedstocks such as bacteria, yeast, and filamentous fungi [50]. These resources are readily available in agriculture-based industries and food processing plants where substantial waste materials are generated.

15.4.2 Process Optimization for Energy Efficiency

Process optimization significantly reduces energy consumption in biosurfactant production by enhancing fermentation conditions, including pH, temperature, and aeration, leading to heightened biosurfactant yields [51]. Transitioning from batch to continuous fermentation methods is also impactful, minimizing energy usage. Strategies encompass maintaining optimal pH levels, controlling temperature, ensuring proper aeration, and supplying nutrients for enhanced yields. Employing automation, optimization techniques, and continuous monitoring reduces resource wastage, fostering sustainable and cost-effective biosurfactant production.

15.4.3 Eco-friendly Extraction and Purification Techniques

Employing eco-friendly extraction and purification techniques contributes to sustainability. These methods avoid the use of hazardous chemicals and lower energy requirements. Examples include supercritical fluid extraction, which deploys non-toxic supercritical CO_2 as a solvent, membrane filtration with selective pore-sized membranes, and adsorption techniques utilizing materials like activated carbon. These practices minimize environmental harm, reduce chemical usage, and yield purer biosurfactants, aligning with green production principles [52].

15.5 HEALTHCARE APPLICATIONS OF BIOSURFACTANTS

Biosurfactants find diverse applications in healthcare, showcasing their versatility and potential. They play a pivotal role in drug delivery systems, where their unique properties aid in enhancing drug solubility and bioavailability. Additionally, biosurfactants contribute to wound healing and infection control by facilitating efficient wound cleaning and promoting antimicrobial activities. These molecules are instrumental in diagnostic applications, assisting in the development of biosensors and diagnostic tools for various medical conditions [53]. Furthermore, biosurfactants are increasingly integrated into personal care products, including skincare and cosmetics, where their gentle, biocompatible nature aligns with the demands of healthcare and personal wellness, underscoring their significance in the healthcare sector. The imperative for greener biosurfactant production in healthcare lies in its potential to mitigate environmental impact and align with sustainable healthcare practices. Implementing eco-friendly production methods not only enhances the overall ecological footprint of healthcare applications but also contributes to a responsible and ethical approach, reinforcing the industry's commitment to environmental responsibility for future generations [54].

15.6 FOOD SECTOR APPLICATIONS

In the food sector, biosurfactants play a multifaceted role, offering sustainable solutions and addressing various challenges. They excel in food emulsification and stabilization, enabling the creation of smooth textures and consistent product quality (Figure 15.3). Additionally, biosurfactants contribute to food preservation by enhancing the solubility of antimicrobial agents, thereby extending the shelf life of perishable goods. Their ability to encapsulate flavors provides an avenue for flavor enhancement and controlled release in food products. Moreover, biosurfactants support the development of biodegradable food packaging materials, aligning with eco-conscious initiatives and minimizing environmental impact across the food supply chain [55]. In these ways, biosurfactants prove invaluable in elevating the quality, safety, and sustainability of food sector applications. The significance of researching and adopting greener production methods for biosurfactants in the food sector lies in fostering sustainability, reducing environmental footprint, and ensuring a responsible and ethical approach that resonates with the evolving demands of eco-conscious consumers and the imperative for a sustainable food supply chain [28]. A study by Liu et al. (2018) demonstrated that incorporating

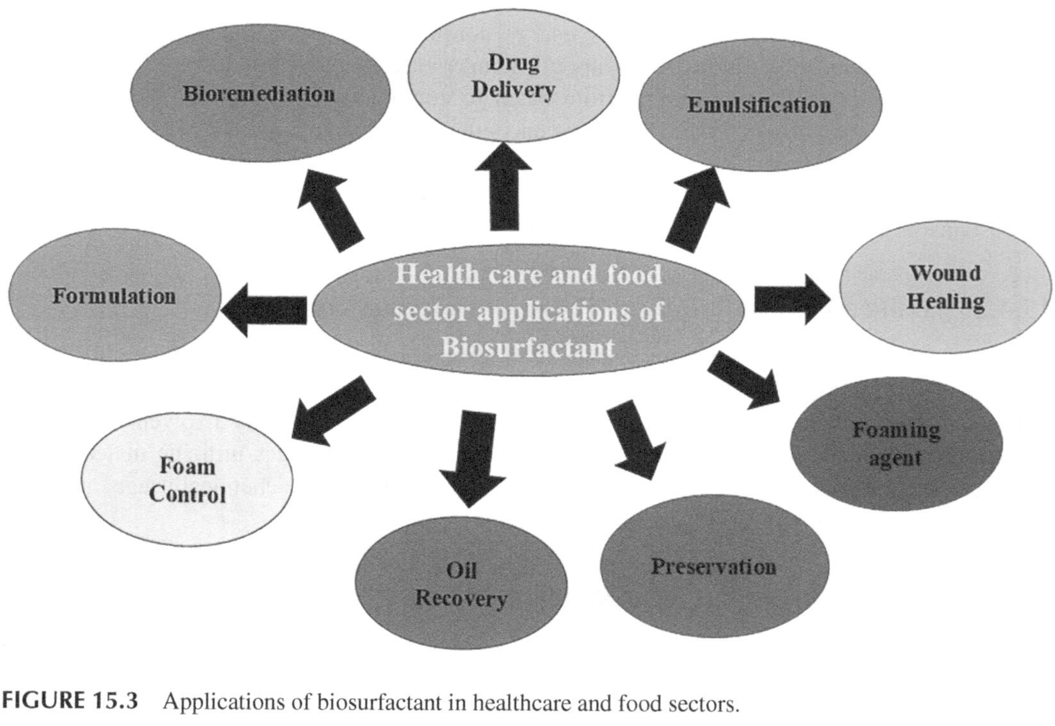

FIGURE 15.3 Applications of biosurfactant in healthcare and food sectors.

rhamnolipids as biosurfactants in food packaging materials improved the solubility and efficacy of antimicrobial agents, thus enhancing the preservation of perishable goods [56]. In a case study by Daverey et al. (2021), the application of sophorolipids in dairy products showed extended shelf life due to their antimicrobial properties, preventing spoilage and enhancing the overall quality and safety of the food [57]. Research conducted by Zhang et al. (2022) highlighted the role of lipopeptides in improving the solubility of natural antimicrobial agents in food matrices, resulting in prolonged shelf life and reduced microbial contamination in various food products [58].

15.7 ENVIRONMENTAL SUSTAINABILITY

Environmental sustainability in biosurfactants is further bolstered by notable advancements in extraction and purification technologies. Innovative methods like supercritical fluid extraction, membrane technology, and biocompatible approaches significantly reduce energy use and chemical consumption during purification processes. These advancements not only enhance biosurfactant purity but also minimize environmental footprint, positioning biosurfactants as eco-friendly solutions for sustainable production practices and addressing critical ecological concerns. Their biodegradability and eco-friendly extraction make them pivotal in wastewater treatment, aiding in pollutant breakdown for cleaner ecosystems and supporting circular economy principles by utilizing agro-industrial waste streams for biosurfactant production [59]. These innovations underline biosurfactants' substantial role in advancing environmental sustainability and responsible resource management.

15.8 REGULATORY AFFAIRS OF BIOSURFACTANTS IN HEALTHCARE AND FOOD-BASED INDUSTRIES

The incorporation of biosurfactants and bioemulsifiers into healthcare and as food additives is pivotal due to their non-toxicity and droplet coalescence properties [60]. Despite lacking defined regulatory recommendations for these additives, adhering to established regulatory frameworks, including acute toxicity assessments and allergy evaluations, is crucial for market commercialization [61]. For instance, glycolipids as a food additive in mayonnaise production displayed a dosage much lower than acute toxicity levels, indicating potential safety [62]. Evaluation methods adhering to OECD (Organisation for Economic Co-operation and Development) guidelines and the establishment of Acceptable Daily Intake (ADI) values support the safety assessment of Biosurfactant and Bioemulsifiers for potential food applications [62, 63].

Several non-pathogenic strains, like Marinobacter, demonstrating biosurfactant production without pathogenicity, expand the scope of biosurfactant biosynthesis for food-related applications, enhancing safety prospects [61]. Commercial endeavors like Jeneil Biotech and Evonik exhibit advancements in natural biosurfactant production, highlighting their potential use in food preservation and other applications [64]. Compliance with policies, alongside the recognition of biosurfactants as Generally Recognized as Safe (GRAS) by regulatory bodies like the Food and Drug Administration (FDA), supports their utilization in food applications [65, 66]. Ultimately, biosurfactants, with their biodegradable composition and low cytotoxicity at low concentrations, present potential for safe adoption and functionality in food formulations, potentially reducing costs associated with their implementation.

15.9 LIFE CYCLE ASSESSMENT (LCA) OF BIOSURFACTANTS

Life Cycle Assessment involves a systematic evaluation of the environmental impacts associated with the entire life cycle of these bioproducts. This assessment typically includes the stages of raw material extraction, production processes, transportation, product use, and disposal or end-of-life treatment. LCA aims to quantify environmental burdens such as energy consumption, greenhouse gas emissions, water and resource use, and potential toxicity throughout these stages, providing insights into the overall environmental footprint of biosurfactants. The assessment allows for informed decision-making, enabling researchers, industries, and policymakers to identify areas for improvement and implement strategies to enhance the environmental sustainability of biosurfactant production and usage [67, 68].

A study by Briem et al. consolidates and reviews limited, non-comparable LCA research on biosurfactants, emphasizing the need for more comprehensive assessments for environmental understanding and mitigation [68]. The three scrutinized studies presented Life Cycle Assessment (LCA) outcomes of various microbial-derived biosurfactants using experimental data [67]. While Aru and Ikechukwu's study did not specify the surfactant type, Baccile et al. and Kopsahelis et al. delved into sophorolipids and rhamnolipids extensively from a life cycle perspective. Kopsahelis et al. and Aru and Ikechukwu adopted a gate-to-gate approach, whereas Baccile et al. appraised the application of the investigated biosurfactant in hand-washing detergent, comparing it with conventional surfactants. Kopsahelis et al. highlighted the paucity of similar studies in current literature, a fact mirrored by the absence of references to other studies in the articles by Baccile et al. and Aru and Ikechukwu [69–71].

15.10 CHALLENGES AND FUTURE PERSPECTIVES

The challenges and future perspectives of biosurfactant utilization encompass several key dimensions. Scaling up production while maintaining cost-effectiveness remains a critical challenge, necessitating innovative strategies to optimize large-scale biosurfactant manufacturing processes. Achieving regulatory approvals and safety assessments is pivotal for broader industrial adoption, demanding rigorous evaluation to ensure biosurfactant safety and compliance with established standards [72]. Regulatory gaps in standardized toxicity testing and clear labeling pose challenges, necessitating comprehensive safety assessments for medical and personal care applications. The seamless integration of biosurfactants into existing industrial processes poses a practical challenge, requiring adaptations and technological innovations to harness their full potential within established systems. Looking ahead, the development of multifunctional biosurfactants with enhanced properties holds great promise, opening doors to broader and more diverse applications, thereby propelling biosurfactant research into an exciting and dynamic future.

15.11 CONCLUSION

The exploration of greener biosurfactant production methods and their diverse applications in healthcare and the food sector illuminates a path toward sustainable innovation. These surface-active compounds, harnessed from renewable resources and synthesized through eco-friendly techniques, hold immense potential for revolutionizing multiple industries [13]. In healthcare, biosurfactants promise to redefine drug delivery systems, enhance wound healing, enable diagnostic advancements, and elevate personal care products, ushering in a new era of improved well-being. Meanwhile, the food sector benefits from biosurfactants through improved emulsification and stabilization, food preservation, flavor encapsulation, and the development of biodegradable packaging, contributing to both quality and sustainability. Their biodegradable nature, pollution mitigation capabilities, and alignment with circular economy principles underscore their significance in environmental sustainability and offer efficient cleaning performance. Although challenges exist in terms of scaling up production, regulatory approvals, and integration into existing processes, the future holds exciting possibilities with multifunctional biosurfactants poised to enhance a wide array of applications in the market [73]. As we journey toward a greener and more sustainable future, biosurfactants stand as versatile allies, offering innovative solutions that bridge the realms of healthcare and the food industry, ultimately benefiting both humanity and the planet.

REFERENCES

[1] Gayathiri E, Prakash P, Karmegam N, Varjani S, Awasthi MK, Ravindran B. (2022) 'Biosurfactants: Potential and eco-friendly material for sustainable agriculture and environmental safety—A review'; *Agronomy*, 12, 662. DOI: 10.3390/agronomy12030662

[2] Biniarz P, Łukaszewicz M. (2017) 'Direct quantification of lipopeptide biosurfactants in biological samples via HPLC and UPLC-MS requires sample modification with an organic solvent'; *Appl Microbiol Biotechnol*, 101, 4747–4759. DOI: 10.1007/s00253-017-8272-y

[3] Sharma J, Sundar D, Srivastava P. (2021) 'Biosurfactants: Potential agents for controlling cellular communication, motility and antagonism'; *Front Mol Biosci*, 8, 727070. DOI: 10.3389/fmolb.2021.727070

[4] Mgbechidinma CL, Akan OD, Zhang C, Huang M, et al. (2022) 'Integration of green economy concepts for sustainable biosurfactant production - A review'; *Bioresour Technol*, 364, 128021.

[5] Begum W, Saha B, Mandal U. (2023) 'A comprehensive review on production of biosurfactants by biodegradation of waste carbohydrate feedstocks: an approach towards sustainable development'; *RSC Adv*, 13, 25599–25615. DOI: 10.1039/D3RA05051C

[6] Gutnick DL, Bach H. (2017) 'Biosurfactants'. In *Reference module in life sciences*. Elsevier. DOI: 10.1016/B978-0-12-809633-8.09184-6

[7] Singh A, Van Hamme JD, Ward OP. (2007) 'Surfactants in microbiology and biotechnology: Part 2. Application aspects'; *Biotechnol Adv*, 25(1), 99–121, Epub 2006 October 28. PMID: 17156965. DOI: 10.1016/j.biotechadv.2006.10.004

[8] Begum W, Saha B, Mandal U. (2023) 'A comprehensive review on production of biosurfactants by biodegradation of waste carbohydrate feedstocks: an approach towards sustainable development'; *Royal Society of Chemistry Adv*, 13, 25599.

[9] Sansarode D, Sahasrabudhe S. (2018) 'Biosurfactant: classification, properties and recent application in cosmetic'; *J. Emerg Technol Innov Res*, 5(10), 160–167.

[10] Eras-Muñoz E, Farré A, Sánchez A, Font X, Gea T. (2022) 'Microbial biosurfactants: a review of recent environmental applications'; *Bioengineered*, 13(5), 12365–12391. DOI: 10.1080/21655979.2022.2074621

[11] Nagtode VS, Cardoza C, Yasin HKA, Mali SN, Tambe SM, Roy P, Singh K, Goel A, Amin PD, Thorat BR, Cruz JN, Pratap AP. (2023) 'Green surfactants (biosurfactants): A petroleum-free substitute for sustainability—comparison, applications, market, and future prospects'; *ACS Omega*, 8(13), 11674–11699. Published online March 24, 2023. DOI: 10.1021/acsomega.3c00591

[12] Pardhi DS, Panchal RR, Raval VH, Joshi RG, Poczai P, Almalki WH, Rajput KN. (2022) 'Microbial surfactants: A journey from fundamentals to recent advances'; *Front Microbiol*, 13, 982603. DOI: 10.3389/fmicb.2022.982603

[13] Vigneshwaran C, Jerold M, Vasantharaj K, Sivasubramanian V. (2016) 'A review on biosurfactants and its environmental applications'; *J Environ Sci Toxicol Food Technol*, 10(8), 152–160, ISSN: 2319-2399. DOI: 10.9790/2402-100802152160

[14] Singh V. (2012) 'Biosurfactant – Isolation, production, purification & significance'; *Int J Sci Res Publ*, 2(7), July 2012.

[15] Kosaric N. (1992) 'Biosurfactants in industry'; *Pure & Appl. Chem*, 64(11), 1731–1737.

[16] Silva IA, Veras BO, Ribeiro BG, Aguiar JS, Campos Guerra JM, Luna JM, Sarubbo LA. (2020) 'Production of cupcake-like dessert containing microbial biosurfactant as an emulsifier'; *PeerJ*, 1–23. DOI: 10.7717/peerj.9064

[17] Ji F, Li L, Ma S, Wang J, Bao Y. (2016) 'Production of rhamnolipids with a high specificity by *Pseudomonas aeruginosa* M408 isolated from petroleum-contaminated soil using olive oil as sole carbon source'; *Ann Microbiol*, 66, 1145–1156. DOI: 10.1007/s13213-016-1203-9

[18] Rita de Cássia F, Almeida DG, Meira HM, Silva EJ, Farias CBB, Rufino RD, Luna JM, Sarubbo LA. (2017) 'Production and characterization of a new biosurfactant from *Pseudomonas cepacia* grown in low-cost fermentative medium and its application in the oil industry'; *Biocatal Agric Biotechnol*, 12, 206–215. DOI: 10.1016/J.BCAB.2017.09.004

[19] Kaskatepe SYB. (2016) 'Rhamnolipid biosurfactants produced by Pseudomonas species, *Arch Biol Technol*, 60786, 10.1590/1678-4324-2016160786

[20] Vera ECS, de Azevedo PODS, Domínguez JM, de Souza Oliveira RP. (2018) 'Optimization of biosurfactant and bacteriocin-like inhibitory substance (BLIS) production by *Lactococcus lactis* CECT-4434 from agroindustrial waste'; *Biochem Eng J*, 133, 168–178. DOI: 10.1016/j.bej.2018.02.011

[21] Durval IJB, Mendonça AHR, Rocha IV, Luna JM, Rufino RD, Converti A, Sarubbo LA. (2020) 'Production, characterization, evaluation and toxicity assessment of a Bacillus cereus UCP 1615 biosurfactant for marine oil spills bioremediation'; *Mar Pollut Bull*, 157, 111357. DOI: 10.1016/j.marpolbul.2020.111357

[22] Verma R, Sharma S, Kundu LM, Pandey LM. (2020) 'Experimental investigation of molasses as a sole nutrient for the production of an alternative metabolite biosurfactant'; *J Water Process Eng*, 38, 101632. DOI: 10.1016/J.JWPE.2020.101632

[23] Narendra Kumar P, Swapna TH, Khan MY, Reddy G, Hameeda B. (2017) 'Statistical optimization of antifungal iturin A production from *Bacillus amyloliquefaciens* RHNK22 using agro-industrial wastes'; *Saudi J Biol Sci*, 24, 1722–1740. DOI: 10.1016/J.SJBS.2015.09.014

[24] Freitas de Oliveira DW, Lima França IW, Nogueira Félix AK, Lima Martins JJ, Aparecida Giro ME, Melo VMM, Gonçalves LRB. (2013) 'Kinetic study of biosurfactant production by *Bacillus subtilis* LAMI005 grown in clarified cashew apple juice'; *Colloids Surf B Biointerfaces*, 101, 34–43. DOI: 10.1016/J.COLSURFB.2012.06.011

[25] Vigneshwaran C, Sivasubramanian V, Vasantharaja K, Krishnan N, Jerold M. (2018) 'Potential of *Brevibacillus sp.* AVN 13 isolated from crude oil contaminated soil for biosurfactant production and its optimization studies'; *J Environ Chem Eng*, 6, 4347–4356. DOI: 10.1016/j.jece.2018.06.036

[26] Nazaret GUSTC, Zanutto CP, Maass D, Souza AAU, Arruda SM. (2021) 'Bioconversion of low-cost brewery waste to biosurfactant: an improvement of surfactin production by culture medium optimization'; *Biochem Eng J*. DOI: 10.1016/j.bej.2021.108124

[27] de Oliveira JG, Garcia-Cruz CH. (2013) 'Properties of a biosurfactant produced by *Bacillus pumilus* using vinasse and waste frying oil as alternative carbon sources'; *Braz Arch Biol Technol*, 56, 155–160. DOI: 10.1590/S1516-89132013000100020

[28] Sarubbo LA, Silva MGC, Durval IJB, Bezerra KGO, Ribeiro BG, Silva IA, Twigg MS, Banat IM. (2022) '*Biosurfactants: Production, properties, applications, trends, and general perspectives*'; *Biochem Eng J*, 181, 108377, April 2022. DOI: 10.1016/j.bej.2022.108377

[29] Jimoh AA, Senbadejo TY, Adeleke R et al. (2021) 'Development and genetic engineering of hyper-producing microbial strains for improved synthesis of biosurfactants'; *Mol Biotechnol* 63, 267–288. DOI: 10.1007/s12033-021-00302-1

[30] Salek K, Euston SR. (2019) 'Sustainable microbial biosurfactants and bioemulsifiers for commercial exploitation'; *Process Biochem*, 85, 143–155. DOI: 10.1016/j.procbio.2019.06.027

[31] Ambaye TG, Vaccari M, Prasad S, Rtimi S. (2021) 'Preparation, characterization and application of biosurfactant in various industries: a critical review on progress, challenges and perspectives'; *Environ Technol Innov*, 24, 102090. DOI: 10.1016/j.eti.2021.102090

[32] Hu F, Liu Y, Li S. (2019) 'Rational strain improvement for surfactin production: enhancing the yield and generating novel structures'; *Microb Cell Fact*, 18, 42. DOI: 10.1186/s12934-019-1089-x

[33] Wu Q, Zhi Y, Xu Y. (2019) 'Systematically engineering the biosynthesis of a green biosurfactant surfactin by *Bacillus subtilis* 168'; *Metab Eng*, 52, 87–97. DOI: 10.1016/j.ymben.2018.11.004

[34] Singh P, Patil Y, Rale V. (2019) 'Biosurfactant production: emerging trends and promising strategies'; *J Appl Microbiol*, 126, 2–13. DOI: 10.1111/jam.14057

[35] Ramani K, Jain SC, Mandal AB, Sekaran G. (2012) 'Microbial induced lipoprotein biosurfactant from slaughter house lipid waste and its application to the removal of metal ions from aqueous solution'; *Colloids Surf B: Biointerfaces*, 97, 254–263.

[36] Patil SB, Sawant YP, Kamble LH, Raorane CJ. (2016) 'Primary screening of actinomycetes in prospects 'with biosurfactant production from animal fat waste'; *Int J CurrMicrobiol Appl Sci*, 5, 92–97.

[37] Sellami M, Khlifi A, Frikha F, Miled N, Belbahri L, Rebah FB. (2016) 'Agro-industrial waste based growth media optimization for biosurfactant production by *Aneurinibacillusmigulanus*'; *J Microbiol Biotechnol Food Sc*, 5, 578–583.

[38] Chooklin CS, Maneerat S, Saimmai A. (2014) 'Utilization of banana peel as a novel substrate for biosurfactant production by *Halobacteriaceae archaeon* AS65'; *Appl Biochem Biotechnol*, 173, 624–645.

[39] Sharma D, Ansari MJ, Gupta S, Al Ghamdi A, Pruthi P, Pruthi V. (2015) 'Structural characterization and antimicrobial activity of a biosurfactant obtained from *Bacillus pumilus*DSVP18 grown on potato peels'; *Jundishapur J Microbiol*, 8, e21257.

[40] Kumar AP, Janardhan A, Viswanath B, Monika K, Jung JY, Narasimha G. (2016) 'Evaluation of orange peel for biosurfactant production by *Bacillus licheniformis* and their ability to degrade naphthalene and crude oil'; *Biotech*, 6, 1–10.

[41] Lins AB, Bione AM, Silva TC, De Souza DG, Campos-Takaki GM. (2016) 'Low-cost production of biosurfactant by *Cunninghamellaphaeospora* using agro-industrial wastes'. In *Microbes in the spotlight: Recent progress in the understanding of beneficial and harmful microorganisms*, ed. Mendez Vilas A, pp. 339–343. Brown Walker Press.

[42] Rubio-Ribeaux D, da Silva Andrade RF, da Silva GS, de Holanda RA, Pele MA, Nunes P, Campos-Takaki GM. (2017) 'Promising biosurfactant produced by a new *Candida tropicalis* UCP 1613 strain using substrates from renewable-resources'; *Afr J Microbiol Res*, 11, 981–991.

[43] Magalhaes ERB, Silva FL, Sousa MADSB, DosSantos ES. (2018) 'Use of different agro industrial waste and produced water for biosurfactant production'; *Biosci, Biotechnol Res Asia*, 15, 17–26.

[44] Radzuan MN, Banat IM, Winterburn J. (2017) 'Production and characterization of rhamnolipid using palm oil agricultural refinery waste'; *Biores Technol*, 225, 99–105.

[45] Raheb J, Hajipour MJ. (2011) 'The stable rhamnolipid biosurfactant production in genetically engineered *Pseudomonas* strain reduced energy consumption in biodesulfurization'; *Energy Sources, Part A: Recovery Util Environ Eff*, 33, 2113–2121.

[46] Souza AF, Rodriguez DM, Ribeaux DR, Luna MA, Lima e Silva TA, Andrade RFS, Campos-Takaki GM. (2016) 'Waste soybean oil and corn steep liquor as economic substrates for bioemulsifier and biodiesel production by *Candida lipolytica* UCP 0998'; *Int J Mol Sci*, 17, 1608.

[47] Kiran GS, Sabarathnam B, Thajuddin N, Selvin J. (2014) 'Production of glycolipid biosurfactant from sponge-associated marine actinobacterium *Brachybacteriumparaconglomeratum* MSA21'; *J Surfactant Deterg*, 17, 531–542.

[48] Li J, Deng M, Wang Y, Chen W. (2016) 'Production and characteristics of biosurfactant produced by *Bacillus pseudomycoides* BS6 utilizing soybean oil waste'; *Int Biodet Biodegrad*, 112, 72–79.

[49] Moya-Ramírez I, Vaz DA, Banat IM, Marchant R, Alameda EJ, Garcia-Roman M. (2016) 'Hydrolysis of olive mill waste to enhance rhamnolipids and surfactin production'; *Biores Technol*, 205, 1–6.

[50] Gaur VK, Sharma P, Sirohi R, Varjani S, Taherzadeh MJ, Chang JS, Ng HY, Wong JWC, Kim S-H. (2022) 'Production of biosurfactants from agro-industrial waste and waste cooking oil in a circular bioeconomy: An overview'; *Bioresour Technol*, 343, ISSN 0960-8524. DOI: 10.1016/j.biortech.2021.126059

[51] Zoua C, Wanga M, Xingb Y, Lana G, Gea T et al. (2014) 'Characterization and optimization of biosurfactants produced by *Acinetobacter baylyi* ZJ2 isolated from crude oil-contaminated soil sample toward microbial enhanced oil recovery applications'; *Biochem Eng J*, 90, 49–58.

[52] Nadarajah N, Singh A, Ward OP. (2002) 'De-emulsification of petroleum oil emulsion by a mixed bacterial culture'; *Process Biochem*, 37, 1135–1141.

[53] Markande AR, Patel D, Varjani S. (2021) 'A review on biosurfactants: properties, applications and current developments'; *Bioresour Technol*, 330, ISSN 0960-8524 DOI: 10.1016/j.biortech.2021.124963

[54] Manga EB, Celik PA, Cabuk A, Banat IM. (2021) 'Biosurfactants: Opportunities for the development of a sustainable future'; *Curr Opin Colloid Interface Sci*, 56, 101514, ISSN 1359-0294. DOI: 10.1016/j.cocis.2021.101514

[55] Nadarajah N, Singh A, Ward OP. (2002) 'De-emulsification of petroleum oil emulsion by a mixed bacterial culture'; *Process Biochem*, 37, 1135–1141.

[56] Liu G, Zhong H, Yang X, Liu Y, Shao B, Liu Z. (2018) 'Advances in applications of rhamnolipids biosurfactant in environmental remediation: A review'; *Biotechnol' Bioeng*, 115(4), 796–814, PMID: 29240227. DOI: 10.1002/bit.26517

[57] Daverey A, Dutta K, Joshi S, Daverey A. (2021) 'Sophorolipid: a glycolipid biosurfactant as a potential therapeutic agent against COVID-19'; *Bioengineered*, 12(2), 9550–9560, PMID: 34709115. PMCID: PMC8810061. DOI: 10.1080/21655979.2021.1997261

[58] Zhang B, Xu L, Ding J, Wang M, Ge R, Zhao H, Zhang B, Fan J. (2022) 'Natural antimicrobial lipopeptides secreted by *Bacillus spp.* and their application in food preservation, a critical review'; *Trends Food Sci Technol*, 127, 26–37. DOI: 10.1016/j.tifs.2022.06.009

[59] Sánchez C. (2022) 'A review of the role of biosurfactants in the biodegradation of hydrophobic organopollutants: prod'uction, mode of action, biosynthesis and applications'; *World J Microbiol Biotechnol*, 38, 216. DOI: 10.1007/s11274-022-03401-6

[60] Sharma D, Singh D, Sukhbir-Singh GM, Karamchandani BM, Aseri GK, Banat IM, Satpute SK. (2023) 'Biosurfactants: Forthcomings and regulatory affairs in food-based industries'; *Molecules*, 28(6), 2823. DOI: 10.3390/molecules28062823

[61] Kralova I, & Sjöblom J. (2009). 'Surfactants used in food industry: A review'; *J Disper Sci Technol*, 30(9), 1363–1383. DOI: 10.1080/01932690902735561

[62] Campos JM, Stamford TL, Rufino RD, Luna JM, Stamford TCM, Sarubbo LA. (2015). 'Formulation of mayonnaise with the addition of a bioemulsifier isolated from *Candida utilis*'; *Toxicol Rep*, 2, 1164–1170. DOI: 10.1016/j.toxrep.2015.08.009

[63] Tripathi L, Twigg MS, Zompra A, Salek K, Irorere VU, Gutierrez T, Spyroulias GA, Marchant R, Banat IM. (2016) 'Biosynthesis of rhamnolipid by a Marinobacter species expands the paradigm of biosurfactant synthesis to a new genus of the marine microflora'; *Microb Cell Fact*, 18, 164. DOI: 10.1186/s12934-019-1216-8

[64] Jeniel Biotech Inc. *Home Page*. Available online: https://www.jeneilbiotech.com (accessed on 27 February 2023).

[65] Marchant R, Banat IM. (2012) Biosurfactants: A sustainable replacement for chemical surfactants? *Biotechnology Letters*, 34, 1597–1600.

[66] Sharma V, Singh D, Manzoor M, Banpurkar AG, Satpute SK, Sharma D. (2022) 'Characterization and cytotoxicity assessment of biosurfactant derived from *Lactobacillus pentosus* NCIM 2912'; *Braz J Microbiol*, 53, 327–340.

[67] Kopsahelis A, Kourmentza K, Zafiri C, Kornaros M. (2018) 'Gate-to-gate life cycle assessment (LCA) of biosurfactants and bioplasticizers production via biotechnological exploitation of fats and waste oils'; *J Chem Technol Biotechnol*, 93(10). DOI: 10.1002/jctb.5633

[68] Briem, A-K, Bippus L, Oraby A, Noll P, Zibek S, Albrecht S. (2022) 'Environmental impacts of biosurfactants from a life cycle perspective: a systematic literature review'; *Adv Biochem Engin/Biotechnol*, 181, 235–269, PMID: 35318488. DOI: 10.1007/10_2021_194

[69] Aru OO, Ikechukwu NEO. (2018) 'Life cycle assessment of the environmental impact of biosurfactant production from oil waste by a Diculture of *Azotobacter vinelandii* and *Pseudomonas sp*'; *J Bioremed Biodegr*, 9. DOI: 10.4172/2155-6199.1000435

[70] Baccile N, Babonneau F, Banat IM, et al. (2017) 'Development of a cradle-to-grave approach for acetylated acidic sophorolipid biosurfactants'; *ACS Sustain Chem Eng*, 5, 1186–1198. DOI: 10.1021/acssuschemeng.6b02570

[71] Kopsahelis A, Kourmentza C, Zafiri C, et al. (2018) 'Gate-to-gate life cycle assessment of biosurfactants and bioplasticizers production via biotechnological exploitation of fats and waste oils'; *J Chem Technol Biotechnol*, 6, 426. DOI: 10.1002/jctb.5633

[72] Ambaye TG, Vaccari M, Prasad S, Rtimi S. (2021) 'Preparation, characterization and application of biosurfactant in various industries: A critical review on progress, challenges and perspectives'; *Environ Technol Innov*, 24, 102090, ISSN 2352-1864. DOI: 10.1016/j.eti.2021.102090

[73] Research and Markets. (2023) 'Biosurfactants Global Market Report', 2023.

16 Microalgae Lipid Extraction for Biodiesel Production

Focus on Harvesting, Pre-Treatment, and Transesterification Processes

Michael Rahul Soosai, S. Sathish, L. Muruganandam, G. S. Nirmala, and I. Ganesh Moorthy

16.1 INTRODUCTION

It is estimated that nearly 80% of our energy demand is currently fulfilled by the burning of fossil fuels[1]. The need for energy is increasing day by day, and it is expected that by 2030, the global energy requirement will be 50% more than what is consumed today[2].

The excessive use of fossil fuels has caused some severe damage to the atmosphere, like climate change and pollution. This situation has led the global communities to divert their attention to the use of clean, environmentally friendly fuel[3]. Thus, in this regard, biofuel is expected to play a crucial role. Biodiesel, which is carbon neutral, produces less pollution, has the potential to be produced in bulk quantities, and is one of the best contenders of biofuels. Vegetable oil, animal fat, and waste cooking oil are the feedstock used in the commercial use of biodiesel. Also, microalgae, which can be easily cultivated, are competing as a good feedstock for the production of biodiesel. Microalgae are the most abundant autotrophic microorganisms of plant life. Microalgae are made up of lipid, protein, and carbohydrate. Microalgae lipid generally consists of carboxylic acid, and its composition consists of higher molecular weight compounds like C14 to C26. Microalgae can accumulate a large quantity of lipid (50–70% lipid per dry biomass), and also it can also produce 58,700 L per hectare annually. The other benefits of using microalgae are that, if it is cultivated in wastewater, it can treat the water by consuming the dangerous nutrients in wastewater. Carbon dioxide is a major constituent of greenhouse gas emissions. Moreover, microalgae require CO_2 for photosynthesis; thus, they contribute to a decrease in pollution level during their growth[4–6].

Production of biodiesel from microalgae faces some bottleneck problem before it can compete with commercial diesel obtained from fossil fuels. The major criteria are the availability, cost, and the quality of fuel. The production of biodiesel from microalgae passes through various steps like cultivation, harvesting, lipid extraction, pre-treatment, transesterification, and biodiesel property analysis. The efficiency of the process and quantity of biomass produced in each step play a vital role in determining the cost of the biodiesel[7–9].

Lipid extraction accounts for nearly 90% total energy input. Pre-treatment is generally done to disrupt the cell wall to increase lipid extraction. The cell wall composition of macromolecules, toughness, and its structure determine the nature of pre-treatment[10, 11].

There are many methods available for cultivation. The harvesting efficiency is the quantity of biomass recovered to the amount of energy spent in harvesting the biomass. In recent years, the number of studies done on the transesterification has increased exponentially. The key variable involved in the transesterification reaction is the reaction time, catalyst dosage, temperature, and alcohol-to-oil molar ratio. The ideal case to reduce the cost of biodiesel in the market is to develop

DOI: 10.1201/9781003732907-16

a process that yields a higher percentage of biodiesel. The process should be developed to handle a large quantity of lipid for transesterification. Preferably, the process should proceed at a lower temperature, in a short time, with a low alcohol-to-lipid molar ratio, and using a small catalyst dosage. Various studies are also conducted to increase the number of runs for the reuse of the catalyst at a higher efficiency. The implementation of this technique will decrease the operating cost in biodiesel production[12–15].

In this chapter, a comprehensive analysis and insight are provided on the recent studies made in the field of harvesting, pre-treatment, and transesterification methods employed for microalgae lipids. Also, the biodiesel standards and the quality associated with biodiesel are discussed.

16.2 MICROALGAE HARVESTING TECHNOLOGIES

For the mass production of biodiesel from microalgae, effective harvesting of the biodiesel is needed. The algal harvesting cost is generally high and depends on the value of the desired product, size, and density of the microalgae biomass. The harvesting is done either by thickening or bulk harvesting (Figure 16.1). The method employed is centrifugation, flocculation (Auto flocculation, Chemical coagulation – Inorganic coagulants/Organic flocculants/Combined flocculation, Electrolytic process), filtration and screening, gravity sedimentation, Flotation (Dissolved air flotation, Dispersed air flotation, ozone floatation, electrolytic floatation), and electrolysis[4].

Using gravitational force, heavy and large-sized algal cells can be harvested through the process of sedimentation[16]. Centrifugation can achieve about 80–90% efficiency in harvesting with about 2–3 min of operation (Figure 16.2). Centrifugation process is reliable, efficient, but it is energy-intensive, requires high operational cost, and damages the cell wall during biomass recovery or separation processes.[4, 17, 18]

Another method by which harvesting is done is filtration. In this process, algal cells are passed through a membrane or screen of some fixed mesh size or pore size. Reverse osmosis, microfiltration, ultra-filtration, dead-end filtration, vacuum filtration, and membrane filtration are some methods employed for harvesting. Some of the challenges faced in the use of filtration as a harvesting method are high operating and capital costs, high energy requirement, and biofouling[16, 19, 20]. Gas/air bubbles are used for floating algal cells on the top of the liquid surface through a simple process called flotation. Dissolved air flotation, electro flotation, ozone flotation, and dispersed flotation are the different types of flotation used by researchers. In ozone flotation, the algal biomass releases protein by disrupting cell walls, and this protein then acts as bio-flocculants to facilitate the process[21]. Microalgal cells are small, have a density greater than water, and are negatively charged. This cell can be attached to flocculants or coagulants for harvesting through an energy-efficient and

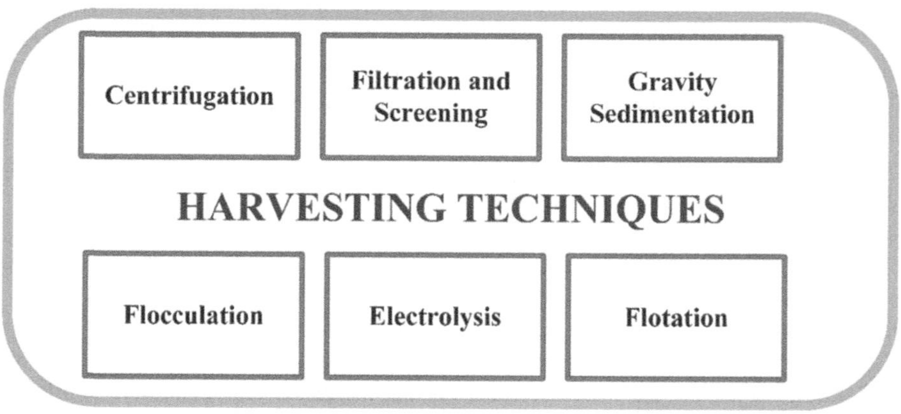

FIGURE 16.1 Various harvesting methods employed for microalgae.

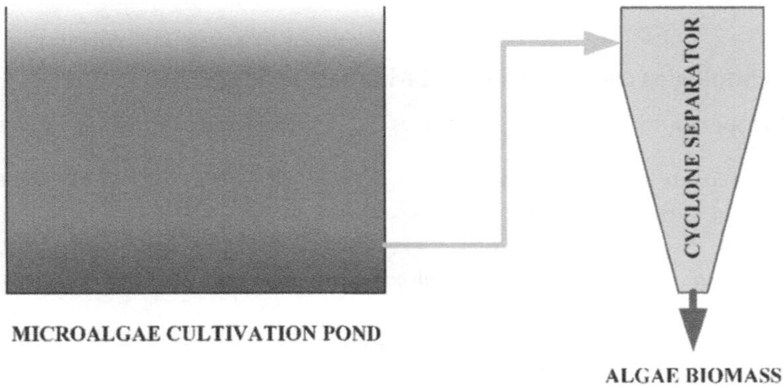

FIGURE 16.2 Algal harvesting using cyclone separator.

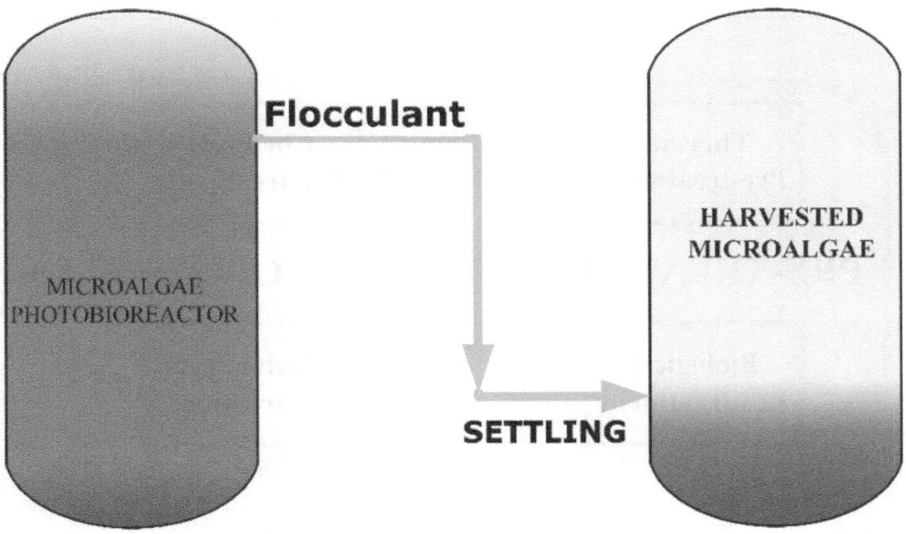

FIGURE 16.3 Algal harvesting using flocculation.

simple process of flocculation (Figure 16.3)[22]. Also, there are various types of electrolytic harvesting, where ions formed in the process attach with a negatively charged algal cell for separating the cells. In one such process called electrophoresis, the zeta potential of microalgal cells is used in an electric field for harvesting[21]. Table 16.1 gives a glimpse of the harvesting technique employed for various algal species for large volumes.

16.3 METHODS OF PRE-TREATMENT FOR MICROALGAE FEEDSTOCK

The microalgae biomass obtained from harvesting the culture may or may not contain water content in it, depending on the methods employed[24]. A complex cell wall structure protects microalgae, which are composed of cellulose and hemicellulose. This cell wall poses a challenge in the extraction of lipid. Thus, pre-treatment of microalgae biomass to disrupt the cell wall becomes an important step for the production of biodiesel[25, 26]. There are many processes[10] available for the pre-treatment (Figure 16.4):

TABLE 16.1
Harvesting Technologies used for Different Microalgae Species[23]

S. No.	Microalgal Species	Harvesting Techniques	Harvesting Efficiency (%)	Treated Volume (L)
1.	*Chaetoceros muelleri*	Centrifugation	90	550
2.	*Chaetoceros calcitrans*	pH increment (Autoflocculation)	≥80	10–1000
3.	*Chaetoceros sp.*	Foam fractionation	90	220
4.	*Chlorella muelleri*	pH increment (Autoflocculation)	≥80	10–1000
5.	*Chlorella sp.*	Foam flotation	≥92	10–10.2
6.	*Desmodesmus sp.*	Gravity sedimentation coupled with filtration	80	200
7.	*Rhodomonas salina*	pH increment (Autoflocculation)	≥80	10–1000
8.	*Scenedesmus sp.*	Flocculation	≥96	1000
9.	*Tetraselmis suecica*	pH increment (Autoflocculation)	≥80	10–1000
10.	*Thalassiosira pseudonana*	pH increment (Autoflocculation)	≥80	10–1000

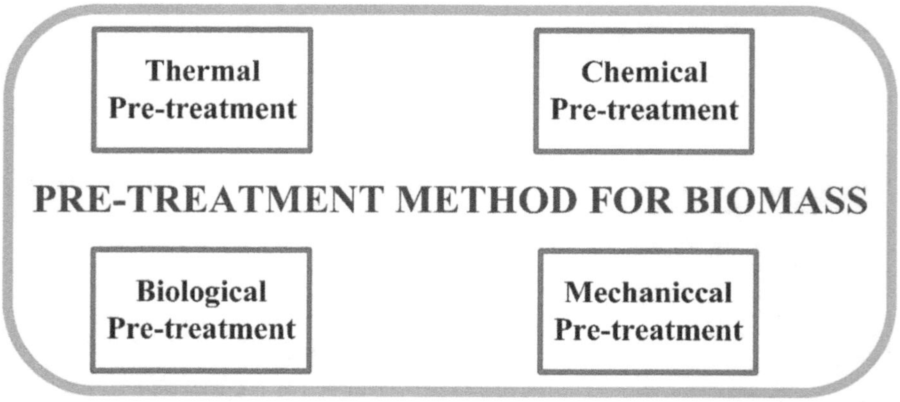

FIGURE 16.4 Various pre-treatment methods employed for microalgae.

- Thermal Pre-treatment method, which can be further classified as hydrothermal, thermal with steam explosion
- Chemical Pre-treatment method in acidic or alkaline medium
- Biological Pre-treatment method using enzyme
- Mechanical Pre-treatment method, which can be carried out using microwave, ultrasound, or blending.

These methods can be employed individually or can be grouped. For the extraction of total hydrocarbons from the mass cultivation of microalgae, Ashokkumar et al.[27] employed various pre-treatment methods for his process. For 15 min, at 15 Kpa pressure, biomass was autoclaved at a temperature of 121°C; for 15 min, at a constant temperature of 50°C ± 1, Ultra-sonication using an ultra-sonic probe at 24 kHz; for 15 min, at 500W and 2455 MHz, the algal biomass was placed in the microwave, and for a holding time of 2 h, the algal biomass was stored at −20°C in a deep freezer for deep freezing pre-treatment.

There are various parameters that are to be considered before employing a proper pre-treatment method. This parameter includes fitness for large-scale use, the cell wall's toughness, lipid product contamination (toxicity), reusability, time and cost, and maintenance[28].

In the thermal pre-treatment method, heat is used to disrupt the cell wall and increase lipid recovery.

Magota et al.[29], for their experiment on the thermal pre-treatment method, introduced 300 mg of the algal sample in 200 mL of water. This sample was placed in a vessel filled with nitrogen gas; this solution was heated in the range of 50–90°C for a varying amount of time. An increase in the holding time of the recovery of hydrocarbons from some strain of *Botryococcus braunii* increases effectively, and for 2.5 min, the recovery can be between 74.9% and 94.9% for a thermal pre-treatment temperature between 80–90°C for various strains. Similar experiments were performed by Kita et al.[30] on *Botryococcus braunii* biomass and established that with an algal-to-water ratio as low as 1.5 g/L at 85°C, nearly 90% hydrocarbon can be obtained, making the production of biodiesel through microalgae more economical.

Yoo et al.[31] used chemical pre-treatment for algal biomass, protic acid, and NaOH with mild heating, which led to the hydrolysis of starch and cellulose as the treatment resulted in breaking of cell wall chemical linkage. Chemical pre-treatment has some serious outcomes as the use of acid or base could directly affect the purity of the downstream product; also, the chemicals used could lead to the corrosion of reactors. Biological pre-treatment of biomass mostly employs the use of enzymes like lysozyme, β-glucuronidase, sulfatase, pectinase, laminarinase, and chitinase. The enzyme can easily break the cell wall and aid in further processing of biomass to biofuels. Still, they are quite expensive and there is a problem related to their recovery or separation; thus, biological pre-treatment is not viable for large-scale processes[32].

Mechanical pre-treatment has the advantage of simplicity, less possibility of contaminating lipid products, and the process is not species-specific, but the process needs high energy demand for its operation[28]. Ultrasonication employs high vibration stress to disrupt the microalgae cell, thus yielding high oil content for biodiesel production[33]. Dejoye et al.[34] used mechanical pre-treatment for extracting oil from *Chlorella vulgaris* biomass and observed that, if the extraction of oil from the biomass using supercritical carbon dioxide was preceded by microwave pre-treatment, the yield of oil increases as the electromagnetic radiation provides immense thermal energy for cell wall disruption.

Ozonation is a process in which the harvesting technique is combined with the pre-treatment method, as during the harvesting, the algal cell wall is subjected to disruption, which eventually results in increased lipid production[35]. Ozoflotation is one such process in which foaming achieved by the use of ozone could result in 98% of harvest[36], and the resulting lipid recovery improves significantly[37]. Table 16.2 represents the various pre-treatment methods used for various algal species.

TABLE 16.2
Pre-treatment Method Employed for Various Algal Species[10]

S. No.	Microalgae Species	Pre-treatment Method	Lipid Content (%)
1.	*Botryococcus sp.*	Microwaves	28.6
2.	*Chlorella sp.*	Sonication	20.1
3.	*Chlorella vulgaris*	Grinding	29
4.	*Chlorella vulgaris*	Microwaves	11
5.	*Nostoc sp.*	Sonication	18.2
6.	*Scenedesmus sp.*	Microwaves	11.5
7.	*Schizochytrium sp.* S31	Osmotic shock	48.7
8.	*Thraustochytrium sp.* AMCQS5-5	Osmotic shock	29.1
9.	*Tolypothrix sp.*	Sonication	14

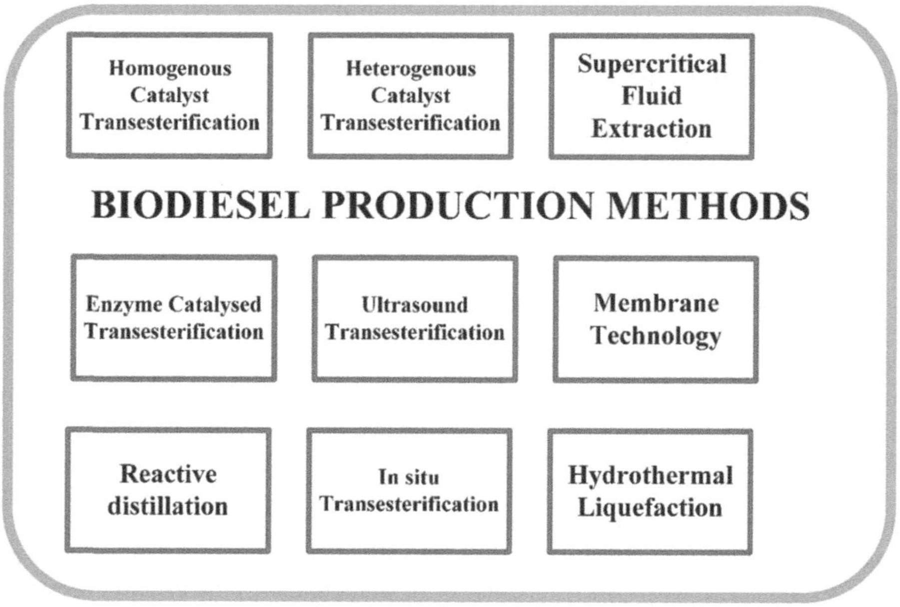

FIGURE 16.5 Various methods employed for converting microalgae lipid to biodiesel.

16.4 BIODIESEL PRODUCTION

In recent years, many processes (Figure 16.5) have been developed for converting the microalgae lipid or entire biomass into biodiesel. Transesterification, being simple and fast, is widely considered for the production of biodiesel. The key parameter which has to be particularly considered for the production of biodiesel are the molar ratio of alcohol to oil, reaction time, and temperature required, along with the catalyst dosage required.

16.4.1 Use of Homogenous Catalyst for Transesterification

Acid and alkali catalysts are used in liquid form in this method. The carbonyl group in triglycerides is protonized by the acid catalyst. The protonated carbon is subsequently attacked by alcohol, creating a tetrahedral intermediate. The major concern in the homogenous-base catalyzed reaction is the formation of nucleophilic alkoxide from alcohol, meant for attacking the electrophilic part of the carbonyl group of the triglycerides. Triglyceride collapses in three stages: first, an intermediate tetrahedral is formed, then the intermediate tetrahedral which is unstable disintegrates into diglyceride ion and fatty acid ester, and finally, the catalyst is recovered by proton transfer. Every fatty acid ester present undergoes these three mechanisms, and at the end, glycerol and three fatty acid esters are formed. In biodiesel production industries, this method involving the use of homogenous catalysts has been used primarily. Focus is now moved toward the use of heterogeneous catalyst[15]. Table 16.3 gives a glimpse of various transesterification processes (Figure 16.6) used by researchers to convert algal biomass into biodiesel.

16.4.1.1 Esterification Process by Using Acid Catalyst

Biomass with high free fatty acids favors this method involving any of the sulfuric, hydrochloric, sulfonic, and phosphoric acids catalysts. Based on the fatty acids present in the feedstock, this method has the advantage of producing bespoke biofuel; this method is very sensitive to water presence. Presence of in this method causes ester to form soap in alkaline conditions. Overall, this method prefers a high oil-to-alcohol ratio, at low to moderate pressure and temperature, along with

TABLE 16.3
Transesterification Process Employed for Algal Species

S. No.	Algae Species	Method	Reaction Media/Catalyst	Biodiesel Yield (%)	Reference
1	*Acutodesmus obliquus*	Transesterification with heterogeneous catalyst	Methanol, calcium oxide catalyst from waste eggshell	86.41	54
2	*Botryococcus sp.*	Ultrasound-assisted transesterification with heterogeneous catalyst	Novozyme CAL-B (*Candida antarctica* lipase B) immobilized on Celite, dimethyl carbonate, ultrasonicated 6 h at 40°C	88	55
3	*Chaetoceros gracilis* (wet)	Transesterification with homogenous catalyst	Methanol, chloroform, sulfuric acid catalyst, 20 min at 80°C	84 (FAME)	56
4	*Chlorella protothecoides*	Supercritical Transesterification	Methanol	90.8	57
5	*Chlorella pyrenoidosa*	Microwave-assisted transesterification with heterogeneous catalyst	Methanol, chloroform, sulfonated graphene oxide, microwave irradiated for 40 min at 90°C.	84.6	58
6	*Chlorella pyrenoidosa* (wet)	Microwave-assisted Transesterification homogenous catalyst	Methanol, chloroform, sulfuric acid catalyst, Microwave 30 min	10.51	59
7	*Chlorella sp.*	Ultrasound-assisted Transesterification homogenous catalyst	Ultrasonic power 137 W, sulfuric acid catalyst	81.2	60
8	*Chlorella sp.*	Transesterification with heterogeneous catalyst	Methanol, lithium hydroxide (LiOH) pumice catalyst	47 (FAME)	61
9	*Chlorella sp.*	Supercritical Transesterification	Methanol: algae (10:1) at supercritical conditions	45.62 wt.%	62
10	*Chlorella vulgaris* CCAP (211/19)	Supercritical Transesterification	Hexane	7.06 ± 1.03	63
11	*Nannchloropsis gaditana*	Transesterification with heterogeneous catalyst	lipase catalyst from *Candida antarctica*, t-butanol	99.5 (FAME)	64
12	*Nannochloropsis* (CCMP1776)	Supercritical Transesterification	Methanol	85.75	65
13	*Nannochloropsis gaditana*	vacuum distillation refining	Water, hexane, methanol, sulfuric acid catalyst	85.5 ± 2.6 (FAME)	66
14	*Nannochloropsis gaditana*	Transesterification with homogenous catalyst	Sulfuric acid catalyst	97.8 wt.% (FAEE)	67
15	*Nannochloropsis gaditana*	Transesterification with homogenous catalyst	1, 2-dichloroethane, ethanol, heated at 185.08°C for 3 h	92 (FAEE)	68
16	*Nannochloropsis gaditana*	Transesterification with homogenous catalyst	Methanol, chloroform, sulfuric acid catalyst, 95°C for 120 min	90	69
17	*Nannochloropsis gaditana*	Supercritical Transesterification	Supercritical methanol	45.8 (FAME)	70
18	*Nannochloropsis gaditana*(CCMP-1775)	Supercritical Transesterification	Methanol	59.28	71
19	*Nannochloropsis oceanica* (wet)	Transesterification with homogenous catalyst	Methanol, chloroform, sulfuric acid catalyst,	91 (FAME)	72
20	*Nannochloropsis oculata*	Supercritical Transesterification	Ethanol: algae (9:1) at supercritical conditions	67 (FAEE)	73
21	*Nannochloropsis salina*	Supercritical Transesterification	Ethanol: algae (9:1) at supercritical conditions	65 (FAEE)	74

(Continued)

TABLE 16.3 (Continued)

S. No.	Algae Species	Method	Reaction Media/Catalyst	Biodiesel Yield (%)	Reference
22	*Nannochloropsis sp.*	Transesterification with homogenous catalyst	Methanol, n-hexane, heated for 4 h at 60°C	90.9	68
23	*Nannochloropsis sp.*	Transesterification with heterogeneous catalyst	Methanol: methylene dichloride (3:1), Mg-Zr solid catalyst	28	75
24	*Nannochloropsis sp.*	Supercritical Transesterification	Ethanol: algae (9:1) at supercritical conditions	21.79 wt.%	62
25	*Nannochloropsis sp.* (dry)	Microwave-assisted transesterification with heterogeneous catalyst	Methanol, chloroform, strontium oxide catalyst, microwave assistance for 5 min at 60°C	37.1	76
26	*Nannochloropsis sp.* (wet)	Microwave-assisted Transesterification homogenous catalyst	Water, methanol, and sodium hydroxide catalyst, microwave-assisted for 10 min at 50°C	75	77
27	*Nannochlorpsis sp.*	Supercritical Transesterification	Supercritical CO_2	62	78
28	*Scenedesmus obliquus*	Transesterification with heterogeneous catalyst	Hexane, methanol, chromium-aluminum catalyst	98.28	79

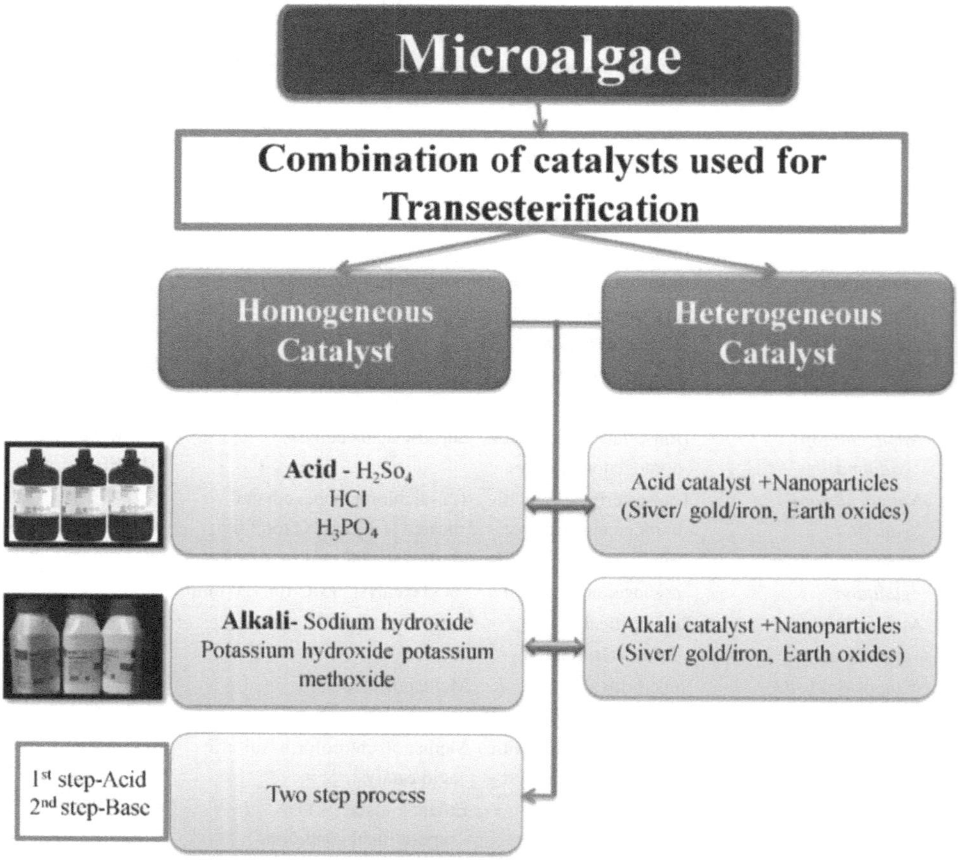

FIGURE 16.6 Various methods for biodiesel production.

a high concentration of catalyst. Still, in the end, a larger amount of funds is to be spent in the separation and purification process[38, 39]. The high viscosity and free fatty acid content present in crude lipid extracted from *Chlorella vulgaris* require high methanol and catalyst consumption. Lam et al.[40] overcame this problem by using a co-solvent tetrahydrofuran (THF), in the process involving sulfuric acid (H_2SO_4) as a catalyst. Attaining 95% (FAME) content, the reaction time needed is 3 h, at a temperature of 60°C, with a catalyst concentration of 21% by weight, and methanol-to-THF-to-lipid molar ratio needed is 60:15:1.

16.4.1.2 Transesterification Process Using Alkali Catalysts

This process, which is commonly used in laboratory, pilot, and industrial-scale levels, implies the use of either one of the alkaline metal hydroxides and alkoxides, otherwise sodium or potassium carbonates. The catalyst is formed very economically and with low free fatty acid feedstocks that show very high performance. By using this method, in very short reaction times, a large amount of methyl esters is formed from triglycerides. Some of the demerits of this reaction include the problematic revival of glycerol, treatment of alkaline wastewater, separation of catalyst from the product, and undesired side reactions due to low selectivity[41].

16.4.1.3 Transesterification Process by Two Steps

Biomass with high free fatty acids (FFAs) favors this process. If the FFA levels range from a mass fraction of 0.5% to less than 3% of the oil, then the alkaline catalysts cannot directly catalyze the transesterification of oils. Acid is catalyzed, FFAs are converted to fatty acid methyl ester (FAME) by esterification, followed by alkali-catalyzed transesterification[42]. A two-step transesterification process was used on algal oil from *Enteromorpha compressa* containing high FFA by Suganya et al.[43]. The acid esterification was employed with a 12:1 methanol-to-oil ratio, 1.5% sulfuric acid for 90 min, 60°C at 400 rpm to reduce the FFA from 6.3% to 0.34%. Further, the base transesterification was carried out for 70 min with a 9:1 methanol-to-oil ratio, 1% sodium hydroxide at 600 rpm and 60°C to yield 90.6% biodiesel.

16.4.2 Use of Heterogeneous Catalyst for Transesterification of Vegetable Oil

The high cost of raw materials and processing expenditure causes the production cost of biodiesel to be high. Using heterogeneous (solid) catalysts minimizes processing and material cost as it can be reused and recycled several times with better separation of the final product; thus, it proves to be superior to the homogenous catalyst. A heterogeneous catalyst has the property of both acid and basic nature, which is appropriate for the transesterification of triglycerides, as it primarily contains magnesium and aluminum oxide. The reaction mechanism using a solid catalyst is similar to that of using homogenous catalysis. Based on the catalytic temperature, solid catalysts can be divided into low-temperature catalysts and high-temperature catalysts. Solid catalyst eliminates the time-consuming and costly step to separate and recover the spent catalyst; also, the need for wastewater treatment is greatly reduced as compared with the use of a homogenous catalyst. Glycerol formed as a by-product is 98% pure, and methyl ester formed with a purity of over 99%. The presence of water or FFAs does not affect the reaction steps during transesterification, as heterogeneous catalysts can easily be changed to embrace desired catalyst properties[44].

16.4.2.1 Transesterification Using Heterogeneous Acid

The heterogeneous acid catalyst is environmentally friendly as it is less toxic and less corrosive. While dealing with biomass, it can simultaneously catalyze both transesterification and esterification reactions. A heterogeneous acid catalyst has a variety of acid sites with different strengths of Bronsted or Lewis acidity, as compared to the homogenous acid catalysts, so it is preferred at the industrial level. Solid acid catalysts, which have sufficient acid site strength, such as Nafion-NR50, sulfated zirconia, and tungstate zirconia, were selected to catalyze biodiesel-forming transesterification. For converting acidic oils into biodiesel using sulfur-based heterogeneous catalysts shows higher tolerance to FFA and water than the basic homogenous catalysts. Some sulfated salts and heteropoly acid catalysts are present in the high-temperature range, as they do with low-temperature heterogeneous acid catalysts[45].

16.4.2.2 Transesterification Using Heterogeneous Alkali

Transesterification using heterogeneous alkali has the advantage of straightforward production and purification processes, less amount of basic wastewater needed, a decrease in process equipment needed, and a good catalyst recovery rate. During transesterification of soybean oil with methanol, FAME yield by using a heterogeneous alkali catalyst was found to have almost the same activity as a homogenous NaOH catalyst under optimized reaction conditions. Heterogeneous alkali catalyst has low tolerance to FFA and water in raw materials, so they are not used for direct processing of crude oils with a high acid number[46].

16.4.2.3 Heterogeneous Catalysts from Biomass Materials

The cost-effective and environmentally friendly process can be made possible from heterogeneous catalysts which are made from renewable feedstocks or biomass materials that show catalytic properties which is suitable for biodiesel production. The preparation of existing acid and solid base heterogeneous solid catalysts which is used for the production of biodiesel is complicated and lengthy for its synthesis and requires great knowledge in this process. The catalyst derived from natural and biological materials like eggshell, limestone calcite, cuttlebone, dolomite, and hydroxyapatite contains calcium material. By using this catalyst, the transesterification reaction can yield about 95% and 98% biodiesel. Before transesterification, these materials are calcinated to form calcium oxides. The incomplete carbonization of carbohydrate materials, trailed by the sulfonation, results in solid catalysts, which have a high density of active SO_3H sites. These are capable of converting high free fatty acids (FFAs) contained waste cooking oils to biodiesel, and this catalyst, even after successively reusing after 50 cycles, shows an extraordinarily high proportion of its original catalytic activity with superb operational stability. Thus, naturally obtained solid catalysts are recyclable, eco-friendly, and are highly suited for the production of biodiesel from transesterification reactions, then the heterogeneous catalyst derived from a chemical origin counterpart, thus gaining their term as eco-friendly[47].

16.4.3 Method Involving Supercritical Fluid

At the industrial level, biodiesel is produced using a catalyst which necessitates the treatment of free fatty acids and the triglycerides in separate reaction stages. The other drawback includes consumption of catalyst, separating the catalyst from the product mixture, low purity of glycerine, as well as the harmful effect of water present in the mixture and the wastewater produced. If biodiesel were to be produced at supercritical conditions most, this disadvantage seems to vanish. At supercritical conditions, without the use of a catalyst, both esterification of FFA and transesterification of triglycerides can be achieved. This method is suitable for any type of raw material, like the one which is tedious to treat, involving oils high in free fatty acids or animal fats. The pressure, temperature, and alcohol-to-oil ratio are the key parameters identified; mostly ethanol and methanol are the preferred alcohol types. The operating pressure is preferred to be around 10–30 MPa, and the temperature to be 280–400°C, with a high alcohol-to-triglycerides molar ratio needed for methanol 42:1[48].

16.4.3.1 Raw Materials for Supercritical Extraction

Due to the vast amount of data available for *Jatropha curcas* oil, its abundance, as well as its high free fatty acid content, makes it well-suited for research. In these processing techniques, it is easier to evaluate and compare new techniques against well-known processes. Waste vegetable oil, a range of other non-edible oils, rice bran oil, corn oil, palm oil, coffee oil, microalgae, soybean, tea seed, grape seed, and rubber seed oil are other oils of interest for the supercritical fluid method. The cost of the raw material determines the cost of biodiesel, and also the FFA contents of the non-edible oil do not hamper the process by the supercritical method.

Using animal fat as a potential source for biodiesel production solves the problem of waste management associated with meat production. Also, it does not compete with food; being cheap, it provides the option of producing low-fat meat from chicken, lamb, and leather. Using supercritical technology to combine both the extraction stage and the reaction stage results in reducing the processing costs. The problem associated with the production of biodiesel using the enzymatic process from animal fat is the high melting point of the oil, which is very close to the denaturize temperature of the enzyme. Using supercritical CO_2 as a solvent for triglycerides offers processing temperatures that are within the optimum conditions for lipase[15].

16.4.3.2 Supercritical Processes

In the reaction stage of biodiesel, supercritical fluid technology has been more extensively used rather than the extraction of oil. Various types of supercritical processes are mentioned below.

16.4.3.2.1 Supercritical Methods: Type of Alcohol Used

Due to the good process for optimization condition to be achieved and the effects of free fatty acid content, methanol is the most favored alcohol for research. FFA content and water do not hinder the process using supercritical methanolysis. Ethanol, being completely renewable and derived from fossil-type sources, is the other preferred alcohol; additionally, the properties of cloud point and pour point of ethyl esters are better than those of methyl esters. The factor of temperature is superior in ethanolysis than in methanolysis. Still, methanolysis has a higher optimum yield than over ethanolysis, making it more favorable, and also the emulsion formed by the reaction mixture using ethanol makes the separation of the product very difficult[49].

16.4.3.2.2 Supercritical Methods: Use of Catalyst

Methanolysis, which is a noncatalytic process, requires high temperature and pressure to undergo a reaction. Using carbon dioxide as an enzyme for supercritical processes can bring the reaction rates close to ambient temperatures. With the help of cesium-doped heteropoly acid catalyst near the critical point of methanol, high conversions with moderate temperatures can be achieved with supercritical transesterification. Research has been carried out using heterogeneous inorganic catalysts or by using co-solvents (carbon dioxide and propane)[50].

16.4.4 Transesterification Process Using Enzyme Catalysts

Some of the problems faced by using the conventional catalyst are high energy requirement, wastewater treatment, catalyst removal from product, and feedstock pre-treatment, so research has been carried out to use enzymes as catalysts as an environmentally friendly means for the production of biodiesel. Enzymes refer to the biocatalyst called lipase which could be extracellular lipases and intracellular lipases. For intracellular lipases, separation processes or enzyme purification are not needed as they are deployed in the whole-cell form. Also, they can be immobilized on several materials for reusability and enzyme stability. Biocatalyst from enzymes can handle different triglycerides having FFA from 0.5% to 80%. Also, the glycerol recovered is of high grade and easier as compared to the alkaline process. Biocatalyst has the property of high efficiency and yield; the amount of methyl ester produced is more without any side reaction involved. Research has revealed that at a temperature of around 30–40°C, all the FFA present in waste oil and fats are completely transformed into alkyl esters using *Rhizopus oryzae* lipase. At present, biocatalyst is not being used at an industrial level as it takes more time to complete the reaction as compared to a reaction carried out using a base catalyst. Also, the production cost of lipase catalyst is higher, so its reuse and regeneration are limited[51, 52].

16.4.5 Transesterification Using Ultrasound

In transesterification, ultrasonication provides the mechanical energy needed for intensive mixing between the oil and alcohol phases by providing cavitation of bubbles near the phase boundary which leads to an increase in temperature. The formation of microjets, as a result of the production of biodiesel by ultrasounds, does not require mixing or heating of the system. Hence, the speed of the reaction increases and so does the effective molar ratio of alcohol to oil, and biodiesel is formed from lipids, with less energy consumption as compared to the other traditional mixing methods. The success of the production of biodiesel using ultrasonic irradiation is mainly dependent on the irradiation frequency, which controls the efficiency of cavitation, the mass transfer rate between reacting mixtures, and the rate of reaction. For a reaction time of about 50 min, a conversion of 99% was recorded for the transesterification of canola oil (*Brassica campestris*) with methanol to fatty acid methyl esters. The shortcoming of this process is that, for the downstream processing or purification process, a large amount of catalyst is to be used. This condition tends to lead to a side reaction involving the formation of unwanted soap. As a result, the biodiesel formed is to be separated from the alcohol phase by washing with water to remove the unused catalyst, soap, and glycerine, forming a significant amount of wastewater which is not such an environmentally friendly process[53].

16.4.6 Biodiesel Production Using Membrane Technology

Membrane reactor technology executes integration of membrane separation and chemical reaction simultaneously at close propinquity. This principle can be successfully applied to the reversible chemical reaction, like the transesterification of lipids. In terms of thermal stability, inorganic membranes are better than polymeric organic membranes. Research has revealed that oil droplets present in the reactor are bigger than the various pore sizes tested. Thus, biodiesel, as it gets formed, can be removed simultaneously, so the purity of biodiesel obtained is higher, thus eliminating side reactions. Though this method proves economical in terms of separation and recycling, the problem of downstream purification in regards to FAME-rich phase still exists[99].

16.4.7 Biodiesel Production Via Reactive Distillation

In the production of biodiesel, the end product of the reaction is mainly manipulated by chemical equilibrium. For this reason, a promising multifunctional reactor is used in a chemical reaction. Also, thermodynamic separation is combined in a single unit that can be successively applied to the production of biodiesel. For feeds containing a high amount of FFA, this reactive distillation process can be applied. A reactive core zone, stripping and rectifying section are the main components of a reactive distillation column (RDC). The number of trays needed for the stripping and rectifying section depends on the extent of purity needed for the product formed, which could also be guaranteed by adding an extra flash and decanter. Being more volatile, methanol and water will be dominant in the rectifying section, while the fatty ester and acid will be dominant in the stripping section. Various researches have been undertaken to optimize the reaction conditions, using a catalyst, dual reactive distillation to thermally coupled distillation columns, heat pinch, and process simulation like using free fatty acids as feedstock. This process does not need excess alcohol and has high productivity, short reaction time, and low capital cost due to the small size of RDC. An additional separation unit is not needed, and also, besides owing to the use of a solid catalyst, it does not need a separate separation process. In general, the energy consumption of the total process is very low as compared to the other conventional process[100]. A typical layout of RDC for producing biodiesel from microalgae is shown in Figure 16.7.

16.4.8 In-Situ Transesterification Process

The in-situ transesterification, also known as reactive extraction, is an economical method of producing biodiesel. The fatty acid alkyl esters are produced directly from the microalgal biomass, thus avoiding the solvent extraction step; hence, the overall process becomes much simpler and

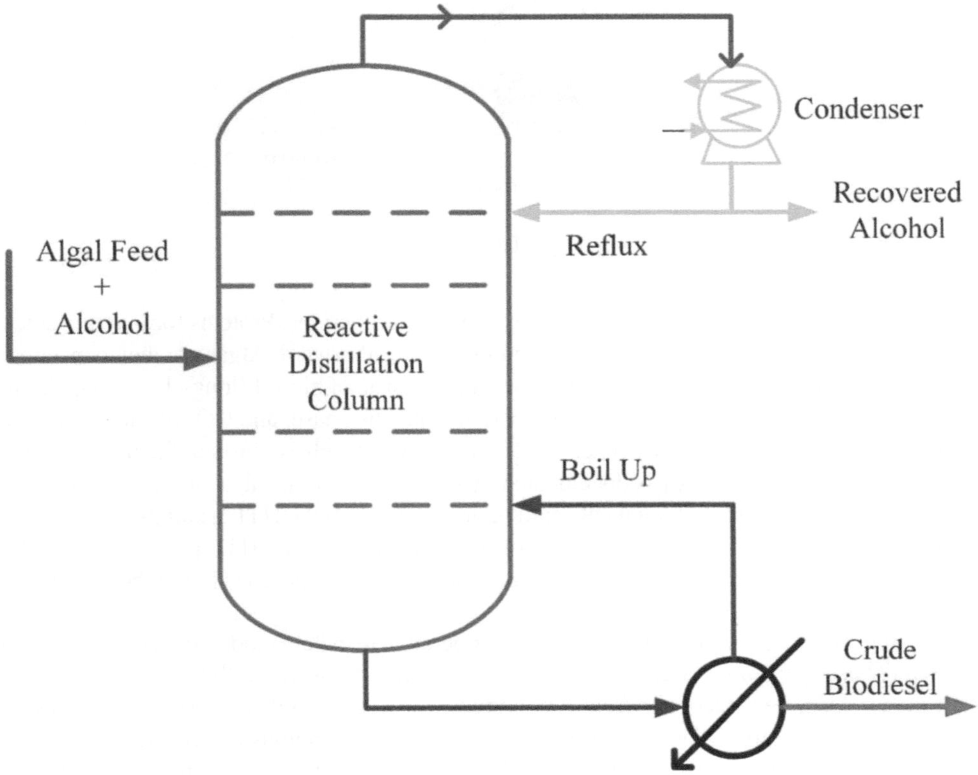

FIGURE 16.7 Reactive distillation column for producing biodiesel from microalgae.

cost-effective. This process results in better biodiesel yield as compared to the conventional process, and also the waste by-product formed is much eliminated in the in-situ transesterification process[101]. Tsigie et al.[102] used *Chlorella vulgaris* to produce biodiesel without the use of any catalyst at the subcritical condition and produced 0.29 g FAME per g dry biomass using 20 mL of methanol at 175°C with 5 g of wet biomass stirred for 4 h. This gain is far better than 0.20 g FAME per g dry biomass produced using the conventional transesterification process. Orta et al.[103] observed that for the process to be more economical, the alcohol consumption should be reduced or recycled. For *Chlorella vulgaris* feedstock, with a low yield alkaline catalyst acts rapidly than an acid catalyst. According to the process conversion rate required, the algal biomass available, and the catalyst separation required, the nature of the catalyst required has to be selected. For making the overall process economical, Ehimen et al.[101] suggested that the variables that affect the cost parameter of the process are the biomass moisture content, reaction time, temperature, and the alcohol volume required, along with the energy required for process mixing. However, an increase in alcohol volume and reaction temperature helps in better production of biodiesel, but for 15 g of Chlorella biomass, the fatty acid methyl ester produced did not show any significant yield for an alcohol volume of greater than 60 mL.

16.4.9 Hydrothermal Liquefaction (HTL)

The hydrothermal liquefaction is carried out to convert algal biomass into a slurry which consists of a solvent, at sufficient pressure and moderate temperature, aimed to retain the solvent in its liquid phase. In this process, the drying of algae is not needed, and, in some research, water is used as the solvent, which is advantageous in many ways, as water is the main constituent in the algal growth media[104]. In the HTL processes (Figure 16.8), through various steps, first due to the applied pressure and temperature, the cell wall and membrane get disrupted, and then the organic compounds in the cell membrane and cell walls like lipids, proteins, and carbohydrates react to depolymerization and hydrolysis.

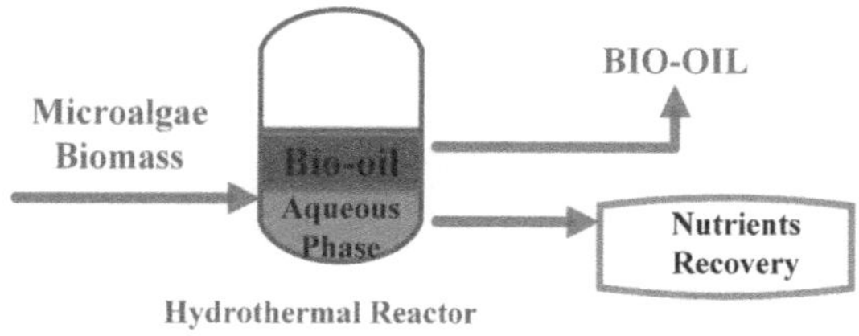

FIGURE 16.8 Hydrothermal liquefaction for producing bio-oil.

Carbohydrates get converted to monosaccharides and polysaccharides. Proteins form amino acids and peptides, and lipids get converted to glycerol and fatty acids afterward. Algae are then converted into products in gas, liquid, and solid residue. This product consists of bio-oil (long-chain nonpolar molecules), while polar molecules of short-chain forms an aqueous phase, and CO_2 formed is released to the atmosphere[82]. The bio-oil can be separated from the aqueous phase through the use of an organic solvent. About 30–75% efficiency is documented by researchers for the algae to bio-oil process[105].

The high moisture content of algal cells and their small size make HTL a suitable process for the transition of algal cells to fuel. There are various methods in which the HTL process can be utilized for converting algal cells to fuel[106]. These methods include noncatalytic HTL and the use of homogenous and heterogeneous catalysts.

HTL using non-catalytic operation is mostly operated in batch mode for its operation. Still, Elliott et al.[104] used the temperature of 623 K and a moderate pressure of 20 MPa to operate HTL in a batch mode. Also, heat recovery from a continuous process would increase HTL process' efficiency[107]. An alkali catalyst like Na_2CO_3 is mostly used in homogenous catalyst HTL process. The use of alkali catalyst has several advantages, such as inhibiting degradation of monomers in biomass by raising the pH and avoiding char and tar formation[42]. In the process of HTL using a homogenous catalyst, the lipid and proteins are converted to bio-oil. In contrast, the carbohydrates are converted to bio-oil with the help of a catalyst (Na_2CO_3). Still, the problem faced in the use of homogenous catalysts is the poor separation or recovery of the catalyst after the process[85].

Heterogeneous catalyst provides a great alternative for the problem as the catalyst could be recovered easily using a simple method such as filtration after the HTL process, but the solid residue does not only contain catalyst as well as other by-products formed during the reaction.

Thus, regeneration of the catalyst becomes a grave problem[108]. Table 16.4 gives an overview of the HTL technique used for different algal species.

TABLE 16.4
HTL Process Employed for Various Microalgae Species

S. No.	Algae Species	Method	Reaction Media	Catalyst	Bio-oil Yield (wt.%)	Reference
1	*Botryococcus braunii*	Homogenous catalysis	Water	Na_2CO_3	64	80
2	*Chlorella pyrenoidosa*	Noncatalytic direct liquefaction	Water	-	39.4	81
3	*Chlorella pyrenoidosa*	Noncatalytic direct liquefaction	Water	-	39.4	82
4	*Chlorella pyrenoidosa*	Noncatalytic direct liquefaction	Ethanol	-	64.6	83
5	*Chlorella pyrenoidosa*	Heterogeneous catalysis	Ethanol	HZSM-5, Raney-Ni	71.3	84
6	*Chlorella vulgaris*	Homogenous catalysis	Water	HCOOH	27	85

(Continued)

TABLE 16.4 (Continued)

S. No.	Algae Species	Method	Reaction Media	Catalyst	Bio-oil Yield (wt.%)	Reference
7	*Chlorella vulgaris,*	Homogenous catalysis	Water	Na_2CO_3, HCOOH	35.8	85
8	*Desmodesmus sp.*	Noncatalytic direct liquefaction	Water	-	49	86
9	*Dunaliella salina*	Heterogeneous catalysis	Water	Ni/REHY	72	87
10	*Dunaliella tertiolecta*	Noncatalytic direct liquefaction	Water	-	36.9	88
11	*Dunaliella tertiolecta*	Homogenous catalysis	Water	Na_2CO_3	43.8	89
12	*Dunaliella tertiolecta*	Homogenous catalysis	Water	Na_2CO_3	25.8	88
13	*Enteromorpha prolifera*	Homogenous catalysis	Water	Na_2CO_3	23	90
14	*Microcystis viridis*	Homogenous catalysis	Water	Na_2CO_3	33	91
15	*Nannochloropsis occulata*	Homogenous catalysis	Water	Na_2CO_3, HCOOH	34.3	85
16	*Nannochloropsis occulta*	Homogenous catalysis	Water	HCOOH	26	85
17	*Nannochloropsis salina*	Noncatalytic direct liquefaction	Water	-	46	92
18	*Nannochloropsis sp*	Noncatalytic direct liquefaction	Water	-	43	93
19	*Nannochloropsis sp.*	Noncatalytic direct liquefaction	Water	-	66	94
20	*Porphyridium creuntum*	Homogenous catalysis	Water	HCOOH	27.1	85
21	*Porphyridium cruentum*	Homogenous catalysis	Water	Na_2CO_3, HCOOH	20	42
22	*Sargassum polycystum*	Microwave-assisted Homogenous catalysis	Ethylene glycol	H_2SO_4	87.7	95
23	*Spirulina platensis*	Noncatalytic direct liquefaction	Water	-	39.9	96
24	*Spirulina platensis*	Noncatalytic direct liquefaction	Water	-	41	97
25	*Spirulina platensis*	Homogenous catalysis	Water	Na_2CO_3	51.6	98
26	*Spirulina platensis*	Homogenous catalysis	Water	$Ca_3(PO_4)_2$	34.5	98
27	*Spirulina platensis*	Homogenous catalysis	Water	NiO	30.2	98

16.5 PROPERTIES OF BIODIESEL

There are various properties which are considered important for the biodiesel to be deemed fit for considering it as a viable alternative to commercially available diesel from crude oil. Some of the important properties are discussed below.

The ability to flow is associated with any fuel, as it determines the proper functioning of spray atomization and fuel injection equipment at low temperatures. Due to the large chemical structure and large molecular mass, the kinematic viscosity of biodiesel is 10–15 times relatively greater than that of diesel obtained from fossil fuels. At low temperature, it becomes very viscous and sometimes even solidifies; consequently, the volume flow and injection spray characteristics in the engine are thus compromised.

Weight per unit volume is called density, and if the density of the component is compared with water, it is known as relative density. EN ISO 3675/12185 and ASTM D1298 are the standards used for measuring density at a temperature of about 15–20°C. For calculating viscosity and flow properties, mass-to-volume conversion, and judging homogeneity of biodiesel tanks, the property of relative density must be known. When exposed to flame or spark, the temperature at which the fuel ignites is called the flashpoint of that fuel. Biodiesel flashpoint is more than that of diesel derived from fossil fuel in the prescribed limit, and hence, it is safe for storage, handling, and transport. When the fuel has cooled, the temperature at which wax crystals first become visible is the cloud point, and the lowest temperature at which the flow of fuel is possible is called the pour point. The knowledge of the property of cloud point is particularly useful as partial or full solidification of fuel may block the fuel line. The low-temperature performance of fuel is determined by the property called the cold filter plugging point (CFPP). The ability of the fuel to auto-ignite quickly after being injected into the engine is determined by the cetane number (CN). A high CN value indicates high ignition quality. In the combustion chamber, a high VN number indicates a short time between ignition and fuel injection. The quality of biodiesel is determined by the oxidation of biodiesel fuel. Potential reactivity with air and the degree of oxidation are determined by oxidation stability. The property of oxidation stability necessitates the need for antioxidants. The lubrication property of biodiesel, which helps to increase the engine life and improves the effective brake power by reducing friction, is better with biodiesel as compared with diesel derived from fossil fuel.

The amount of FFA present in a fresh fuel sample is the neutralization number or acid number. High FFA content gives rise to a high acid number which leads to corrosion in the fuel supply system of an engine. The term Total Glycerine notes the amount of triglyceride that remains unconverted into methyl esters. The amount of monoglycerides, triglycerides, glycerine, and diglycerides, which is the result of the incomplete reaction, is used for calculating total glycerol. The total glycerine, if present above the specified value of fuel, may cause deposition on the piston, valve, and nozzle. In biodiesel, water may be present either as suspended water droplets or in undissolved form. Diesel obtained from fossil fuel may contain a maximum of 50 ppm of water, but biodiesel may contain a maximum of about 1500 ppm of water content. Dirt particles, suspended rust, or insoluble compounds present in fuel are the main ingredients of sediments. Water content present in fuel causes corrosion in the form of rust, and also water may become acidic with time and may attack the fuel storage tank. Secondly, water may promote the growth of yeast, fungi, and bacteria at its interference with fuel; some of these organisms may convert sulfur into sulfuric acid, which causes further corrosion of fuel storage tank[2]. Table 16.5 compares biodiesel properties obtained from various algal species to the standard prescribed for biodiesel.

TABLE 16.5
Comparison between Biodiesel Standards and the Biodiesel Obtained from Various Algal Species

S. No.	Fuel Properties	Units	USA (ASTM 6751-12)[109]	Europe (EN 214)[110]	*Chlorella sp.*[111]	*Crypthecodinium cohnii*[112]	*Nannochloropsis sp.*[113]
1	Acid value	mg KOH/g	0.50 max	0.50 max	0.37	0.14	-
2	Net calorific value	MJ/kg	n/a	n/a	37.06	39.86	40.72
3	Cetane number	–	47 min	51.0 min	-	46.5	-
4	Cloud point	°C	Report on request	report on request	-	16	7
5	Density @15°C	kg/m³	Report	860–900	883.6	912	869
6	Diglyceride	%mass	n/a	0.20 max	0.122	-	-
7	Ester content	%mass	n/a	96.5 min	97.42	-	-
8	Flashpoint	°C	130 min	120 min	179	95	144
9	Free glycerol	%mass	0.020 max	0.020 max	-	-	-

(Continued)

TABLE 16.5 (Continued)

S. No.	Fuel Properties	Units	USA (ASTM 6751-12)[109]	Europe (EN 214)[110]	*Chlorella sp.*[111]	*Crypthecodinium cohnii*[112]	*Nannochloropsis sp.*[113]
10	Iodine number	gI_2/100 g	n/a	120 max	97.12	-	-
11	Methanol	%mass	0.2 max	0.20 max	-	-	-
12	Monoglyceride	%mass	n/a	0.80 max	0.792	-	-
13	Oxidation stability	h@110°C	3 min	6.0 min	6.76	-	-
14	Phosphorus	mg/kg	10 max	10.0 max	-	-	-
15	Sulphur	mg/kg	15 max	10.0 max	8.1	7.5	-
16	Total glycerol	%mass	0.240 max	0.25 max	-	-	-
17	Triglyceride	%mass	n/a	0.20 max	0.197	-	-
18	Viscosity @4°C	mm^2/s	1.9–6.0	3.5–5.0	4.73	5.06	4.19
19	Water & sediment	%vol	0.05 max	0.05 max	-	0	-

16.6 CONCLUSION

Our current civilization is driven by energy obtained from fossil fuels, which has caused severe stress to the environment. Energy requirement is increasing daily, and biodiesel from microalgae as an alternate source of fuel has been researched extensively in recent decades. Despite this extensive research, an ideal solution pertaining to the solution to this problem has not yet been discovered. There are some key bottleneck issues related to this problem that hinder the establishment of large-scale production of biodiesel from microalgae. Most of the research is performed at a small scale in the laboratory. More research has to be focused on the production of large-scale biodiesel. The yield of biodiesel from microalgae is mainly a species-specific process. Harvesting is an energy-consuming process, and the need to develop a more efficient harvesting process has to be considered. Proper pre-treatment of biomass will relax the stress on operating costs. Due to the low cost and simplicity, transesterification is regarded as the best process for the production of biodiesel. Numerous studies have been reported on the development of various high biodiesel yield transesterification processes. The cost analysis and scale-up study of such a process will increase the feasibility approach of such a process for large-scale production of biodiesel. Some of the qualities of biodiesel are discussed in this chapter. ASTM 6751-3 (American Standards for Testing Materials) and EN 14214 (European Union) are the international standards prescribed for maintaining the quality of biodiesel. The produced biodiesel which meets the international standard can be easily vouched for commercial use.

REFERENCES

1. Hallenbeck P. Biological hydrogen production; fundamentals and limiting processes. *International Journal of Hydrogen Energy*. 2002;27(11–12):1185–1193. doi:10.1016/s0360-3199(02)00131-3
2. Atabani AE, Silitonga AS, Badruddin IA, Mahlia TMI, Masjuki HH, Mekhilef S. A comprehensive review on biodiesel as an alternative energy resource and its characteristics. *Renewable and Sustainable Energy Chapters*. 2012;16(4):2070–2093. doi:10.1016/j.rser.2012.01.003
3. Chisti Y. Biodiesel from microalgae. *Biotechnology Advances*. May–June 2007;25(3):294–306. doi:10.1016/j.biotechadv.2007.02.001
4. Chen CY, Yeh KL, Aisyah R, Lee DJ, Chang JS. Cultivation, photobioreactor design and harvesting of microalgae for biodiesel production: a critical review. *Bioresource Technology*. January 2011;102(1):71–81. doi:10.1016/j.biortech.2010.06.159
5. Gouveia L, Oliveira AC. Microalgae as a raw material for biofuels production. *Journal of Industrial Microbiology and Biotechnology*. February 2009;36(2):269–274. doi:10.1007/s10295-008-0495-6
6. Shi F, Wang P, Duan Y, Link D, Morreale B. Recent developments in the production of liquid fuels via catalytic conversion of microalgae: experiments and simulations. *RSC Advances*. 2012;2(26):9727–9747.

7. Sharma YC, Singh B, Korstad J. A critical review on recent methods used for economically viable and eco-friendly development of microalgae as a potential feedstock for synthesis of biodiesel. *Green Chemistry*. 2011;13(11):2993–3006.
8. Ahmad AL, Yasin NHM, Derek CJC, Lim JK. Microalgae as a sustainable energy source for biodiesel production: A review. *Renewable and Sustainable Energy Reviews*. 2011;15(1):584–593. doi:10.1016/j.rser.2010.09.018
9. Williams PJlB, Laurens LM. Microalgae as biodiesel & biomass feedstocks: review & analysis of the biochemistry, energetics & economics. *Energy & Environmental Science*. 2010;3(5):554–590.
10. Kadir WNA, Lam MK, Uemura Y, Lim JW, Lee KT. Harvesting and pre-treatment of microalgae cultivated in wastewater for biodiesel production: a review. *Energy Conversion and Management*. 2018;171:1416–1429.
11. Lardon L, Helias A, Sialve B, Steyer JP, Bernard O. Life-cycle assessment of biodiesel production from microalgae. *Environmental Science & Technology*. September 1 2009;43(17):6475–6481. doi:10.1021/es900705j
12. Deshmukh S, Kumar R, Bala K. Microalgae biodiesel: A review on oil extraction, fatty acid composition, properties and effect on engine performance and emissions. *Fuel Processing Technology*. 2019;191:232–247.
13. Mata TM, Martins AA, Caetano NS. Microalgae for biodiesel production and other applications: a review. *Renewable and Sustainable Energy Reviews*. 2010;14(1):217–232.
14. Akubude VC, Nwaigwe KN, Dintwa E. Production of biodiesel from microalgae via nanocatalyzed transesterification process: A review. *Materials Science for Energy Technologies*. 2019;2(2):216–225. doi:10.1016/j.mset.2018.12.006
15. Aransiola EF, Ojumu TV, Oyekola OO, Madzimbamuto TF, Ikhu-Omoregbe DIO. A review of current technology for biodiesel production: State of the art. *Biomass and Bioenergy*. 2014;61:276–297. doi:10.1016/j.biombioe.2013.11.014
16. Cheah WY, Ling TC, Show PL, Juan JC, Chang J-S, Lee D-J. Cultivation in wastewaters for energy: A microalgae platform. *Applied Energy*. 2016;179:609–625. doi:10.1016/j.apenergy.2016.07.015
17. Uduman N, Qi Y, Danquah MK, Forde GM, Hoadley A. Dewatering of microalgal cultures: A major bottleneck to algae-based fuels. *Journal of Renewable and Sustainable Energy*. 2010;2(1). doi:10.1063/1.3294480
18. Teixeira CMLL, Kirsten FV, Teixeira PCN. Evaluation of Moringa oleifera seed flour as a flocculating agent for potential biodiesel producer microalgae. *Journal of Applied Phycology*. 2012;24(3):557–563. doi:10.1007/s10811-011-9773-1
19. Shekhar M, Shriwastav A, Bose P, Hameed S. Microfiltration of algae: Impact of algal species, backwashing mode and duration of filtration cycle. *Algal Research*. 2017;23:104–112. doi:10.1016/j.algal.2017.01.013
20. Rashid N, Ur Rehman MS, Sadiq M, Mahmood T, Han J-I. Current status, issues and developments in microalgae derived biodiesel production. *Renewable and Sustainable Energy Reviews*. 2014;40:760–778. doi:10.1016/j.rser.2014.07.104
21. Pragya N, Pandey KK, Sahoo P. A review on harvesting, oil extraction and biofuels production technologies from microalgae. *Renewable and Sustainable Energy Reviews*. 2013;24:159–171.
22. Rodolfi L, Chini Zittelli G, Bassi N, et al. Microalgae for oil: strain selection, induction of lipid synthesis and outdoor mass cultivation in a low-cost photobioreactor. *Biotechnology Bioengineering*. January 1 2009;102(1):100–112. doi:10.1002/bit.22033
23. Ndikubwimana T, Zeng X, Murwanashyaka T, et al. Harvesting of freshwater microalgae with microbial bioflocculant: a pilot-scale study. *Biotechnology Biofuels*. 2016;9(1):47. doi:10.1186/s13068-016-0458-5
24. de Souza Silva APF, Costa MC, Lopes AC, et al. Comparison of pre-treatment methods for total lipids extraction from mixed microalgae. *Renewable Energy*. 2014;63:762–766.
25. González-Fernández C, Sialve B, Bernet N, Steyer J-P. Impact of microalgae characteristics on their conversion to biofuel. Part II: Focus on biomethane production. *Biofuels, Bioproducts and Biorefining*. 2012;6(2):205–218. doi:10.1002/bbb.337
26. Johnson LA. Oil recovery from soybeans. In *Soybeans*. In L. A. Johnson, P. J. White, & R. Galloway (Eds.), *Soybeans: Chemistry, production, processing, and utilization*, Elsevier; 2008:331–375.
27. Ashokkumar V, Rengasamy R, Deepalakshmi S, Sivalingam A, Sivakumar P. Mass cultivation of microalgae and extraction of total hydrocarbons: A kinetic and thermodynamic study. *Fuel*. 2014;119:308–312. doi:10.1016/j.fuel.2013.11.062

28. Lee AK, Lewis DM, Ashman PJ. Disruption of microalgal cells for the extraction of lipids for biofuels: Processes and specific energy requirements. *Biomass and Bioenergy*. 2012;46:89–101. doi:10.1016/j.biombioe.2012.06.034
29. Magota A, Saga K, Okada S, Atobe S, Imou K. Effect of thermal pre-treatments on hydrocarbon recovery from Botryococcus braunii. *Bioresource Technology*. November 2012;123:195–198. doi:10.1016/j.biortech.2012.07.095
30. Kita K, Okada S, Sekino H, Imou K, Yokoyama S, Amano T. Thermal pre-treatment of wet microalgae harvest for efficient hydrocarbon recovery. *Applied Energy*. 2010;87(7):2420–2423.
31. Yoo G, Park MS, Yang J-W. Chemical pre-treatment of algal biomass. In A. Pandey, S. Negi, & P. Binod (Eds.) *Pre-treatment of biomass*. Elsevier; 2015:227–258.
32. Gerken HG, Donohoe B, Knoshaug EP. Enzymatic cell wall degradation of Chlorella vulgaris and other microalgae for biofuels production. *Planta*. January 2013;237(1):239–253. doi:10.1007/s00425-012-1765-0
33. Surendhiran D, Vijay M. Effect of various pre-treatment for extracting intracellular lipid from nannochloropsis oculata under nitrogen replete and depleted conditions. *ISRN Chemical Engineering*. 2014;2014:1–9. doi:10.1155/2014/536310
34. Dejoye C, Vian MA, Lumia G, Bouscarle C, Charton F, Chemat F. Combined extraction processes of lipid from Chlorella vulgaris microalgae: microwave prior to supercritical carbon dioxide extraction. *International Journal of Molecular Sciences*. 2011;12(12):9332–9341. doi:10.3390/ijms12129332
35. Zheng H, Gao Z, Yin J, Tang X, Ji X, Huang H. Harvesting of microalgae by flocculation with poly (gamma-glutamic acid). *Bioresource Technology*. May 2012;112:212–220. doi:10.1016/j.biortech.2012.02.086
36. Betzer N, Argaman Y, Kott Y. Effluent treatment and algae recovery by ozone-induced flotation. *Water Research*. 1980;14(8):1003–1009. doi:10.1016/0043-1354(80)90144-x
37. Valeriano González MT, Monje-Ramírez I, Orta Ledesma MT, Gracia Fadrique J, Velásquez-Orta SB. Harvesting microalgae using ozoflotation releases surfactant proteins, facilitates biomass recovery and lipid extraction. *Biomass and Bioenergy*. 2016;95:109–115. doi:10.1016/j.biombioe.2016.09.020
38. Fukuda H, Kondo A, Noda H. Biodiesel fuel production by transesterification of oils. *Journal of Bioscience and Bioengineering*. 2001;92(5):405–416. doi:10.1016/s1389-1723(01)80288-7
39. Maçaira J, Santana A, Recasens F, Angeles Larrayoz M. Biodiesel production using supercritical methanol/carbon dioxide mixtures in a continuous reactor. *Fuel*. 2011;90(6):2280–2288. doi:10.1016/j.fuel.2011.02.017
40. Lam MK, Lee KT. Catalytic transesterification of high viscosity crude microalgae lipid to biodiesel: Effect of co-solvent. *Fuel Processing Technology*. 2013;110:242–248. doi:10.1016/j.fuproc.2012.12.021
41. Frascari D, Zuccaro M, Pinelli D, Paglianti A. A pilot-scale study of alkali-catalyzed sunflower oil transesterification with static mixing and with mechanical agitation. *Energy & Fuels*. 2008;22(3):1493–1501. doi:10.1021/ef700584h
42. Canakci, M., & Van Gerpen, J. (2001). Biodiesel production from oils and fats with high free fatty acids. *Transactions of the ASAE*, 44(6), 1429–1436.
43. Suganya T, Nagendra Gandhi N, Renganathan S. Production of algal biodiesel from marine macroalgae Enteromorpha compressa by two step process: optimization and kinetic study. *Bioresource Technology*. January 2013;128:392–400. doi:10.1016/j.biortech.2012.10.068
44. Endalew AK, Kiros Y, Zanzi R. Heterogeneous catalysis for biodiesel production from Jatropha curcas oil (JCO). *Energy*. 2011;36(5):2693–2700. doi:10.1016/j.energy.2011.02.010
45. Zhang S, Zu YG, Fu YJ, Luo M, Zhang DY, Efferth T. Rapid microwave-assisted transesterification of yellow horn oil to biodiesel using a heteropolyacid solid catalyst. *Bioresource Technology*. February 2010;101(3):931–936. doi:10.1016/j.biortech.2009.08.069
46. Zhang J, Chen S, Yang R, Yan Y. Biodiesel production from vegetable oil using heterogenous acid and alkali catalyst. *Fuel*. 2010;89(10):2939–2944. doi:10.1016/j.fuel.2010.05.009
47. Sharma YC, Singh B, Korstad J. Application of an efficient nonconventional heterogeneous catalyst for biodiesel synthesis from pongamia pinnata oil. *Energy & Fuels*. 2010;24(5):3223–3231. doi:10.1021/ef901514a
48. Al-Zuhair S, Hussein A, Al-Marzouqi AH, Hashim I. Continuous production of biodiesel from fat extracted from lamb meat in supercritical CO2 media. *Biochemical Engineering Journal*. 2012;60:106–110. doi:10.1016/j.bej.2011.10.010
49. Alenezi R, Leeke GA, Winterbottom JM, Santos RCD, Khan AR. Esterification kinetics of free fatty acids with supercritical methanol for biodiesel production. *Energy Conversion and Management*. 2010;51(5):1055–1059. doi:10.1016/j.enconman.2009.12.009

50. Rathore V, Madras G. Synthesis of biodiesel from edible and non-edible oils in supercritical alcohols and enzymatic synthesis in supercritical carbon dioxide. *Fuel*. 2007;86(17–18):2650–2659. doi:10.1016/j.fuel.2007.03.014
51. Gog A, Roman M, Toşa M, Paizs C, Irimie FD. Biodiesel production using enzymatic transesterification – Current state and perspectives. *Renewable Energy*. 2012;39(1):10–16. doi:10.1016/j.renene.2011.08.007
52. Szczęsna Antczak M, Kubiak A, Antczak T, Bielecki S. Enzymatic biodiesel synthesis – Key factors affecting efficiency of the process. *Renewable Energy*. 2009;34(5):1185–1194. doi:10.1016/j.renene.2008.11.013
53. Koh MY, Ghazi TIM. A review of biodiesel production from Jatropha curcas L. oil. *Renewable and Sustainable Energy Reviews*. 2011;15(5):2240–2251. doi:10.1016/j.rser.2011.02.013
54. Pandit PR, Fulekar M. Egg shell waste as heterogeneous nanocatalyst for biodiesel production: optimized by response surface methodology. *Journal of Environmental Management*. 2017;198:319–329.
55. Sivaramakrishnan R, Incharoensakdi A. Direct transesterification of Botryococcus sp. catalyzed by immobilized lipase: ultrasound treatment can reduce reaction time with high yield of methyl ester. *Fuel*. 2017;191:363–370.
56. Wahlen BD, Willis RM, Seefeldt LC. Biodiesel production by simultaneous extraction and conversion of total lipids from microalgae, cyanobacteria, and wild mixed-cultures. *Bioresource Technology*. February 2011;102(3):2724–2730. doi:10.1016/j.biortech.2010.11.026
57. Nan Y, Liu J, Lin R, Tavlarides LL. Production of biodiesel from microalgae oil (Chlorella protothecoides) by non-catalytic transesterification in supercritical methanol and ethanol: Process optimization. *The Journal of Supercritical Fluids*. 2015;97:174–182.
58. Cheng J, Qiu Y, Zhang J, Huang R, Yang W, Fan Z. Conversion of lipids from wet microalgae into biodiesel using sulfonated graphene oxide catalysts. *Bioresource Technology*. 2017;244:569–574.
59. Cheng J, Yu T, Li T, Zhou J, Cen K. Using wet microalgae for direct biodiesel production via microwave irradiation. *Bioresource Technology*. 2013;131:531–535.
60. Karimi M. Exergy-based optimization of direct conversion of microalgae biomass to biodiesel. *Journal of Cleaner Production*. 2017;141:50–55.
61. de Luna MDG, Doliente LMT, Ido AL, Chung T-W. In situ transesterification of Chlorella sp. microalgae using LiOH-pumice catalyst. *Journal of Environmental Chemical Engineering*. 2017;5(3):2830–2835. doi:10.1016/j.jece.2017.05.006
62. Jazzar S, Quesada-Medina J, Olivares-Carrillo P, et al. A whole biodiesel conversion process combining isolation, cultivation and in situ supercritical methanol transesterification of native microalgae. *Bioresource Technology*. August 2015;190:281–288. doi:10.1016/j.biortech.2015.04.097
63. Najafabadi HA, Vossoughi M, Pazuki G. The role of co-solvents in improving the direct transesterification of wet microalgal biomass under supercritical condition. *Bioresource Technology*. 2015;193:90–96.
64. López EN, Medina AR, Cerdán LE, Moreno PAG, Sánchez MDM, Grima EM. Fatty acid methyl ester production from wet microalgal biomass by lipase-catalyzed direct transesterification. *Biomass and Bioenergy*. 2016;93:6–12.
65. Patil PD, Gude VG, Mannarswamy A, et al. Comparison of direct transesterification of algal biomass under supercritical methanol and microwave irradiation conditions. *Fuel*. 2012;97:822–831.
66. Torres S, Acien G, García-Cuadra F, Navia R. Direct transesterification of microalgae biomass and biodiesel refining with vacuum distillation. *Algal Research*. 2017;28:30–38.
67. Park J, Kim B, Chang YK, Lee JW. Wet in situ transesterification of microalgae using ethyl acetate as a co-solvent and reactant. *Bioresource Technology*. 2017;230:8–14.
68. Religia P, Wijanarko A. Utilization of n-hexane as co-solvent to increase biodiesel yield on direct transesterification reaction from marine microalgae. *Procedia Environmental Sciences*. 2015;23:412–420.
69. Kim B, Im H, Lee JW. In situ transesterification of highly wet microalgae using hydrochloric acid. *Bioresource Technology*. 2015;185:421–425.
70. Jazzar S, Olivares-Carrillo P, Pérez de los Ríos A, et al. Direct supercritical methanolysis of wet and dry unwashed marine microalgae (Nannochloropsis gaditana) to biodiesel. *Applied Energy*. 2015;148:210–219.
71. Sitthithanaboon W, Reddy HK, Muppaneni T, et al. Single-step conversion of wet Nannochloropsis gaditana to biodiesel under subcritical methanol conditions. *Fuel*. 2015;147:253–259.
72. Im H, Lee H, Park MS, Yang JW, Lee JW. Concurrent extraction and reaction for the production of biodiesel from wet microalgae. *Bioresource Technology*. 2014;152:534–537. doi:10.1016/j.biortech.2013.11.023

73. Reddy HK, Muppaneni T, Patil PD, et al. Direct conversion of wet algae to crude biodiesel under supercritical ethanol conditions. *Fuel*. 2014;115:720–726.
74. Patil PD, Reddy H, Muppaneni T, et al. In situ ethyl ester production from wet algal biomass under microwave-mediated supercritical ethanol conditions. *Bioresource Technology*. July 2013;139:308–315. doi:10.1016/j.biortech.2013.04.045
75. Li Y, Lian S, Tong D, et al. One-step production of biodiesel from Nannochloropsis sp. on solid base Mg–Zr catalyst. *Applied Energy*. 2011;88(10):3313–3317.
76. Koberg M, Cohen M, Ben-Amotz A, Gedanken A. Bio-diesel production directly from the microalgae biomass of Nannochloropsis by microwave and ultrasound radiation. *Bioresource Technology*. March 2011;102(5):4265–4269. doi:10.1016/j.biortech.2010.12.004
77. Loong TC, Idris A. One step transesterification of biodiesel production using simultaneous cooling and microwave heating. *Journal of Cleaner Production*. 2017;146:57–62.
78. Taher H, Al-Zuhair S, Al-Marzouqi A, Haik Y, Farid M. Growth of microalgae using CO2 enriched air for biodiesel production in supercritical CO2. *Renewable Energy*. 2015;82:61–70.
79. Guldhe A, Moura CV, Singh P, et al. Conversion of microalgal lipids to biodiesel using chromium-aluminum mixed oxide as a heterogeneous solid acid catalyst. *Renewable Energy*. 2017;105:175–182.
80. Dote Y, Sawayama S, Inoue S, Minowa T, Yokoyama S-Y. Recovery of liquid fuel from hydrocarbon-rich microalgae by thermochemical liquefaction. *Fuel*. 1994;73(12):1855–1857.
81. Yu G, Zhang Y, Schideman L, Funk T, Wang Z. Distributions of carbon and nitrogen in the products from hydrothermal liquefaction of low-lipid microalgae. *Energy & Environmental Science*. 2011;4(11):4587–4595.
82. Yu G, Zhang Y, Schideman L, Funk T, Wang Z. Hydrothermal liquefaction of low lipid content microalgae into bio-crude oil. *Transactions of the ASABE*. 2011;54(1):239–246.
83. Duan P, Bai X, Xu Y, et al. Non-catalytic hydropyrolysis of microalgae to produce liquid biofuels. *Bioresource Technology*. May 2013;136:626–634. doi:10.1016/j.biortech.2013.03.050
84. Vardon DR, Sharma BK, Blazina GV, Rajagopalan K, Strathmann TJ. Thermochemical conversion of raw and defatted algal biomass via hydrothermal liquefaction and slow pyrolysis. *Bioresource Technology*. April 2012;109:178–187. doi:10.1016/j.biortech.2012.01.008
85. Biller P, Ross AB. Potential yields and properties of oil from the hydrothermal liquefaction of microalgae with different biochemical content. *Bioresource Technology*. January 2011;102(1):215–225. doi:10.1016/j.biortech.2010.06.028
86. Garcia Alba L, Torri C, Samorì C, et al. Hydrothermal treatment (HTT) of microalgae: evaluation of the process as conversion method in an algae biorefinery concept. *Energy & Fuels*. 2011;26(1): 642–657.
87. Yang C, Jia L, Chen C, Liu G, Fang W. Bio-oil from hydro-liquefaction of Dunaliella salina over Ni/REHY catalyst. *Bioresource Technology*. March 2011;102(6):4580–4584. doi:10.1016/j.biortech.2010.12.111
88. Zou S, Wu Y, Yang M, Li C, Tong J. Bio-oil production from sub- and supercritical water liquefaction of microalgae Dunaliella tertiolecta and related properties. *Energy & Environmental Science*. 2010;3(8):1073–1078. doi:10.1039/c002550j
89. Minowa T, Yokoyama S-Y, Kishimoto M, Okakura T. Oil production from algal cells of Dunaliella tertiolecta by direct thermochemical liquefaction. *Fuel*. 1995;74(12):1735–1738.
90. Zhou D, Zhang L, Zhang S, Fu H, Chen J. Hydrothermal liquefaction of macroalgae Enteromorpha prolifera to bio-oil. *Energy & Fuels*. 2010;24(7):4054–4061.
91. Yang YF, Feng CP, Inamori Y, Maekawa T. Analysis of energy conversion characteristics in liquefaction of algae. *Resources, Conservation and Recycling*. 2004;43(1):21–33. doi:10.1016/j.resconrec.2004.03.003
92. Toor SS, Reddy H, Deng S, et al. Hydrothermal liquefaction of Spirulina and Nannochloropsis salina under subcritical and supercritical water conditions. *Bioresource Technology*. March 2013;131:413–419. doi:10.1016/j.biortech.2012.12.144
93. Brown TM, Duan P, Savage PE. Hydrothermal liquefaction and gasification of Nannochloropsis sp. *Energy & Fuels*. 2010;24(6):3639–3646.
94. Faeth JL, Valdez PJ, Savage PE. Fast hydrothermal liquefaction of Nannochloropsis sp. to produce biocrude. *Energy & Fuels*. 2013;27(3):1391–1398.
95. Guo J, Zhuang Y, Chen L, Liu J, Li D, Ye N. Process optimization for microwave-assisted direct liquefaction of Sargassum polycystum C. Agardh using response surface methodology. *Bioresource Technology*. 2012;120:19–25.

96. Jena U, Das KC, Kastner JR. Effect of operating conditions of thermochemical liquefaction on bio-crude production from Spirulina platensis. *Bioresource Technology*. May 2011;102(10):6221–6229. doi:10.1016/j.biortech.2011.02.057
97. Jena U, Das KC. Comparative evaluation of thermochemical liquefaction and pyrolysis for bio-oil production from microalgae. *Energy & Fuels*. 2011;25(11):5472–5482. doi:10.1021/ef201373m
98. Jena U, Das K, Kastner J. Comparison of the effects of Na2CO3, Ca3 (PO4) 2, and NiO catalysts on the thermochemical liquefaction of microalga Spirulina platensis. *Applied Energy*. 2012;98:368–375.
99. Dube MA, Tremblay AY, Liu J. Biodiesel production using a membrane reactor. *Bioresource Technology*. February 2007;98(3):639–647. doi:10.1016/j.biortech.2006.02.019
100. Kiss AA, Dimian AC, Rothenberg G. Biodiesel by catalytic reactive distillation powered by metal oxides. *Energy & Fuels*. 2008;22(1):598–604. doi:10.1021/ef700265y
101. Ehimen E, Sun Z, Carrington C. Variables affecting the in situ transesterification of microalgae lipids. *Fuel*. 2010;89(3):677–684.
102. Tsigie YA, Huynh LH, Ismadji S, Engida AM, Ju Y-H. In situ biodiesel production from wet Chlorella vulgaris under subcritical condition. *Chemical Engineering Journal*. 2012;213:104–108.
103. Velasquez-Orta SB, Lee JGM, Harvey A. Alkaline in situ transesterification of Chlorella vulgaris. *Fuel*. 2012;94:544–550. doi:10.1016/j.fuel.2011.11.045
104. Elliott DC, Hart TR, Schmidt AJ, et al. Process development for hydrothermal liquefaction of algae feedstocks in a continuous-flow reactor. *Algal Research*. 2013;2(4):445–454. doi:10.1016/j.algal.2013.08.005
105. Pham M, Schideman L, Scott J, Rajagopalan N, Plewa MJ. Chemical and biological characterization of wastewater generated from hydrothermal liquefaction of Spirulina. *Environmental Science & Technology*. February 19 2013;47(4):2131–2138. doi:10.1021/es304532c
106. Jena U, Vaidyanathan N, Chinnasamy S, Das KC. Evaluation of microalgae cultivation using recovered aqueous co-product from thermochemical liquefaction of algal biomass. *Bioresource Technology*. February 2011;102(3):3380–3387. doi:10.1016/j.biortech.2010.09.111
107. Jazrawi C, Biller P, Ross AB, Montoya A, Maschmeyer T, Haynes BS. Pilot plant testing of continuous hydrothermal liquefaction of microalgae. *Algal Research*. 2013;2(3):268–277. doi:10.1016/j.algal.2013.04.006
108. Toor SS, Rosendahl L, Rudolf A. Hydrothermal liquefaction of biomass: A review of subcritical water technologies. *Energy*. 2011;36(5):2328–2342. doi:10.1016/j.energy.2011.03.013
109. Standard A. D6751-12. *Standard Specification for Biodiesel Fuel Blend Stock for Middle Distillate Fuels, ASTM International, West Conshohocken*. 2012.
110. European Committee for Standardization. (2004). *EN 10025-1: Hot rolled products of structural steels - Part 1: General technical delivery conditions* (EN 10025-1:2004). CEN Central Secretariat: Rue de Stassart, 36 B-1050 Brussels, Belgium.
111. Makarevičienė V, Lebedevas S, Rapalis P, Gumbyte M, Skorupskaite V, Žaglinskis J. Performance and emission characteristics of diesel fuel containing microalgae oil methyl esters. *Fuel*. 2014;120:233–239.
112. Islam MA, Rahman MM, Heimann K, et al. Combustion analysis of microalgae methyl ester in a common rail direct injection diesel engine. *Fuel*. 2015;143:351–360. doi:10.1016/j.fuel.2014.11.063
113. Haik Y, Selim MYE, Abdulrehman T. Combustion of algae oil methyl ester in an indirect injection diesel engine. *Energy*. 2011;36(3):1827–1835. doi:10.1016/j.energy.2010.11.017

Index

Pages in **bold** refer to tables.

D

E

F

M

N

O

P

R